Springer Series in Optical Sciences

Volume 169

For further volumes:
http://www.springer.com/series/624

Springer Series in Optical Sciences

The Springer Series in Optical Sciences, under the leadership of Editor-in-Chief William T. Rhodes, Georgia Institute of Technology, USA, provides an expanding selection of research monographs in all major areas of optics: lasers and quantum optics, ultrafast phenomena, optical spectroscopy techniques, optoelectronics, quantum information, information optics, applied laser technology, industrial applications, and other topics of contemporary interest.
With this broad coverage of topics, the series is of use to all research scientists and engineers who need up-to-date reference books.
The editors encourage prospective authors to correspond with them in advance of submitting a manuscript. Submission of manuscripts should be made to the Editor-in-Chief or one of the Editors. See also www.springer.com/series/624

Wolfram Hergert · Thomas Wriedt
Editors

The Mie Theory

Basics and Applications

Springer

Editors
Wolfram Hergert
Martin-Luther-Universität
Halle-Wittenberg
Fakultät für Naturwissenschaften II
Von-Seckendorff-Platz 1
06120 Halle/Saale
Germany

Thomas Wriedt
Stiftung Institut für Werkstofftechnik
Badgasteiner Str. 3
28359 Bremen
Germany

ISSN 0342-4111 ISSN 1556-1534 (electronic)
ISBN 978-3-642-43614-7 ISBN 978-3-642-28738-1 (eBook)
DOI 10.1007/978-3-642-28738-1
Springer Heidelberg New York Dordrecht London

Printed on acid-free paper

Springer is part of Springer Science+Business Media (www.springer.com)

Preface

The small conference we arranged to celebrate the 100th anniversary of Mie's theory was a fascinating event. The two nephews of Gustav Mie were present, one of them presenting a real live account on his uncle. We got in touch with them when one of us (TW) started to redesign a web page on a collection of Mie scattering programs back in 2000. A photo of Gustav Mie scanned from a source at the local university library and some information about his life was included in this web page. This photo can still be found at the ScattPort light scattering information portal [1] and since then the photo has been widely distributed on the Internet. One of the nephews called by phone introducing himself as the caretaker of the scientific inheritance of his uncle and asked for the source of the photo. Apparently, he did not know about this specific photo. Since that time our interest on the history of science in the topic of light scattering arose and we collected papers and other sources of information. Somehow the idea to arrange a conference on the anniversary in 2008 of Gustav Mie's original paper arose. After two attempts we managed to secure some funding to arrange this conference. Scientists from all over the world presented talks on the current state of light scattering by particles including simulations and applications. The current edited volume is a compilation of selected extended versions of papers presented at this conference arranged in Halle (Saale) in 2008. We would like to thank all the authors for their contributions and acknowledge the support of the conference by Deutsche Forschungsgemeinschaft (DFG).

Halle, Bremen, Germany

Wolfram Hergert
Thomas Wriedt

Reference

1. Information on Gustav Mie at ScattPort, http://www.scattport.org/index.php/gustav-mie-special.

Contents

Contributors

Logan K. Ausman Department of Chemistry, Northwestern University, 2145 Sheridan Road, Evanston, IL 60208-311, USA, e-mail: ausmanlk@uwec.edu

Ferdinando Borghese Dipartimento di Fisica della Materia e Ingegneria Elettronica, Università di Messina, Viale Ferdinando Stagno d'Alcontres 31, 98166 Messina, Italy, e-mail: ferdinando.borghese@unime.it

Nils Jørgen Christensen Institut for Informatik og Matematisk Modellering, Danmarks Tekniske Universitet, Richard Petersens Plads, Bygning 321, 2800 Lyngby, Denmark, e-mail: njc@imm.dtu.dk

Paolo Denti Dipartimento di Fisica della Materia e Ingegneria Elettronica, Università di Messina, Viale Ferdinando Stagno d'Alcontres 31, 98166 Messina, Italy, e-mail: paolo.denti@unime.it

Jeppe Revall Frisvad Institut for Informatik og Matematisk Modellering, Danmarks Tekniske Universitet, Richard Petersens Plads, Bygning 321, 2800 Lyngby, Denmark, e-mail: jrf@imm.dtu.dk

Gerard Gouesbet CORIA (Complexe de Recherche Interprofessionnel en Aérothermochimie), Laboratoire d'Energétique des Systèmes et Procédés (LESP), Unité Mixte de Recherche (UMR) 6614 du Centre National de la Recherche Scientifique (CNRS), Université de Rouen et Institut National des Sciences Appliquées de Rouen, B.P. 08, 76131 Mont Sait Aignan Cedex, France, e-mail: gerard.gouesbet@coria.fr

Wolfram Hergert Institute of Physics, Martin Luther University Halle-Wittenberg, Von-Seckendorff-Platz 1, 06120 Halle, Germany, e-mail: wolfram.hergert@physik.uni-halle.de

Henrik Wann Jensen Computer Graphics Laboratory, Computer Science and Engineering, University of California, San Diego, CSE 4116, 9500 Gilman Drive, La Jolla, CA 92093-0404, USA, e-mail: henrik-at-cs.ucsd.edu

Philip Laven Geneva, Switzerland, e-mail: philip@philiplaven.com

Daniel Mackowski Mechanical Engineering Department, Auburn University, Auburn, AL 36849, USA, e-mail: mackodw@auburn.edu

Rosalba Saija Dipartimento di Fisica della Materia e Ingegneria Elettronica, Università di Messina, Viale Ferdinando Stagno d'Alcontres 31, 98166 Messina, Italy, e-mail: rosalba.saija@unime.it

George C. Schatz Department of Chemistry, Northwestern University, 2145 Sheridan Road, Evanston, IL 60208-311, USA, e-mail: schatz@chem.northwestern.edu

Thomas Wriedt Stiftung Institut für Werkstofftechnik, Badgasteiner Str. 3, 28359 Bremen, Germany, e-mail: thw@iwt.uni-bremen.de

Chapter 1
Gustav Mie: From Electromagnetic Scattering to an Electromagnetic View of Matter

Wolfram Hergert

Abstract The year 2008 saw the 100th anniversary of the publication of Gustav Mie's work "Contributions to the optics of turbid media, particularly colloidal metal solutions" in Annalen der Physik. This event was an occasion to express appreciation for his contribution to the theory of scattering of electromagnetic waves. The achievements of Gustav Mie in the framework of the development of an unified field theory and his contributions to the discussion around the formulation of the theory of general relativity have been mentioned many times in other contexts. A detailed description of Gustav Mie's work has, until now, been unavailable. This contribution undertakes the task of presenting a comprehensive account of the life and scientific work of Gustav Mie.

1.1 Introduction

It is generally perceived that the revolutions in physics at the beginning of the twentieth century are marked by the development of quantum physics and the special and general theories of relativity. Planck's introduction of the concept of quantum of action, regarded as the dawn of quantum physics, in December 1900 and Einstein's *annus mirabilis* 1905 mark this development. These milestones of progress in modern physics are often reflected as achievements of theoretical physics, although their starting points were experimental results and questions of those times. The best-known is the discovery of a new kind of radiation by W.C. Röntgen in the year 1895, based on experiments with cathode rays. In 1891,

W. Hergert (✉)
Department of Physics, Martin Luther University Halle-Wittenberg, Halle, Germany
e-mail: wolfram.hergert@physik.uni-halle.de

W. Hergert and T. Wriedt (eds.), *The Mie Theory*, Springer Series in Optical Sciences 169, DOI: 10.1007/978-3-642-28738-1_1,

Fig. 1.1 Gustav Mie (29.9.1868–13.2.1957). [Archives University of Freiburg (UAF)]

Gustav Mie (1868–1957) (Fig. 1.1), then a 23-year-old postdoctoral research fellow, stood at the beginning of his scientific career. Besides the development of modern physics, changes in the structure of the institutions occured at that time. Cahan [1] discussed the institutional revolution in German physics in the years 1865–1914. The professional ethos of physicists at universities changed from pure teaching to research. The typical structure of a physical institute developed at German universities in the second half of the nineteenth century, accompanied by the construction of new and adequate buildings. Rostock and Heidelberg, were Mie studied, received new buildings for the physical institutes only after the turn of the century, whereas during 1890–1891, the buildings in Greifswald, Halle and Freiburg, where he later got a professorship, were ready for occupancy and operation. Gustav Mie's most creative years coincided with the time, when theoretical physics was established as an independent discipline [2].

Mie said of himself about his work:

> As far as I can adjudicate on me, it became my occupation to establish the connection between theoretical and experimental physics, which stood at that time in danger, to diverge more and more. I believe that my most important achievement was the publication of a text book on electricity, in which I, above all, show Maxwell's differential equations, which to an outsider appear quite abstract, are nothing else as the shortest and most exact formulation of experimentally found laws, by which electrical and magnetic fields are linked with one another.[1]

[1] [Mie81], p. 738f.

First, the stages of Mie's scientific career are reviewed. His scientific work is then highlighted in the second section. The connection between experimental and theoretical physics was very important for him. The following discussion will be concentrated on his theoretical work. Classical electrodynamics, field theory, theory of relativity and the quantum theory are his main fields of research. The work on the theory of matter and the contributions to the theory of the scattering of electromagnetic waves are particularly important. Gustav Mie's role as a lecturer along with a description of his disciples is the main theme of the next section. Mie's publications document also his desire, to present modern physical developments to a broader audience. These activities are treated together with his role as a lecturer.

The scientific achievements and conflicts of his long life were also overshadowed by two world wars and the time of the National Socialist movement. In this context, his reflections of social developments will also be discussed, as far as they can be concluded from his diaries, letters and secondary sources. I attempt to list a bibliography of the publications of Gustav Mie, as completely as possible. A bibliography of biographic notes on Mie is also provided.

Biographic details are specified in footnotes for persons who were important to Mie's scientific career and life; otherwise only biographical data is given. The origins of the figures and the photographs which did not originate from the author are explicitly cited.

1.2 Life and Scientific Career

1.2.1 Childhood in Rostock

Gustav Adolf Feodor Wilhelm Ludwig Mie was born on 29 September 1868 in Rostock (Fig. 1.2) to Hans Friedrich Ernst Amandus Mie (1828–1906), a salesman active in the insurance business and his wife Caroline, nee Ziegler (1834–1901). His parents were both children of Prostestant pastors. Many of his Mecklenburgian relatives were Protestant pastors.[2] His uncle, Fedor Mie (1840–1905), worked as a Protestant pastor at St. Petri in Rostock.

Gustav Mie spent his childhood and youth in Rostock. He had three brothers and one sister. His brother Friedrich (1865–1911) became a high school professor in Berlin. His brother Johannes Mie worked as a merchant in Hamburg. His brother Amandus carried on with the family tradition and worked as a Protestant pastor in Scharnebeck.[3] Gustav Mie attended High School at the Great City School (Große Stadtschule) in Rostock (Fig. 1.3), an educational institution rich in tradition, from

[2] The origins of the branched family can be followed up in [3].

[3] H. Spehl, Mie, Gustav Adolf Feodor Wilhelm Ludwig in: [Bio6] p. 186ff., [4], [Mie81], p. 734.

Fig. 1.2 Mie's birthplace (house on the corner), Wokrenter Strasse 35 in Rostock. The house was all but destroyed in the war; however, it was carefully rebuilt in 1981–1982

1877 to 1886.[4] The Christian Protestant family tradition instigated his desire for studies in theology after his Abitur. During high school, his interest shifted from theology to natural sciences. In his memoirs, he wrote about this change:

> It was not until I was a sixth former that I got the idea of becoming a natural scientist and mathematician. Even today, I can recall the blessedness I felt when this idea came to mind and my parents agreed upon it. I have never been happier in my whole life than during those days and it felt like I was wandering around in paradise. The fact that I took the exam in Hebrew for my Abitur in 1886, typical of someone aiming at studying theology, proved how late I made my decision.[5]

However, it was not all plain sailing. Mie's brother Friedrich struggled to leave his career as a Protestant pastor imposed on him by his parents. In the year 1886, Mie received his school leaving certificate (Abitur). In the Abitur certificate (cf. Fig. 1.4), he was assessed to have a commendable behaviour. His diligence and attention was reported as, "in former times [as] insufficient, in the last terms good and very good". He finished the subject Mathematics with the grade "very good". "In Mathematics, he demonstrated ...good knowledge. Solving problems

---

[4] The Great City School was installed in 1580 under the superintendence of the humanist Nathan Chyträus (1543–1598) in the former Dominician monastery St. Johannis. In the years 1864–1867 the new classicistic building was built by the Rostock architect Th.H. Klitzing. Between Easter 1860 and Michaelis 1901, 7 pupils named Mie completed the Great City School [5].

[5] [Mie81], p. 734.

Fig. 1.3 The Great City School in Rostock

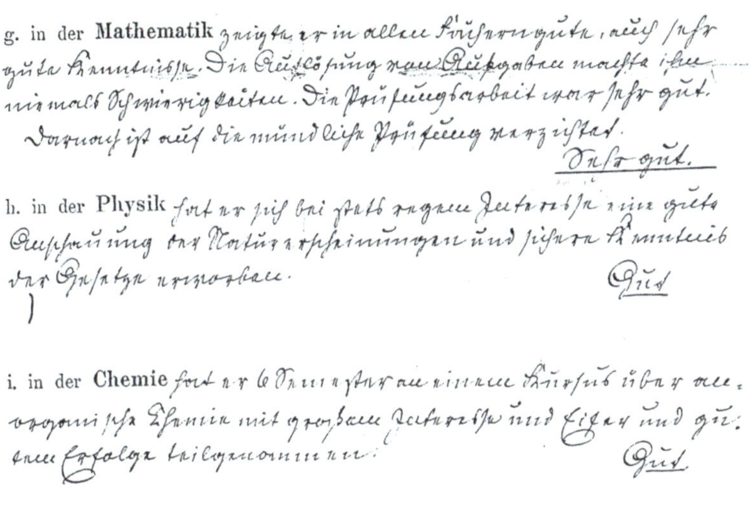

g. in der **Mathematik** zeigte er in allen Teilen gute, auch sehr gute Kenntnisse. Die Auflösung von Aufgaben machte ihm niemals Schwierigkeiten. Die Prüfungsarbeit war sehr gut. Darnach ist auf die mündliche Prüfung verzichtet. Sehr gut.

h. in der **Physik** hat er sich bei stets regem Interesse eine gute Auffassung der Naturerscheinungen und sichere Kenntnis der Gesetze erworben. Gut

i. in der **Chemie** hat er 6 Semester an einem Kursus über anorganische Chemie mit großem Interesse und Eifer und gutem Erfolge teilgenommen. Gut.

Fig. 1.4 Part of Gustav Mie's Abitur certificate. (UAF, Nachlass Mie, C136/1—Zeugnisse)

never caused difficulties for him. The exam results were very good. Therefore, he was excused from the verbal examination".[6] He also achieved a good grade in Physics. The school leaving certificate states: "In Physics he always had keen interest, a good understanding of natural phenomena and acquired a good

[6] Archives University of Freiburg (UAF), Nachlass Mie, C136/1—Zeugnisse.

knowledge of the physical laws".[7] In Chemistry, Hebrew, Greek and History he also achieved good grades, while he was not as successful in the foreign languages English and French, and neither was he very successful in his native German.

1.2.2 Study in Rostock and Heidelberg (1886–1890)

Mie completed his studies at the university in his hometown Rostock as well as in Heidelberg. He first enrolled in Rostock in the winter semester of 1886–1887. After three terms, he moved on to Heidelberg in the summer of 1888, "to become acquainted with something else" as he writes in his memories.[8] It might be, that also the Rostock mathematician Martin Krause (1851–1920), a disciple of Leo Königsberger (1837–1921) the prominent mathematician in Heidelberg at that time, contributed to his decision. The summer term of 1889 saw him back in Rostock. He finished his studies in Heidelberg, where he spent the remaining time from the winter semester 1889–1890 up to the summer semester 1890.

Initially, he placed emphasis on chemistry, geology and mineralogy. In the winter semester of 1886–1887, 18 students were enrolled in chemistry at Rostock University.[9] Chemistry was lectured by Oscar Jacobsen.[10] Chemistry in Rostock had a good reputation at the time. Mie attended lectures on mineralogy and geology given by Eugen Geinitz.[11]

Prior to the beginning of Mie's studies in Rostock, mathematics, physics, geology and astronomy were represented by Hermann Karsten,[12] born into an old Mecklenburgian family, which produced a number of scholars. Karsten was, at the same time, director of the Navigation School. Only late, in contrast to other German universities, did an independent chair for physics become installed in Rostock. The first full professor was Heinrich Matthiessen[13] who started at 1873.

Physics was housed in the same building as chemistry for a long time. As a result of considerable extensions to the university available in the years 1870–1890 a

---

[7] ibid.

[8] [Mie81] p. 735.

[9] [6], p. 1003.

[10] Oscar Jacobsen (1840–1889), study in Kiel, 1865 assistant at the Chemical Laboratory of the University of Kiel, 1868 graduation, 1871 private lecturer, 1873 professorship in Rostock, see also [7].

[11] Eugen Geinitz (1854–1925), study in Dresden and Leipzig, 1876 graduation Leipzig, 1877 habilitation Göttingen, private lecturer in Göttingen and Heidelberg, 1878 professorship in Rostock. Geinitz is the father of the geology of Mecklenburg (Book "Geologie Mecklenburgs").

[12] Hermann Karsten (1809–1877), studies in Bonn, Berlin and in Königsberg with Bessel, 1830 private lecturer in mathematics and mineralogy in Rostock, 1832 extraordinary professor, 1836 professor.

[13] Heinrich Friedrich Ludwig Matthiessen (1830–1906) studies of natural sciences in Kiel with G. Karsten, grammar school teacher in Jever and Husum, 1873 professor in Rostock.

Fig. 1.5 *Left* View of the old Institute of Physics in the year 1885 (still without the front extension). On the back of the original photo Prof. Matthiessen noted the following in August 1885: "The Physical Institute—1885 photo taken by Dr. phil. Mönnich, Privatdozent, and given to the Institute. The persons presented on it, from left to the right, are: Institute servant Maass, cand. König, stud. Black (IInd assistant), cand. Karnatz (assistant) and teacher Klingberg", *Right* Interior of the old Institute of Physics, persons from left to right Dr. Groesser, Prof. Matthiessen, Dr. Brüsch and Maass ([9], pp. 96 and 99.)

building within the yard of the main building became available for physics, which had already served as a chemical laboratory from 1834 to 1844 [8] (see Fig. 1.5). This was the situation the student Gustav Mie encountered at the physical institute.

In the year 1879, the mathematical–physical seminar was established; the first directors were physicist Hermann Matthiessen and mathematician and astronomer Johann Martin Krause.

Paragraph 1 of the statutes of this seminar reads as:

> The mathematical–physical seminar has to give the students suggestions and guidance for independent investigations and free lectures in the field of abstract mathematics and in mathematical physics. Subject to further regulations, the control of this institute will be transferred to the professors of mathematics and physics under the superintendence of the Ministry, Department of Education. Independent from each other they offer topics for small and larger written work and free lectures and give suggestion and guidance to the members for execution of the tasks. The details of the contents and scheduling of the seminar-like exercises are left to the directors.[14]

The directors could nominate three students for one award per term. Mie was awarded in the year 1889.[15] Chemistry and physics, however, quickly slipped into the background of Mie's studies.

In the middle of the nineteenth century, Heidelberg was oustandingly represented in physics, chemistry and mathematics. Chemist Robert Bunsen (1811–1899) taught there from 1852 until retirement in the year 1889. Hermann von Helmholtz (1821–1894) a versatile physiologist and physicist, held the chair for physiology in Heidelberg from 1858 to 1870. Gustav Kirchhoff (1824–1887)

---

[14] [10], p. 223.

[15] [11], p. 828.

Fig. 1.6 *Left* Leo Königsberger (1837–1921), mathematician, *Right* Harry Rosenbusch (1836–1914), mineralogist and petrographer. (Courtesy of UAH)

was a full professor for physics in Heidelberg from 1854 to 1875. His successor was Georg Hermann Quincke (1834–1924). Mathematics was also outstandingly represented in Heidelberg. Following Ludwig Otto Hesse (1811–1874), Leo Königsberger[16] taught for a short period between 1869 and 1874 (Fig. 1.6). Succeeding Immanuel Lazarus Fuchs (1833–1902), Königsberger taught from 1884 to 1914 for the second time in Heidelberg. In 1869, a mathematical–physical seminar was established at Heidelberg university, with Kirchhoff and Königsberger as the first directors.[17] In the year 1890, the subjects physics, chemistry, botany, zoology, mineralogy, mathematics and agriculture were separated from the Philosophical Faculty and combined to form a new Scientific-mathematical Faculty. The physicist Quincke was the first dean and the mathematician Königsberger became his deputy.

In spite of Quincke and Bunsen being outstanding scholars in physics and chemistry, Mie continued to concentrate on mathematics and mineralogy.

[16] Leo Königsberger (1837–1921) Professor in Greifswald, Dresden Vienna and Heidelberg, work on the theory of elliptical and hyperelliptical integrals and complex differential equations, Helmholtz biography, see also [12].

[17] Statutes of the seminar in, Archives of University of Heidelberg (UAH), Fak. -Akte H-IV-102/71, Nr. 78, fol. 75–76, cf. [13]. Statutes of such seminars are regularly printed in "Zeitschrift für mathematischen und naturwissenschaftlichen Unterricht", for example in vol. **5**, 173 (1874) the statutes of the mathematical–physical seminar of Halle University can be found.

One reason for this choice may have been the absence of lectures in theoretical physics. Mie acquired his knowledge in theoretical physics mainly by self-instruction. It may be, due to his engagement in mineralogy, that Mie was not a regular member of the mathematical–physical seminar; only students having mathematics and physics as their main subjects could be regular members.

In Heidelberg, the mathematician Leo Königsberger and the mineralogist and petrographer Rosenbusch[18] (Fig. 1.6) were his academic teachers. Königsberger was a excellent academic teacher. A student reported on his lectures:

> Königsberger performed with virtuoso control of the material rapidly, clearly, dragging the listener along. His fresh, self-confident, strong nature, as well as his kindness and fairness secured him the affection of the academic youth.[19]

Mie completed the summer semester of 1889, once again in Rostock, and concluded his study in the summer of 1890 in Heidelberg. The emphasis of his study was mineralogy and mathematics. His preoccupation with mineralogy was so intense that Harry Rosenbusch, the founder of the systematic petrography, employed him as a junior research assistant. Mie's task was to organise and catalogue the hand piece and thin section collection consisting of 7,000 pieces.

Leo Königsberger remembers his student Mie in his memories. He writes:

> My lecturer activity ... turned out well beyond all expectation, excellent young men, who later occupied many chairs of mathematics, physics, astronomy at German and foreign universities, like Ph. Lenard, M. Wolf, G. Mie, K. Boehm among others, whom I call my scholars.[20]

Max Wolf (1863–1932) later was the director of the National Observatory of Baden in Heidelberg and Phillip Lenard (1862–1947), Nobel Laureate of the year 1905, occupied the physics chair in Heidelberg.

Gustav Mie registered for the state examination of the higher teaching profession in 1890. He had to take the exam at the Technical University of Karlsruhe. He requested the examination in mathematics and physics as major subjects, and took mineralogy and chemistry as minor subjects. He also took an exam in religion through his own volition. His topic in physics was "to deduce, from the laws of induction, the basic principles necessary for the construction of dynamo-electric machines, as well as it's mode of action in a mathematical way". He worked on this topic with such enthusiasm and interest, leaving himself only 24 h to write his report on the mathematical topic. The situation was reflected in the assessment on his mathematical work:

> The written examination in mathematics "On the geodetic lines on the surfaces of second order", even if the developments in the individual points could be more detailed and more

[18] Karl Heinrich Rosenbusch (1836–1914), professor for petrography and mineralogy at the universities of Strasbourg and Heidelberg, 1903 Wollaston medal, cf. also [14, 15].

[19] Report of the student O. Rausenberger in [16], p. 179.

[20] [12] p. 187.

clearly expressed and also, the order and care of elaboration could be better, but overall, the contents and conception of the task can probably be rated sufficient.[21]

This is in contrast to the judgement of Otto Lehmann (1855–1922) with respect to the written examination in physics:

The treatment of the topic: "Mathematical deduction of the the basic principles of construction of dynamo-electric machines from the laws of induction" proves that the candidate applied himself to a level of excellence in particularly difficult theories of electricity and is able to move and independently proceed with full certainty in this area. It is acknowledged that the resulting work is be rated as excellent".[22]

In the summer of 1891, Mie attained a doctorate with the thesis "On the Fundamental Principle of the Existence of Integrals of Partial Differential Equations". In the personal record for the thesis he wrote:

During his study time he [the author] visited the lectures of the following Mr. professors and associate professors: Krause, Matthiessen, Geinitz, Jacobsen, Königsberger, Cantor, Schapira, Rosenbusch, Osann, Goldschmidt, Andrae. To all his admired teachers, most of all Mr. Geh. Rat Königsberger and Mr. Geh. Bergrat Rosenbusch, the author expresses devoted thanks [Mie1].

1.2.3 Karlsruhe: Beginning of the Scientific Career (1892–1902)

After graduation, the pursuit of a scientific career seemed unlikely. Mie found an employment in Dresden at a private school in summer 1892.[23] Anyway, Mie sent a copy of his thesis to Otto Lehmann, his examiner in physics in the state examination, and was offered an assistantship at the Technical University of Karlsruhe. Otto Lehmann[24] was the successor of Heinrich Hertz (1857–1894), who was active from 1885–1889 at the TH Karlsruhe (Fig. 1.7). Hertz had accomplished his innovative experiments on the propagation of electromagnetic waves at that time. His successor Lehmann became famous for the investigation of liquid crystals.

Gustav Mie had to lead the physical practical course[25] and used this in order to gain appropriate knowledge in experimental physics. Furthermore, he had to

---

[21] UAF, Nachlass Mie, C136/1 Zeugnisse.

[22] ibid.

[23] From the copy of testimonials existing in the archives, it was revealed that this refers to the Müller Gelinek Six-form High School. This school had a long tradition and before the transformation into a private school on the 26 April 1819, it was the urban Friedrich August School.

[24] Otto Lehmann (1855–1922), Study Strasbourg, Teacher for Physics, Mathematics and Chemistry in Mulhouse (Alsace), 1883 Lecturer in Physics TH Aachen, 1888 Professor for Electrotechnology at the Polytechnic Institute Dresden, 1889 Ordinarius at Karlsruhe Technical University Fridericiana with teaching assignment for Physics and Electrotechnology.

[25] see Lehmann's report in [17], p. 84.

support Lehmann with the experiment lectures. The large number of students in Karlsruhe and the impression of the lectures of Kundt (1839–1894) during his study in Strasbourg brought Lehmann to dedicate himself to the intensive development of experiment lectures.[26] Regarding the experiments shown in the lectures of Lehmann it is said:

> Also today most experimental physicists prefer the demonstration of physical experiments in the large-scale setups, which could be noticed from all places of the lecture room. This style was consequently pursued by Lehmann. The enormous size of his experimental assemblies and the spryness of the experimental presentation led to largest acceptance and concentration among his students. To demonstrate the forces that could be produced by electric currents, Lehmann, according to the verbal report of a student, catapulted a lab assistant to the lecture room ceiling using current power. Although such experimental setups could not become completely accepted, they surely remained not without influence on the modern presentation of physical experiments. They have been certainly treasured by his students, who reported after the lecture grinning due to "Circus Lehmann".[27]

Lehmann's appointment to an extraordinary professorship of electrotechnology at the Polytechnic Institute in Dresden took place with himself quoted as saying, "without any intention of dedicating myself permanently to electrotechnology".[28] In the year 1895 in Karlsruhe the Electrotechnical Department was created and thus electrotechnology was separated from physics. At this time the technical universities Darmstadt and Hannover were prominent in the training of electrical engineers. Karlsruhe, although one of the latecomers with regard to the institutionalisation of electrotechnology, ascended in 1910 to become the prominent training centre for electrical engineers.[29] In the year 1892, an electrotechnical extraordinary professorship was awarded to August Schleiermacher (1857–1953), a former assistant of H. Hertz. Over three decades, from 1896 to 1927, he represented theoretical electrotechnics as professor ordinarius.[30] There was a friendly and collegial contact between Hertz and his assistant of only a year younger, Schleiermacher, recorded in Hertz's diaries.[31] Schleiermacher, as the first professor at a German technical university, conducted his lectures on theoretical electrotechnics based on Maxwell's theory. However, the appointment of Engelbert Arnold[32] in the year 1894 was considerably more important to the

---

[26] see [17], p. 88f., see also [18].

[27] [19], p. 108, private communication K.A. Turban 1881.

[28] see [17], p. 72.

[29] see: W. König, *Gewinner und Verlierer. Der Stellenwert der einzelnen Hochschulen im Institutionalisierungsprozess der Elektrotechnik in Deutschland 1882 bis 1914*, in [20], p. 171ff.

[30] In 1892 he was appointed as first budgetary professor for electrotechnics in Karlsruhe. On 11 July 1896 Schleiermacher became regular professor for theoretical physics in the electrotechnical department of Engelbert Arnold. In 1926 the chair was rededicated to theoretical electrotechnics after Schleiermacher's retirement. cf. [21].

[31] cf. [21], p. 25.

[32] Engelbert Arnold (1856–1911), 1874–1878 studies in Zürich, 1880–1891 Polytechnic Institute Riga, joint founder of the Russian-Baltic Electrotechnical Factory 1888, 1891 chief electrician of Maschinenfabrik Oerlikon, from 1894 Karlsruhe, see also [22].

Fig. 1.7 Engelbert Arnold (1856–1911), Otto Lehmann (1855–1922). (Archives Karlsruhe Institute of Technology, Karlsruhe)

success of electrotechnology in Karlsruhe (Fig. 1.7). Arnold focussed his work in the area of the construction of high performance, high-speed d.c. machines. This was the area which was treated by Mie in his state examination work. All in all, Mie found a scientific environment in Karlsruhe strongly influenced by the ideas and scientific work of J.C. Maxwell and H. Hertz. He did not follow Lehmann's work on liquid crystals; however, an exchange of letters with Lehmann continued until much later.

In the first Karlsruhe years, Gustav Mie trained himself increasingly in physics, where electrodynamics and their technical application became his preferred field, both in practical experiments and theoretical treatment. Suggestions on how to deal with the problems with the systems of units put forward by Lehmann who was himself preoccupied with such questions at the time.[33] Electrotechnical problems were permanently present due to Arnold's work. Mie also worked on electrotechnical problems in Greifswald, the next institution of his academic career. Interestingly, Mie published work on d.c. machines together with Arnold [Mie8].

With the work "Draft of a general Theory of the Energy Transmission" Gustav Mie attained habilitation on 29 July 1897.[34]

[33] O. Lehmann, *Das absolute Maßsystem*, Verh. des Naturwiss. Vereins zu Karlsruhe **12**, 365 (1897), excerpts in: Zeitschr. für den physikal. und chem. Unterricht **10**, 77 (1897).

[34] University and regional Library (ULB) Halle: Dissertationes polytechn. Karlsruhanae 1896–1899, No. 11, *Entwurf einer allgemeinen Theorie der Energieübertragung* von Dr. Gustav Mie, Habilitationsschrift zur Erlangung der venia legendi für mathematische Physik an der Großherzoglichen Technischen Hochschule in Karlsruhe. (Sonderdruck aus den Sitzungsberichten der K. Akad. der Wiss. Wien [Mie6]).

In the year 1900 Mie was appointed professor with the teaching assignment "Modern Trends and Opinions on Electricity and the Science of the Electrical Waves".[35]

In the scientific life of the university town Karlsruhe scientific associations were very important. Such an association was already created in 1840 by Alexander Braun (1805–1877), the director of the Naturalienkabinett at that time. Starting from 1862, the association arose as "Naturwissenschaftlicher Verein in Karlsruhe" with new statute. The "Naturwissenschaftlicher Verein Karlsruhe e.V." has presently been active for 150 years. The report on the meetings of the years 1888–1895 proves that 12 to 15 meetings per year took place. The mean number of listeners was approximately 30. In average, physical topics have been discussed five times a year. The professors and members of the Physical Institute participated actively in this association since its inception. Heinrich Hertz presented his experiments for the first time to the public during one of the meetings of this association. From 1885 to 1889 Hertz presented eight times. Thus, he lectured on 24 June 1887 on "New Relationship between Light and Electricity"[36] and on 22 February 1889 he gave a lecture on "Relations between Light and Electricity".[37] The original apparatuses used by Hertz were still present in the institute with Lehmann's assumption of office. Lehmann wrote:

> I came into possession of the physical cabinet which belonged to Hertz that contained several intrinsically worthless items, however, amongst them were, from a historical perspective, very precious apparatuses. There are in particular items, which in occasion of the International Electrotechnical Exhibition in Frankfurt a. M. in 1891 occupied an important place ... and will be later relinquished to the German Museum in Munich on its strong request ..."[38]

Otto Lehmann spoke regularly at the association meetings. Lehmann met with the other members for meetings mostly in the physics lecture hall, in order to be able to demonstrate experiments. The "Experimental Lecture on Röntgen's X-rays" presentation by Lehmann was held on 7 February 1896.[39] While Hertz and Lehmann participated regularly and actively in the meetings of the scientific association, Engelbert Arnold, a member of the association too, hardly contributed.

Gustav Mie became a member during the 432th meeting on 16 December 1892. He gave his first lecture on 5 January 1894, entitled "On the Nature of Heat". During his time in Karlsruhe Mie gave seven additional lectures in the meetings of

---

[35] [17], p. 84.

[36] Verh. des Naturwiss. Vereins zu Karlsruhe **10**, pp. 150–51 (1883–1889).

[37] ibid. **11**, pp. 41–43 (1883–1895).

[38] cf. [17], p. 71, in 1913 the Karlsruhe devices were inventoried in the German Museum in Munich. For the history of the devices see also: J.H. Bryant, *Heinrich Hertz's Experiments and Experimental Apparatus: His Discovery of Radio Waves and his Delineation of their Properties*, in [23], p. 39 ff.

[39] Verh. des Naturwiss. Vereins zu Karlsruhe **13**, p. 37 (1895–1900), "The lecture, which was given by the speaker, over 3 weeks on five occasions, for different corporations and associations, is printed in the papers"., see also [Mie4].

the scientific association of Karlsruhe. He spoke in the 488th meeting on 15 May 1896 on the topic of “On the Meaning of Hertz’ Mechanics” [Mie10].

The replication of the Hertz experiments with the original equipments still existing in the institute at the time made a lasting impression on Mie. He also used parts of the original Hertz equipment in the meeting of the scientific association on 16 July 1897 to report on “Telegraphing without Wires, with Experiments”.[40] The work crucial for Gustav Mie’s further career was his publication on the Lecher problem [Mie9]. From this work he was noticed, for the first time, as a theoretical physicist, and not as an electrical engineer.

On 14 March 1901, Gustav Mie and Berta Hess, daughter of Friedrich Hess, district surveyor in Heidelberg, got married. It turned out to be a childless marriage. However, Gustav and Berta Mie spent a long and fulfilled life together. On 3 January 1916 he wrote about himself and his wife:

> For an outsider, it is difficult to understand, how an inherently pedantic and quite scientific person, as I am, can be compatible with a person with an artistic nature and infused with fantasy, such as, like my wife.[41]

1.2.4 Greifswald: The productive Period (1902–1917)

From his theoretical work in Karlsruhe, in particular by the investigation of electrical waves on parallel wire systems [Mie9], Mie had already acquired a certain reputation. The attainment of an extraordinary professorship was thus the next logical step.

At the beginning of the twentieth century, Greifswald was a small Prussian university town at the Baltic Sea, not far away from Rostock, Mie’s native city. The university in Greifswald was founded way back in 1456.[42] In the years 1888–1891 the Physical Institute acquired a new building under the auspices of Professor Anton Oberbeck (1846–1900), well adapted to the needs of physical research of that time (Fig. 1.8). Prior to his full professorship in Greifswald, Oberbeck held a chair for theoretical physics in Halle. The University of Halle was one of the first German universities, establishing such a chair.[43]

The successor of Oberbeck in Greifswald was Franz Richarz (1860–1920). Richarz accepted a professorship at Marburg in 1901. The subsequent regulation paved the way for Mie to Greifswald. Carl König (1859–1936), who did not have a large impact in physics, was a professor extraordinarius in Greifswald since 1899. In the proposal by the faculty for Richarz’s successor, Hallwachs (1859–1922) (Dresden) and Dietrici (1858–1929) (Hannover) have been mentioned, but the

[40] Verh. des Naturwiss. Vereins zu Karlsruhe **13**, S. 98–103 (1895–1900).

[41] UAF, Nachlass Mie, E12/69.

[42] For the history of physics in Greifswald see [25–28].

[43] cf. [2] vol. 2, chapter 15, p. 33 ff.

faculty insisted in transfering the chair to König. In the course of this personnel decision the replacement of the professor extraordinarius should be settled at the same time.[44] The proposal list for the occupation of the extraordinary professorship covered Pringsheim (1881–1963) (Berlin), Elster (1854–1920) (Wolfenbüttel), Wiechert (1861–1928) (Göttingen), Mie (Karlsruhe) and Straubel (1864–1943) (Jena). Mie was set only on fourth place and the judgement over him was however farsighted and lead the Minister to entrust the position to a "coming man":

> Mie's work shows the fact that his mathematical skill is linked also with the ability of physical opinion and important achievements for the future are to be expected from him. His publications are however up to now purely theoretical in nature.[45]

The years in Greifswald were Mie's most scientifically fruitful time. To a considerable degree it fulfilled the expectation that accompanied his appointment. In the year 1905, König was relocated to Gießen. During the Richartz follow-up discussion the budgetary professor extraordinarius was still favoured by the faculty, but now the situation took a different turn. The philosophical faculty reacted to the transfer of König with a three person proposal: Rubens (1865–1922) (Charlottenburg), Dieterici (Hannover), Pringsheim (Berlin). Concerning Mie the faculty explicated:

> The present professor extraordinarius, Professor Mie has, by his scientific achievements and his personal integrity, acquired a respectable position and we would not hesitate to suggest him for the full professorship. However, it is our desire to put more emphasis on the experimental aspects of physics and a more fundamental form of presentation than we expect them to be given by Prof. Mie with his predominantly theoretical capabilities.[46]

The ministry insisted on a more detailed rationale, and in the long run Mie got the position. Both appointment procedures show that theoretical physics began to be established that time. On 17 September 1905, Mie became a full professor under the condition that he represented physics in its entire extent. At the same time he was appointed the director of the Physical Institute and the Astronomical–mathematical Institute. From winter semester 1836–1837 until 1887 the Astronomical–mathematical Institute belonged to the Mathematics department. The Astronomical–mathematical Institute was moved to Physics department during the directorship of A. Oberbeck in the winter semester 1887–1888. Lectures in astronomy were given by W. Ebert (1871–1916) from 1903 till 1905.[47] During Mie's directorship astronomy was abandoned. It was reactivated in 1922.[48] Gustav

---

[44] Archive of the University of Greifswald (UAG), Phil.Fak I290.

[45] ibid.

[46] UAG Phil. Fak. I 298.

[47] cf. A. Schnell, Gestrandet in Wien: Wilhelm Ebert (1871–1916), Acta Hist. Astr. **43**, 318 (2011).

[48] The building of the physical institute also contained an observatory. A modern astrodome on the tower can be seen on Fig. 1.8, cf. H. Kersten, *Astronomie in Greifswald* in [25], p. 51ff.

Fig. 1.8 The Physical Institute in Greifswald (built 1889–1891). A commemorative plaque is posted on the wall of the building. The text reads: "Professor Dr. Gustav Mie (1868–1957) cofounder of a uniform theory of particles and field, Greifswald 1902–1917", see also [24]

Mie was Dean of the Philosophical Faculty in Greifswald in the year 1912–1913, and was elected rector of the university on 15 May 1916 for a one-year period. He began his rectorship with the lecture "Law of Nature and Spirit" [Mie39].

In the year 1916, Mie received an offer to be the successor of Ernst Dorn in Halle. After some hesitation and repeated negotiations with the ministry he accepted the offer in December 1916. As candidates for his successor, he brought Johannes Stark (1874–1957) (Aachen), as well as in second place, *pari loco* Johannes Königsberger (1874–1946) (Freiburg) and Clemens Schäfer (1878–1968) (Breslau) to discussion. Mie had already become acquainted with Stark during his time in Greifswald where Stark was serving as a substitute professor, in the years 1907–1908, for professor extraordinarius Starke (1874–1960), who was later an ordinarius at the RWTH Aachen from 1917 to 1940. The faculty at the end followed Mie's proposal.

Inferred from his diaries, it was a happy time for Mie personally in Greifswald, despite difficulties due to the World War. During one evening on a sailing yacht Stark reported:

> He [Mie] had an assistant named Falkenberg, who possessed a beautiful large sailing yacht with a habitable space. It lay on the Ryck opposite Wieck. Once we spent an amusing evening on it. It had already become dark, when we rose to the deck, in order to go ashore. It followed Guschen, as he was referred to by Mrs. Mie, deep in thought as usual, he continued to rise when he fell into the water which was, fortunately, only as deep as his

height. Despite the unfortunate accident we had to laugh at the sudden disappearance of the great philosopher. We pulled him out rapidly and he laughed, too.[49]

1.2.5 Halle: Intermediate Station (1917–1924)

In 1890, at practically the same time as the Physical Institute in Greifswald, the Physical Institute of Halle University also got a new modern functional building. Also in Halle, special care was taken to set up a building that meets all the requirements of the physical research at this time [30]. Here, Ernst Dorn[50] acquired special merits. As successor of Oberbeck in Halle, he was considerably involved in the construction of the new institute.

After 1870, theoretical physics itself began to be institutionalised in Halle.[51] With Anton Oberbeck, Halle had a professor extraordinarius for theoretical physics starting from 1878. In 1884, Oberbeck received an offer to Karlsruhe, his position was subsequently converted to a personal chair in order to maintain a position for him in Halle. The vacant chair in Karlsruhe was occupied by Heinrich Hertz. Hertz had an offer from Greifswald, but preferred Karlsruhe due to the excellent conditions in comparison to those at the institute in Greifswald. Oberbeck's personal professorship was occupied by Ernst Dorn later on. After the death of the full professor of experimental physics, Carl Hermann Knoblauch,[52] Dorn took over his chair for experimental physics and became the director of the institute. The second position was downgraded to an extraordinary professorship for theoretical physics and transferred to Karl Schmidt.[53] This decision was unfortunate, to that extent, since Schmidt's field of interests had been more technical and applied physics than theoretical physics.

The institute in Halle was larger than the institute in Greifswald, had a higher reputation and even a higher remuneration. Theoretical physics, which was Mie's main focus, at least formally, was already institutionalised in Halle. Advancing this institute must have appeared as an appealing task to Mie, so much so that he accepted the offer from Halle. However, the hearings of the appeal were

---

[49] [29], p. 35.

[50] Ernst Dorn (1848–1916), studies in Königsberg, teacher in Königsberg and Berlin, 1873 habilitation in mathematics Greifswald, 1873 extraordinarius for physics in Breslau, 1881 professor in Darmstadt, 1886 professor in Halle as a successor of Oberbeck, 1895 director of the physical institute, see also [30].

[51] see [2] vol. 2, p. 33ff.

[52] Carl Hermann Knoblauch (1820–1895), studies in Berlin, 1845 cofounder of the Berlin Physical Society, 1848 habilitation in Berlin, 1849 professor in Marburg, 1853 professor in Halle, 1886–1871 rector of the University of Halle, 1878 President of the Leopoldina, see also [30, 31].

[53] Karl Schmidt (1862–1946), studies in Göttingen and Berlin, assistant in Strasbourg and Königsberg, 1889 habilitation Halle, 1895 extraordinary professorship for theoretical physics, 1912 personal tenured professorship.

complicated. The special position held by Schmidt, being the cause for conflicts over the next years, was a contributing factor.

During the regulation of the successor to Dorn, the faculty first mentioned Peter Debye (1884–1966), then Max von Laue (1879–1960) and in third place, Gustav Mie. As criteria for the selection the faculty stated:

> Conserving the existing ordinaries is of outmost importance for the faculty. In addition, we would like to engage a scientific personality capable of arranging the academic instruction together with the present full professor in a harmonic and versatile manner. Another decisive aspect is that the former contrast between experimental and theoretical research has vanished and the present full professor's effort has developed in the more experimental direction. Therefore the faculty agreed on personalities with focus on theoretical aspects without excluding experimental work.[54]

Debye and Laue were not to be attracted for Halle. Corresponding to the qualities of Mie the faculty noted:

> As a distant third choice following these two researchers we propose Gustav Mie, Greifswald. ...The number of the Mie's papers is not as large as those of the two physicists selected first, nevertheless the treated problems and the kind of their solution marks Mie as a researcher of outstanding gift, deep thoroughness and great originality.[55]

A large portion of the documents in the archive of Halle University from Mie's time in Halle reveals his conflicts with Karl Schmidt. Schmidt, who had large support within the faculty and from the chancellor, steadily tried to increase his influence in the institute. Mie's plans for the modernisation of the institute in Halle were thwarted by Schmidt.

During his time in Halle, Mie actively participated in the sessions of the so-called "Naturforschende Gesellschaft". This society was founded already in 1779. On 20 June 1918 he spoke about "The study of the physics of the inner atom by X-rays" and on 8 July 1919 about "The decay of atoms during radioactive processes".[56] Other scientific societies were also active in Halle. In 1917 the so-called "Hallescher Verband für die Erforschung der Mitteldeutschen Bodenschätze und Ihrer Verwertung" was founded. Mie was also interested in such activities. He was a member of this society until his departure to Freiburg. On 28 September 1917 he wrote:

> On Wednesday, there was a great occasion: Foundation of Halle's Federation for the Study of the Mineral Resources. I hope that this will result in my increased contact with the industrial circles, because I want to gain much knowledge of the technology. That's my largest desire.[57]

On 7 February 1917 Mie became member of the German Academy of the Natural Scientists Leopoldina. The photo (cf. Fig. 1.9) was filed along with his member documents.

---

[54] Archive of Halle University (UAH), PA11453: Peronalakte Mie 1916–1918.

[55] ibid.

[56] cf. Mitteilungen der Naturforschenden Gesellschaft zu Halle **5**, 53f. (1920), [32].

[57] UAF, Nachlass Mie, E12/71.

Fig. 1.9 Gustav Mie in 1917, when he became a member of the Leopoldina. (Courtesy of Archive of the Leopoldina, member documents Gustav Mie, member number 3412)

Although his theory of matter was conceived in Greifswald, the discussion and exchange of letters about field theory and general relativity with Einstein mainly occurred during his time in Halle. Mie's time in Halle was scientifically fruitful, although he could not realize all his ideas for the development of the institute. On the other hand, Halle was intellectually exciting for him. Mie and his wife regularly visited the theatre and concerts. He visited the art gallery near the institute building, but did not get acquainted with expressionism, a main collection area of the gallery. He met also "old Rostock guys" like the philologist K. Brockmann (1868–1956) and the botanist G.H. Karsten (1863–1937). Both professors also received their Abiturs from the Great City School in Rostock. While in Freiburg, on 26 October 1924, Mie wrote about that time in Halle:

> I think back at Halle, as a time of my highest mental stimulation, in contrast to the enormously rich industries and crowds of dissatisfied workers living in darkness, to the mental contrasts of men of the industry, completely oriented on power and will, and the religious powers, before with which I had never become acquainted. It was the most interesting, spirited interplay of brightness and darkness, although somewhat though on

> the nerves at the time. Now, I withdrew myself here into peace, because I still want to muster my strength, in order to produce something.[58]

1.2.6 Freiburg: Ordinarius and Retirement in Baden (1924–1957)

Gustav Mie did not fully assimilate with Halle, probably due to the conflicts with K. Schmidt. The political riots at the beginning of the 1920s in central Germany surely was another reason for him to look for a position at a respected university. His positive memories of the first years of his scientific career in Karlsruhe and his wife's family ties made a return to Baden seemed reasonable. In a letter to Sommerfeld he wrote:

> I have to say that I am looking forward not only to leaving here, but also to being at the beautiful Freiburg institute and returning to the beautiful Baden country, with which I am so familiar.[59]

In 1895, Franz Himstedt[60] became a full professor in Freiburg succeeding Emil Warburg,[61] who went to Berlin. Wolfgang Gaede's[62] development of pumps was of crucial importance to Himstedt's work on channel rays, alpha-rays and X-rays. As with the appellate procedures in Greifswald and Halle, Mie was not the faculty's favoured applicant, however, the reasons in this case were different. In the report of the selection to the Faculty of Sciences of the University of Freiburg from 25 April 1923 J. Zenneck (1871–1959) (Munich), Chr. Füchtbauer (1877–1959) (Rostock), Cl. Schäfer (Marburg) and W. Gaede (Karlsruhe) have been mentioned.[63] The list finally handed to the ministry included: Zenneck, Schäfer and Füchtbauer. Zenneck rejected the offer, and thus the ministry allowed the faculty to supplement the list. In a letter of the faculty to the rector and senate for submission to the ministry, now Mie was mentioned *primo loco*. It was written:

> We make use of the opportunity granted by the ministry, because it came to our attention, that Professor Gustav Mie (Halle), one of our most prominent physicists, may be inclined

[58] UAF, Nachlass Mie, E12/72.

[59] UAF, Nachlass Mie, E12/37 Nr.18.

[60] Franz Himstedt (1852–1933), studies in Göttingen, graduation and habilitation University of Göttingen, 1878 private lecturer Göttingen, 1880 private lecturer Freiburg, 1889 successor of Röntgen in Gießen, 1895 professor in Freiburg.

[61] Emil Warburg (1846–1931), studies in Heidelberg and Berlin, 1867 graduation, 1870 habilitation, 1872 Strasbourg, 1876 professor in Freiburg, 1894 professor in Berlin, 1905 director of the Physikalisch-Technische Reichsanstalt.

[62] Wolfgang Gaede (1878–1945), studies in Freiburg, 1901 graduation, 1909 habilitation in Freiburg, 1913 professor in Freiburg, 1919 professor in Karlsruhe.

[63] UAF, Generalakten, Bestand B1/1287.

> to accept the offer from Freiburg. We did not mention him before, because we considered it unlikely that he might be willing to accept our offer, however, it seems like he did not get assimilated in Halle and would like to come to South Germany. Prof. M. Wien, Jena, writes about him: "Among the living German physicists, I consider him among those of highest rank, not only as a theoretician, but also as an experimentalist". Not only is he an outstanding researcher, but also a very successful teacher, as proven by numerous experimental and theoretical theses accomplished under his supervision. His style of presentation is plain and clear. Mie is a subtle, highly educated man and a pleasant, kind colleague.[64,65]

Wilhelm Kast (1898–1980), Mie's assistant in Halle followed him to Freiburg.

Mie's years in Freiburg were associated with a number of honours. In 1925, his first station in his scientific career celebrated the 100th anniversary of the foundation. On this occasion Schleiermacher initiated the award of the honorary title "Doctor Engineer" to Mie. Schleiermacher pointed out in his report: "...his habilitation about the general validity of Poynting's theorem of energy flux, prepared in Karlsruhe, showed the direction of research, pointing to general and basic questions".[66] Gaede added in his report: "This outstanding and profound scientist stands out from other physicists by having a lot of sympathy for technical issues and proven this on several occasions".[67] On his 70th birthday the issue 3 of volume 425 of "Annalen der Physik" (1938) was dedicated to Mie; it also contained papers by his disciples. In 1938, G. Mie and A. Sommerfeld became honorary members of the German Physical Society.

Mie was awarded with the Goethe-Medal for Art and Science on his 75th birthday in November 1943. The medal was given to artists, scientists and others as authorised by Hindenburg. In 1934, Adolf Hitler, in his position as German Head of State, took over the authority for awarding this medal. Kast, Mie's former assistant in Halle and Freiburg started the activities to effect this award.[68] One of the main arguments of the referees is the seminal influence of Mie's textbook [Mie37].

The building of the Physical Institute in Freiburg was destroyed during the Second World War. In honour of the eminent physicist, the new building was named "Gustav Mie Haus".[69]

Apart from the obligations at the institute, Mie took part in the scientific life of the university town Freiburg, as he usually did. As in Karlsruhe and Halle, Freiburg also had a lively scientific society, so-called the "Naturforschende

[64] ibid.

[65] Dissertations with purely theoretical content supervised by Mie could not be found.

[66] UAK, Best. 27059, Sig. 1, 385.

[67] ibid.

[68] cf. UAF, B24/2448.

[69] In honour of Mie's time in Halle the new main physics lecture hall of Halle University was named "Gustav-Mie-Hörsaal". Also, a so-called Mie prize is awarded to excellent students of physics in Halle.

Gesellschaft zu Freiburg im Breisgau", founded 1821. On 16 December 1925 Mie spoke about fundamentals of the quantum theory [Mie62].

Mie was active right to his very last semester. He retired in June 1935. On 1 February 1935, Mie wrote to Sommerfeld:

> ...In this winter semester I gave lectures for the gentlemen of my institute and of the physical-chemical institute about your handbook article about "Electron Theory of Metals". Eventually I will continue in the next semester, to speak also about the further development of the theory corresponding to the article of Bethe. I guess, we spoke about that in Pyrmont. I have had a lot of fun with this matter.[70]

After retirement, epistemological and theological questions became the focus of his attention. He published several papers on these subjects.[71] [Mie78, Mie80, Mie83] In honour of Mie's 85th birthday, a colloquium was organised by his former students and assistants on the 1 November 1953. Berta Mie died in 1954 and Gustav Mie passed away on 13 February 1957 at the ripe old age of 89.

1.3 Scientific Fields of Activity

The name of Gustav Mie is surely most frequently mentioned in connection with the term "Mie scattering" and "Mie effect". His contributions to the development of field theory and general theory of relativity are acknowledged in many reviews [33, 34]. Other scientific contributions of Gustav Mie are still of great interest.

In a certain manner, Mie's scientific papers are standard examples in the framework of the sociology of scientific knowledge. Corresponding to R.K. Merton the following thesis holds: "all scientific discoveries are, in principle, multiples, including those that, on the surface, appear to be singletons".[72] Stigler's law of eponymy states that the naming of an effect, a law, an equation rarely follows the name of the discoverer.[73] The scientific community has to be prepared also for the breakthrough of a new theory. "If an early valid statement of a theory falls on deaf ears, and a later restatement is accepted by the science, this is surely proof that the science accepts ideas only when they fit into the then-current state of science".[74] Mie scattering can be discussed in such a context. In the *Dictionary of the named Effects and Laws in Chemistry, Physics und Mathematics*, Mie does not appear as associated with Mie scattering, instead, the "Mie-Grüneisen equation of state" is cited.[75]

---

70 Letter of Mie to Sommerfeld from 1 Februaray 1935, UAF, E12/37, Nr.23.

71 The value of those articles should not be discussed here. It was a subject of controversy among the editors of the planned Mie edition Höhnl and Plötze. cf. UAF, Nachlass Hönl E14/2, see also review to Mie's "Naturwissenschaften und Theologie", Die Naturwissenschaften **20**, 566 (1932).

72 [35], p. 298.

73 Stigler's law of eponymy. [36], cf. also [37].

74 [38], p. 146.

75 [39], p. 212.

Table 1.1 Distribution of Gustav Mie's publications on the stations of his scientific career and on his scientific fields of activity

Period	Station	Publications
1892–1902	Karlsruhe	12
1903–1918	Greifswald	33
1919–1924	Halle	14
1925–1936	Freiburg	21
1937–1950	Retirement	7
Field of activity		Publications
Field theory, theory of relativity		17
Electrodynamics, electrotechnics		10
X-rays		8
Thermodynamics, statistics		7
Quantum physics		6
Mathematics		4
Scattering theory, optics		3
Textbook /textbook contributions		4
Philosophy, religion		7
Popular science		8
Biographic contributions		4
Miscellanea		9

A list of Mie's publications is given as completely as possible in the appendix. Table 1.1 gives an overview of the distribution of his publications at the institutions of his scientific career. Furthermore, the attempt was made to sort his publications corresponding to the subject. Table 1.1 clearly shows that the Greifswald years have been the most successful period. The papers concerning scattering theory [Mie25] as well as the three publications concerning the theory of matter [Mie32, Mie33, Mie36] were written in Greifswald. Also, the beginning of the discussion with Einstein about the basics of the theory of general relativity dates back to that time [34].

A revision of Mie's publications shows that he appeared as the coauthor of an article only on a few occasions. In Karlsruhe, he published a paper with E. Arnold [Mie8] and in Halle with J. Hergweg [Mie53]. Two papers in Freiburg were about roentgenographic investigations of polymeric formaldehyde. The investigations were conducted in collaboration with Hermann Staudinger (1881–1965), later a Nobel Laureate in Chemistry [Mie68, Mie69]. The subject of the investigation was the proof of existence of the fibroid structure. This verification was fundamental for later work of Carothers (1896–1937) to develop fully synthetic fibres. In 1931, he published a paper with E. Frankenberger about the measurement of refractive indices [Mie73].

Some of Mie's areas of scientific interest will be discussed in more detail now. At the beginning, some of his early and nowadays somewhat unknown publications will be considered. His work on electrotechnics and electrodynamics, field theory and the theory of relativity and of course on scattering theory will be discussed in more detail.

1.3.1 Transport of Energy and Thermodynamics

Two publications will be mentioned at first, dealing with problems of fundamental nature. The findings of those papers became common property during the development of physics. Therefore the papers are rarely cited nowadays.

Gustav Mie's paper, produced during his habilitation, "Draft of a General Theory of the Energy Transmission" [Mie6] is one of his less known publications. The development of Maxwell's theory and especially Poynting's work had shown that the transport of energy is not restricted to electric conductors. This changed point of view led to the reconsideration of the term energy. In particular H. Hertz (1857–1894) called for additional investigations in this field.[76] Mie's habilitation was one step in this direction. Max Jammer (1905–2010) writes about this work:

> Mie showed, in this publication, whose importance for the development of modern physics can hardly be overestimated, that not only a fluid under pressure transports energy, transferring it in relation to pressure and velocity, but that every deformation of an elastic body causes a current of energy, that can be exactly determined. ...However, Mie was not able to draw the final conclusion. Matter and energy were still two different aspects of physical reality for him.[77]

It should be mentioned that this work is recently discussed in the framework of problems of scientific education in schools [41].

Methods for a direct investigation of the interaction of atoms were still a long way from being feasible at the beginning of the twentieth century. Therefore, the influence of interatomic interactions on macroscopic properties was studied starting from a number of assumptions. This is also the way of thinking presented by Gustav Mie in his paper "On the kinetic gas theory of single atomic bodies" [Mie13]. This paper is a fundamental contribution to the microscopic theory of solids. Eduard Grüneisen (1877–1949) wrote in his article "State of Solids"[78] for the "Handbuch der Physik":

> In his fundamental work about the kinetic theory of single atomic bodies G. Mie introduced the assumption, that the potential of attractive as well as repulsive forces between two atoms, is inversely proportional to the distance of the atoms. Hence, the potential energy of two atoms at distance ρ is given by

$$\phi = -\frac{a}{\rho^m} + \frac{b}{\rho^n}, \tag{1.1}$$

[76] H. Hertz, *Untersuchungen über die Ausbreitung der elektrischen Kraft*, Gesammelte Werke, Leipzig 1894, Bd. 2, S. 234 und S. 294 and G. Helm, *Die Energetik nach ihrer geschichtlichen Entwicklung*, Leipzig 1898.

[77] [40], p. 187f.

[78] [42], E. Grüneisen: Zustand des festen Körpers, p. 1 ff.

where the first term represents the attractive and the second term the repulsive force. a and b are constants.[79]

Mie considered [Mie13] mainly single atomic gases, but Grüneisen [43] extended the field of application of the potential to solids. Both authors did not speculate about the origin of the forces. Experimental results for the coefficients m and n are given for example by Fürth [44]. Usually the Mie potential is written in the form

$$V(r) = D_0\left\{\frac{n}{m-n}\left(\frac{\sigma}{r}\right)^m - \frac{m}{m-n}\left(\frac{\sigma}{r}\right)^n\right\}. \tag{1.2}$$

The Mie potential is a precursor of the Lennard–Jones potential, which fixes the exponents to $m = 12, n = 6$. The Lennard–Jones potential is written as

$$V = 4\varepsilon\left\{\left(\frac{\sigma}{r}\right)^{12} - \left(\frac{\sigma}{r}\right)^6\right\} = \varepsilon\left\{\left(\frac{r_m}{r}\right)^{12} - 2\left(\frac{r_m}{r}\right)^6\right\}. \tag{1.3}$$

The 12–6 form was introduced by Lennard–Jones (1894–1954) in 1924 [45–47]. The Mie potential witnessed a renaissance as a model potential during the last years. The solution of the Schrödinger equation with a Mie potential was discussed [48–52]. The potential (1.2) was used in [53] to describe the interaction of He with a Cu (110) surface, in relation to He scattering. Thermoelastic properties have also been investigated using such a potential [54, 55]. Barakat et al. [56] used the Mie potential to calculate the binding energy of metallic nanoparticles. Potentials of the form (1.2, 1.3) are named sporadically also Mie-Lennard–Jones potentials.[80] The Mie-Grüneisen equation of state follows from the assumptions of the power function form of the interaction and the corresponding potential.[81]

1.3.2 Electrotechnics and Electrodynamics

Electrotechnics and electrodynamics were Mie's preferred areas of interest during his time in Karlsruhe as well as at the beginning of his Greifswald years. His investigation of electromagnetic waves along a wire are connected to the work of Heinrich Hertz. The impression Hertz's work had on the young Gustav Mie whilst in Karlsruhe was very significant.

Ernst Lecher (1856–1926) published his paper on the investigation of a system of parallel wires in 1890 [58]. Arnold Sommerfeld (1868–1951) published his theoretical investigation about the propagation of electromagnetic waves along a

[79] ibid, p. 11. Grüneisen refers in the discussion on Mie's paper [Mie13] and his own article published considerably later [43].

[80] z.B. W.C. de Markus, *Planetary Interiors*, S. 441 in [57].

[81] s. [42], p. 22, [Mie13], [43].

wire in 1899 [59]. In the year 1900, Mie's article containing the complete theoretical solution of the Lecher problem appeared [Mie9]. Mie gave a complete mathematical analysis, a style we encounter later in his treatment of the scattering problem.

Sommerfeld answered on the consignation of the article on the parallel wire system by Mie:

> Dear colleague, Many thanks for the consignment of your treatise. It was, as you can imagine, exceptionally interesting for me. I am personally not only glad about the scientific progress, but also that my wire waves got such a competent and thorough reader. ...You have done an ample piece of work![82]

Mie answered immediately:

> Dear Professor, I have been very glad about your friendly letter and your appreciative words. In fact, I finished the laborious task, because I was almost completely preoccupied by the task for 3/4 of a year, insofar as when lectures or practical course work did not require my attention. Hence, it is in a certain manner a bit regrettable, that the Fourier series as a result of complex calculations can be truncated in case of air wires almost always after the first term....[83]

This particular investigation established Mie's reputation as a theoretical physicist. During the years 1900 and 1901, Mie exchanged ideas on wire waves with Sommerfeld several times. Soon after, there was to be a further development in this area. Mie's solution was rigorous and was based on Maxwell's theory, but due to the series expansions used by Mie, the results appeared in the form of confusing formulas. Sommerfeld wrote: "But—if you want to open a lock, you have to use a key and not a hatchet. The power series expansion [in Mie's paper] appears in this case as a hatchet, not as a proper key" [60]. D. Hondros, a Ph.D student of Sommerfeld's, investigated in his theses a further development of the theory, starting from considerations of G. Gentile (1875–1944) about diffraction by a pair of parallel wires [61, 62]. Mie did not come back to his investigations of 1900. It seems that he was not very interested in following the development in this field.

Motivation by Engelbert Arnold played an important role for Mie in Karlsruhe and at the beginning of his time in Greifswald by way of occupying himself with practical electrotechnical problems. Arnold contributed substantially to the development of direct current machines. The commutation was one of the main problems in the development of more effective direct current machines. The question came into the focus of several researchers and a couple of publications appeared practically at the same time. Concerning the value of their own publication, written together with Gustav Mie, Arnold explicated:

> Due to the thankworthy cooperation of Dr. Mie the complicated solution of differential equation (3) was successfully presented in a neat manner. The graphical representation of the conversion of energy during the short circuit period leads to results, different from the

[82] Letter of Sommerfeld to Mie from 1 July 1900, UAF, E12/37 Nr.1.

[83] Letter of Mie to Sommerfeld from 9 July 1900, UAF, E12/37 Nr.2.

conceptions used so far for the processes during the short circuit. Therefore the publication of this work is still of interest.[84]

Apart from the development and construction of generators and motors, the rating and improvement of distribution networks was an important issue of that time.[85] Mie studied heat conduction in stranded cables, a problem important for the dimensioning of such cables [Mie17, Mie19].

1.3.3 Scattering Theory

Mie's article "Contributions to the optics of turbid media, particularly colloidal metal solutions" [Mie25], published in the year 1908, belongs to the so-called "Sleeping beauties and wallflowers", i.e. papers that received much-delayed appreciation [64, 65]. It was not until 1945 that this investigation received a broad appreciation and application. The development of computer engineering, especially, advanced the application of Mie's theory. This seminal work is not only important with respect to its scientific content. The commonly used terminology, Mie scattering, does not accommodate the history of Mie's investigation.[86] This work of Gustav Mie seems to be a singular point in his scientific opus. Most of the biographical notes about Mie do not keep track of the genesis and context of this paper on scattering theory. Both points of view will be discussed here in more detail.

The young professor extraordinarius Mie had to represent theoretical physics. The Ph.D students at the institute Karl Degen, Erich Lischner and later also Walter Steubing[87] visited the lectures of Mie. Under supervision of the ordinarius König, experimental investigations on colloidal metal solutions were performed. Degen and Lischner defended their theses connected to such investigations in May 1903 [66–68]. The topic of Degen's Ph.D was the experimental investigation of the dielectric properties of a magnesium sol. Lischner investigated the elliptical polarisation of light, reflected from pigment solutions. One special example of his investigation was fuchsin, a triphenylmethane pigment dissolved in alcohol. Mie was familiar with those investigations of course. At the same time, H. Siedentopf (1872–1940) and R. Zsigmondy (1865–1929) developed the so called ultramicroscopy, a version of dark field microscopy, to uncover submicroscopic particles with a size of a few nanometer by means of their scattered light [69]. Zsigmondy's interest in colloids was motivated by his work on coloured glasses with the glass

[84] cf. [Mie8], p. 97.

[85] cf. [63], p. 132ff.

[86] Instead of Mie theory, occasionally the terms Lorenz-Mie theory or Lorenz-Mie-Debye theory are used.

[87] Walter Steubing (1885–1965), Ph.D 1908 Greifswald, TH Aachen (Stark), 1927 professor for applied physics Breslau, 1948 professor in Hamburg.

factory Schott und Genossen in Jena. It was possible to correlate colour changes with particle sizes and particle forms by means of ultramicroscopy. In 1905, Zsigmondy published a review on the ultramicroscopic investigations on colloids [70]. Zsigmondy wrote:

> A series of examples should be used to demonstrate, how by an incremental fragmentation of a solid—the metallic gold—the properties of the same will change. The fragmentation of the solid should be continued as far as possible; if possible down to molecular dimensions. On the other hand, size and properties of the single particles, obtained as a result of the fragmentation, should be determined exactly.[88]

Chapters 10 and 11 of Zsigmondy's book discuss the connection between particle size and colour, and also the colour change of colloidal gold. Zsigmondy sums up:

> Based on this fact, also supported by investigations of gold ruby glass, we have to denote every attempt to calculate the particle size in case of metal hydrosols from the light absorption as untimely.[89]

In 1906, the journal "Zeitschrift für Chemie und Industrie der Kolloide" was founded.[90] The introduction pointed to Zsigmondy's book [70] and stated, "...that we are at the doorstep of a new great area, where science has to develop and industry to exploit—this is the colloids".[91] In 1902, Mie studied Planck's dispersion theory [71, 72]. This can be concluded from the exchange of letters with Max Planck (1858–1947).[92] The reasons that motivate the investigations of Steubing were revealed clearly in the report "The optical properties of colloidal gold solutions" [Mie23] published in "Zeitschrift für Chemie und Industrie der Kolloide" in November 1907. The topic of Steubing's and Mie's investigation was perfectly suitable for the new journal. Part of the reason why this journal was chosen was probably due to its fast publication turnaround time. This article was a report on a talk given by Mie on 18 September 1907 on the 79th "Versammlung Deutscher Naturforscher und Ärzte" in Dresden.[93] During the discussion of the talk Siedentopf pointed out a paper by J. C. Maxwell Garnett (1880–1958) [73] unknown to Mie that time. Siedentopf also discussed the possibility extending the calculations to elliptical particles.[94] Walter Steubing defended his thesis on 16 December 1907 in Greifswald [74]. In 1908, a corresponding publication appeared in "Annalen der Physik" [74]. Steubing himself did not come back to the topic of his Ph.D work in his future work. He started his scientific career with Johannes

---

88 [70], preface.

89 ibid., p. 112.

90 vol. **1** (1906)–vol. **12** (1913) Zeitschrift für Chemie und Industrie der Kolloide, vol. **13** (1913)–vol. **179** (1961) Kolloid-Zeitschrift, vol. **180** (1962)–vol. **251** (1973) Kolloid-Zeitschrift & Zeitschrift für Polymere, than Colloid & Polymer Science.

91 Zeitschrift für Chemie und Industrie der Kolloide, **1** (1906).

92 UAF: E012/38, exchange of letters Mie-Planck, Nachlass Gustav Mie.

93 see also the publication in "Berichte der Dt. Physikal. Gesellschaft" [Mie24].

94 see also "Verhandlungen der Gesellschaft Deutscher Naturforscher und Ärzte", **79**, 40 (1907).

Stark (1874–1957) in Aachen and published three papers with him on canal rays from 1908 to 1910.

The first communications on the work done in Greifswald appeared at the end of the year 1907. Mie sent his complete paper to Wilhelm Wien, editor of "Annalen der Physik", in January 1908. Mie wrote:

> Dear colleague,
>
> I have enclosed within an article of theoretical work on the optics of turbid media, to request publication in the Annalen. Perhaps, I will send more soon, but hopefully with less voluminous amount of material. ...Eventually it will be possible to meet you at Easter. Maybe I intend to start with some kind of snow sport, because I see that I need some physical fitness. If it is not possible to meet you in Würzburg, I would go to the Bavarian mountains for a few days.[95]

Wilhelm Wien responded:

> Dear colleague,
>
> I confirm the receipt of your work on optics of turbid media, which discusses a really relevant topic. I would be very happy to welcome you here and to discuss a number of things. At the end of the winter semester I have to go to Mittenwald immediately to stay there for several weeks. I have to advise you strongly to start skiing. ...Because, as always, when a number of theoreticians gather together, it is also a time for scientific conversation.[96]

The submission of the paper was practically also the starting point for Mie's participation in the winter sports meetings in Mittenwald.

The publication of the article [Mie25] marked the conclusion of Mie's work on this topic. This was often considered as exceptional in retrospect. It is not uncommon for a scientist to keep working on the same topic for years or decades to become a specialist in a narrow field. Mie's behaviour can be understood from his style of work and the development of physics till that time. In his work on the Lecher problem, a complete mathematical solution of a certain physical problem was given. Also, in case of the seminal contribution to scattering theory, a complete mathematical solution for spherical particles was given. An extension to spheroids, when the expected difficulties in the numerical evaluation were taken into consideration, did not look very promising. Mie was also busy with his "Lehrbuch der Elektrizität und des Magnetismus", which appeared 1910. His book was dedicated to the electromagnetic program and in its intention exceeded the solution of a single problem. Mie himself wrote in the preface:

> It is without any doubt, that the actual foundation of the whole physics is electricity. Electricity fills the place, that was occupied 100 years ago by mechanics. Mechanics itself as well as optics becomes more and more a part of electricity.[97]

---

[95] UAF: E01248, 16, exchange of letters Mie-Wien, Nachlass Gustav Mie.

[96] UAF: E01248, 13, exchange of letters Mie-Wien, Nachlass Gustav Mie.

[97] [Mie27], Vorwort, S. VII.

Table 1.2 Summarizing timetable to the development of scattering theory, based on [77]

Year	Author	Ref.	Achievement, Remark
1863	A. Clebsch	[78]	Elastic wave equation
1871	Lord Rayleigh	[79]	"Rayleigh" scattering
1872	Lord Rayleigh	[80]	Scattering of sound by a sphere
1881	Lord Rayleigh	[81]	Scattering by a dielectric cylinder
1881	H. Lamb	[82]	Elastic waves
1890	L.V. Lorenz	[83]	Results analogous to [Mie25, 84]
1893	J.J. Thomson	[85]	Scattering by a conducting sphere
1899/1900	G.W. Walker	[86, 87]	Scattering by a sphere
1908	G. Mie	[Mie25]	Complete *reference solution*
1909	P. Debye	[84]	Debye potentials
1910	J.W.N. Nicholson	[88]	Scattering by a sphere
1915	H. Bateman	[89]	Complete review
1917	A.J. Proudman et al.	[90]	Numerical results to [91]
1920	G.N. Watson	[92]	Theory of Bessel functions
1920	T.J. Bromwich	[91]	Complete solution (started in 1899)

Einsteins *annus mirabilis* was three years prior in 1908. The theory of special relativity was intensively discussed and Mie found more excitement thinking about a new foundation of a theory of matter instead of puzzling over details of a special problem. A sign of such an attitude was his participation in the congress of natural scientists in Salzburg in 1909. Einstein took part in such a large congress for the first time. Mie participated in the discussions on talks about basic questions in electromagnetism and the theory of special relativity. Contributions to the discussions on other topics are not documented.[98] This demonstrated that his main interest was already focussed on topics other than scattering theory.

The historical development of scattering theory will be now summarised.[99]

Table 1.2 lists a few milestones in the development of the scattering theory, that were mainly, as it were, precursors to Mie's paper. Clebsch (1833–1872) [78] investigated boundary value problems in elastic media, especially of sound waves impinging on spherical surfaces. Clebsch's work is fundamental from several points of view. It had already contained work on potentials that were later named Debye potentials. Clebsch's treatise was published before Maxwell's theory. It presented the theory in component form as opposed to the modern notation of vector analysis. This might be the reason why the work of Clebsch is only rarely cited. The work of Lord Rayleigh (1842–1919) on the exact solution of the scattering problem of a dielectric cylinder [81], published in 1881, is based on Maxwell's theory. Lamb (1849–1934) [82] solved the vectorial wave equation with methods, used also by Clebsch, and provideed a basis for a series of further publications.

[98] see Verhandlungen der Gesellschaft Deutscher Naturforscher und Ärzte, **80** (1908), **81** (1909).

[99] The remarks are based on the the reviews of Kerker [75] and Logan [76, 77]. For a detailed discussion see especially [77].

Fig. 1.10 Ludvig Lorenz (1829–1891). (Portrait from [93])

Ludvig Lorenz[100] (Fig. 1.10) is mainly known as the "second Lorenz" from the Lorentz-Lorenz formula and, incidentally, the so-called Lorenz gauge is often erroneously attributed to H.A. Lorentz.[101] Lorenz's work from 1890 [83] is based on the work of Clebsch and contains the exact solution of the scattering problem for the sphere. The Lorenz theory is not based on Maxwell's theory, but based on the exact solution of the same boundary problem. This work has remained practically unknown. As already mentioned, Mie was pointed to a paper by Maxwell Garnett at the conference of natural scientists in Dresden 1907. Mie cites Maxwell Garnett [73] in his paper, but also an older publication of L. Lorenz [96]. In contrast to Mie, to Maxwell Garnett, the work of Lorenz from 1890 was known. He cited the Lorenz paper [83] together with publications of Lord Rayleigh and H. Hertz in the first part of his own publication. Therefore, Mie could have found a hint to the work by Lorenz. However, Lorenz's original work in the Danish language was most likely not known to him. Nicholson (1881–1955) [88] investigated

[100] Ludvig Lorenz (1829–1891), studies in Kopenhagen, 1852 diploma as chemical engineer, September 1858- Juli 1859 studies in Paris, 1866 teacher at the academy of military sciences in Kopenhagen, 1887 working as a free scientist, funded by Carlsberg, 1876 professor, 1887 doctor honoris causa of Uppsala University (see also [94, 95]).

[101] see for instance H. Klingbeil, Elektromagnetische Feldtheorie, Teubner-Verlag Stuttgart, 2003; W. Nolting, Grundkurs Theoretische Physik 3—Elektrodynamik, Springer-Verlag, Berlin 2002.

the light scattering from a metallic sphere and compared the results with those from geometrical optics. He also developed a series of asymptotic formulae for Bessel functions, necessary for the discussion of the scattering on large spheres. The same formulae have been developed earlier by Lorenz and were part of his 1890 work. The publications of Mie [Mie25] and Debye [97] remained largely unknown in England as exemplified by the Nicholson paper [88]. This lack of knowledge of the work of others is in strong contrast to the complete overview, given by Bateman for the first time [89]. The bilateral appreciation and knowledge of scientific work disembogued at the beginning of Word War I in appeals, that depressed the atmosphere in the international scientific community for a long time (see also Chap. 5). Finally, the work of Bromwich (1875–1929) and his co-workers Proudman, Dodson and Kennedy should be mentioned. Bromwich published his paper in 1920 [91], but the exact solution was already known to him in 1899. Proudman et al. [90] performed detailed numerical calculations, published in 1917.

Logan summarised the distinctive feature of Mie's contribution, which resulted in the paper to be generally accepted as a reference, in an apposite manner:

> What Mie did that was different from the work of previous writers was to set out on an ambitious computing program. He made calculations which involved summing the first several partial waves, in order to be able to obtain numerical results for spheres, which were too large to qualify for the criterion to be small enough for only the first partial wave (Rayleigh scattering) to be important. Mie's paper is complete in itself, and it served as the basis for much of the work which has been done in this field since its publication. ...Mies paper caught the attention of his and later generations because he employed the results of his very *thorough* paper to study a very interesting practical problem, and others, who came after him, found that his paper provided them with a *good* and a *complete*, guide to follow in their applied work in light scattering.[102]

In 1904 Mie's booklet "Molecules, Atoms, World Ether" appeared [Mie18]. It was a result of lectures from a summer school in Greifswald in 1903. He discussed the properties of turbid media, using atmosphere, milk and also gold colloids as examples. Further editions appeared in 1907 and 1911. The final sentences of the part about turbid media are identical in the first two editions:

> The small oscillators in the two colloidal gold specimen, that I have shown, have the special property of the absorption of certain types of light strongly, especially the green color. Therefore the red color prevails in the passing light.[103]

In the third edition from 1911 his intensive study of the subject is reflected in the following reformulation of that passage:

> That the colloidal metal solutions show totally different colors than the usual turbid media is related to the special optical properties of metals. They show with respect to the different types of light a very strong selective behavior, certain colors are strongly absorbed, whereas other colors are strongly reflected. In the red gold solutions, for example, a very strong absorption of green light by the gold particles occurs, but for blue solutions one obtains mostly an abnormally strong lateral emission of yellow-red light.

---

[102] Logan [77], p. 9 and footnote 16.

[103] [Mie18], p. 57 1st edition, p. 61 2nd edition.

> The theory of Lord Rayleigh, which we have already discussed, is valid only for particles of such substances, which do not strongly absorb or reflect, that means substances which would appear as a form of a coarse powder that is white or only weakly colored. Most substances belong to this class.[104]

1.3.4 Theory of Matter and Theory of Relativity

Even today, Gustav Mie's publication "Contributions to the optics of turbid media, particularly colloidal metal solutions" is the basis of many practical applications, but his contributions to the theory of matter are interesting only in a historical context. His three publications "Grundlagen einer Theorie der Materie" [Mie32, Mie33, Mie36] have had strong influence on the development of field theory at that time. Mie's contributions in this area are appreciated and discussed in-depth in a series of reviews and books.[105]

Up to his contribution to scattering theory, Mie's scientific work was much influenced by Maxwell's theory. He contributed to the extension of the theory and, propagation by way of his textbook [Mie29]. Therefore it is not surprising, that Mie is seen, from a historical point of view, as a proponent of the electromagnetic field program. The unified description of electric, magnetic and optic phenomena based on Maxwell's theory as well as the success of electron theory of H.A. Lorentz (1853–1928) provided the electromagnetic field program a dominant role. The program was formulated mainly by E. Wiechert (1861–1928) and J. Larmor (1857–1942) in the 1890s and was at the turn of the century the dominant concept. Significant contributions made by, to name only a few, M. Abraham (1875–1922), W. Wien (1864–1928), A. Sommerfeld (1868–1951), J.J. Thomson (1856–1940), H.A. Lorentz (1853–1928) and H. Poincaré (1854–1912). Vizgin states:

> The radical form of the electromagnetic program advanced by Wiechert in 1894 and somewhat later by Larmor declared the ether to be primary reality; its excited states gave the charged particles (electrons), and the origin of their mass was explained on the basis of the concept of electromagnetic mass developed in the 1880s and 1890s, primarily by British scientists (J.J. Thompson, Heaviside, and also, somewhat later, Searle and Morton). It was assumed that the laws of Newtonian mechanics could be deduced from the equations of the electromagnetic field.[106]

The persistent result of this effort was the introduction of the field concept and the attempt to unify physical phenomena on this basis. Mie's substantial work with the field concept and the attempt to develop a new theory of matter in the framework of the electromagnetic program is rather connected with the end of this development. Late highlights in this sense are the unified theories of Hilbert

---

[104] [Mie18], p. 66 3rd edition.

[105] see Vizgin [98], Kohl [33], Renn (Ed.) [99, 100], Corry [101].

[106] [98], p. 7.

(1862–1943) in 1915 and Weyl (1885–1955) in 1917. They tried to develop a unified theory of the electromagnetic field and the gravitational field based on the the theory of general relativity. Mie's work stimulated the attempts of Hilbert and Weyl and also later work on nonlinear field theories of Born (1882–1970) and Infeld (1898–1968) [102]. The dawn of the 1920s was characterised by the boom of the quantum mechanical program. Most of the physicists preferred to work in this framework, because it was easier to relate the theoretical developments to the real physical processes.

Mie published his theory of matter in three articles in the years 1912 and 1913 [Mie32, Mie33, Mie36]. It was a very ambitious program to unify electromagnetism, gravitation and the first ideas of quantum mechanics. Mie's aim was to finally get a theory which describes the existence and properties of the electron and later on also of atoms and molecules, which produces the atomic spectra and also delivers the field equations of gravitation.

Mie's work was connected with the discussion around the theory of general relativity. Einstein emphasised, that the extensions he aimed for should be independent of any assumptions about the constitution of matter. In contrast to this point of view, Mie was searching for a theory of matter based on an electromagnetic field concept. He wanted to start from a set of equations describing the state and dynamical evolution of the ether. Elementary particles, as an inherent part of the theory, should manifest themselves as so-called knots of the ether. To explain a stable electron in electromagnetic terms (charged sphere) an additional attractive force, the so called Poincaré stress, is necessary. Mie made a different choice. He considered nonlinear field equations. Maxwell's equations are expected to be the weak field limit of the nonlinear field equations. He started from a world function and got the field equations by means of Hamilton's variational principle. In contrast to Einstein, the solution of the gravitational problem would have been a bonus of his unified field theory.

A series of notes in his diary reflect his struggle with mathematical problems but also his pleasure in thinking about the solution of physical problems.

> In the last week I was submerged in a fog of mathematical formulae and calculations. They concerned the effect of the gravitational field on the internal structure of an electron. I believed that I had found a beautiful entry into the dark region by the Born transformation and with the introduction of bipolar coordinates and I hoped to head for a light soon. However, I immediately came into a sea of fog, drove around in it wildly back and forth, because I thought, once I could get to the border of the fog by chance, I could then slip free.[107]

At the end Mie missed his grandiose goal, but his formalism has had a large influence on the further development of unified field theories. A sign of the importance of Mie's work is that Weyl discusses the theory in his famous book "Raum-Zeit-Materie".[108]

---

[107] UAF, Nachlass Mie, E12/68, diary note from 28 October 1915.

[108] See [103] and also [104].

Hilbert included Mies's work in his field theory and mentioned this in his publication at the very beginning. Hilbert wrote:

> The tremendous problems stated by Einstein and his subtle methods conceived to their solution as well as the profound ideas and original conceptualisations whereby Mie formulates his electrodynamics, opened new routes for the investigation of the foundations of physics. I want in the following—in terms of the axiomatic method—essentially starting from two simple axioms, to develop a new system of fundamental equations of physics, which are of ideal beauty. I believe that the equation contains the solution of the problems of Einstein and Mie simultaneously.[109]

Mie was very excited about the progress made by Hilbert. After the receipt of Hilbert's publication he noted:

> Yesterday a separate print of Hilbert's paper "The Fundamentals of Physics" arrived. In this essay he combines my theory of matter with the principle of general relativity, which Einstein aims at, but only Hilbert really fulfils. It is a completely wonderful work! With what a strong fist this Mathematician strikes hard! He smashes the largest obstacles to smithereens. My boldest dreams are exceeded. I believe that it will not take long, before we really get the true world function and will be able to create matter like god in a mathematical way. I just wrote a letter to Hilbert. I am extremely excited about this event.[110]

Mie was also invited to the Wolfskehl lectures[111] in 1917. In 1915, Einstein was invited to give Wolfskehl lectures and M.v. Smoluchovski was the invitee in 1916. Other renowned invitees have been Poincaré, Lorentz, Sommerfeld, Planck, Debye, Born and others. Mie's Wolfskehl lectures appeared in print in 1917 [Mie43, Mie44, Mie45]. After finishing the corrections in the proof sheets, Mie commented:

> I have now completed the last proof-sheet of my Einsteiniade and soon, the third and final lecture will also be printed. I cannot help myself, I found it enormously interesting reading.[112]

An important place for discussion and controversy was the regular meeting of the "Gesellschaft deutscher Naturforscher und Ärzte". Mie and Einstein visited the meetings regularly. They presented their own contributions or commented on contributions from others during the discussion. A conflict between Mie and Einstein appeared at the meeting in Vienna in 1913 (see Fig. 1.11). Einstein presented an overview, "To the state of the art of the gravitational problem". Einstein discussed his theory, developed together with Marcel Großmann, and also

---

[109] D. Hilbert, *Die Grundlagen der Physik. Erste Mitteilung. Vorgelegt in der Sitzung vom 20. November 1915.* in "Nachrichten von der Königlichen Gesellschaft der Wissenschaften zu Göttingen. Mathematisch-physikalische Klasse", Heft 5, 395–407 (1915).

[110] UAF, Nachlass Mie, E12/69, diary note from 13 February 1916.

[111] Paul Wolfskehl (1856–1906) was an industrialist with mathematical interest. He bequeathed 100000 Mark to the Royal Academy of Sciences of Göttingen to be given to that person, who will proof Fermat's Theorem during the next 100 years. (Wolfskehl prize) The interest of the prize money was used to organize the Wolfskehl lectures.

[112] UAF, Nachlass Mie, E12/71, diary note from 8 December 1917.

Fig. 1.11 "Naturforscherversammlung" in Vienna, 21 to 28 September 1913. (Courtesy of Techniseum Mannheim)

included attempts, especially Nordström's theory, in his talk. The discussion was led mainly by Mie and Einstein. Mie was angry that Einstein failed to mention his theory. A controversy documented in the Physikalische Zeitschrift was the result. In an exchange of letters from 1917 to 1919 both scientists discussed their different points of view. Illy summarised the exchange of letters and discussion between Mie and Einstein:

> The case of Einstein and Mie underlines a rather sceptical thesis: there is no perfect understanding of each other. Theories change in the hands of scientists, taking part in a relay race called science with each member of the team leaving the fingerprints of his commitments and beliefs on it. These fingerprints are often nonrational and incomprehensible to others. But this is the way science always grows...[113]

---

[113] [105], p. 256.

The “Naturforscherversammlung” in 1920 took place in Bad Nauheim and was dominated by controversies about Einstein's theory of relativity and the dispute between Einstein and Lenard (1862–1947). Mie spoke about “The electrical field of a charged particle rotating around a center of gravitation”. Mie appeared in the discussion as a critical promoter of the theory of general relativity.[114]

1.4 Gustav Mie as University Teacher

1.4.1 Lectures

Gustav Mie was full professor for experimental physics and director of the institutes in Greifswald, Halle and Freiburg. The obligation which accompanied with these positions was to cover teaching in experimental physics. A check of the university calendars of the corresponding years demonstrates that Mie did not regard teaching as an annoying obligation. On 2 November 1915, he wrote in his diaries with reference to the beginning of the practical course of the students of medicine and the difficult times during World War I:

> In such a situation one is automatically kind and attentive to the young people and immediately forgets to play the role of the schoolmaster, which is often, for me, so inexpressibly distastefull. Yesterday, I took to being a teacher with great pleasure, and by the way, there were also three ladies sprucing up the otherwise so martial audience. ...Then, I had my first lecture in this term ...and it was a great pleasure for me to tell the young people, as an introduction to electricity, a little bit about my old friend, the world ether to teach them respect.[115]

As an example, the canon of the courses that Mie lectured in Freiburg from winter semester of 1925–1926 until the winter semester of 1934–1935 is given in Table 1.3. The courses remained almost unchanged during the whole period. Starting from the summer semester 1930 the half-day courses *Advanced Practical Course I and II*, where Mie lectured the first part together with Kast (1896–1980), replaced *Physical Exercises for advanced Students*. The workload from the lectures was high and, apart from the large experimental physics lecture, also included the work in practical courses.

Theoretical physics was lectured by Emil Cohn[116] and Johann Georg Königsberger.[117] Courses in theoretical physics were only rarely given by Mie.

---

[114] For a summary see of the meeting see Weyl [106].

[115] UAF, Nachlass Mie, E12/69, diary note from 2 November 1915.

[116] Emil Cohn (1854–1944), 1879 Dr. phil. in Strassburg, assistant of A. Kundt, 1884 habilitation in theoretical physics, 1919 honorary professor in Rostock, 1920–1935 honorary professor in Freiburg, 1938 together with Gans, Graetz, Jaffé, Kaufmann and other physicists of Jewish descent withdrew out of DPG, 1939 emigrating to Switzerland.

[117] J.G. Königsberger (1874–1946), 1904–1936 professor of physics at the University of Freiburg, pioneer of semiconductor physics, coined 1911 together with Weiss the term “semiconductor” (sun of the mathematician Leo Königsberger).

Table 1.3 Courses of Gustav Mie at the University of Freiburg from the WS 25/26 up to the WS 34/35

	Lecture	Amount
SS	Experimental Physics I (Mechanics, Acoustics, Thermodynamics)	5 h/week
	Mathematical Additions to Experimental Physics	2 h/week
	Physical Exercises for Students of Natural Sciences	6 h/Wweek
	Small Physical Practical Course for Physicians and Pharmacists	3 h/week
	Physical Exercises for advanced students	Daily
	Guidance for independent work	Daily
	Physical Colloquium	2 h/week
WS	Experimental Physics II (Electricity and Optics)	5 h/week
	Mathematical Additions to Experimental Physics	2h/week
	Physical Exercises for Students the Natural Sciences	6 h/week
	Small Physical Practical Course for Physicians and Pharmacists	3 h/week
	Physical Exercises for advanced students	Daily
	Guidance for independent work	Daily
	Physical Colloquium	2 h/week

see university calendars on-line in FREIMORE - FREIburger Multimedia Object REpository `http://freimore.uni-freiburg.de`

Occasionally, instead of the *the Mathematical Additions to Experimental Physics*, lectures on quantum mechanics were offered. In the summer semester of 1927 *The Schrödinger Mechanics*[118] was taught and *Quantum Mechanics* was offered in the summer semesters of 1929 and 1931. The lecture of the summer semester 1927 continued into the winter term of 1927–1928 with the lecture *The Schrödinger Mechanics II.* This was followed by the lecture *The Dirac Theory of the Electron* in the winter semester of 1929–1930.

The courses in Greifswald and Halle mainly followed the structure in Table 1.3. From 1911 to 1914, Mie also offered theoretical lectures in Greifswald, for example *Theory of Relativity* (WS12/13) and *Theory of the Quantum of Action* (SS13). Due to the difficulties with K.E.F. Schmidt, Mie did not offer lectures in theoretical physics in Halle.

Regarding the structure and effect of Mie's lectures, Kast wrote:

> It had always been his concern that the field representation should be introduced at high school level, not only to university students. Just as he did not place the topic of electric forces but electric field pattern at the beginning of the 2nd part of his basic lecture, he did not begin the first part with the laws of motion but with the weight of bodies as a consequence of the gravitational field of earth, taking his audience as far as the theory of relativity. Therefore, his lectures were not easy, particularly as he did not allow simplification that would have affected the rigour in the content of the presentation. By this however, the lectures attained a degree of coherence that deeply impressed those who were willing to follow.[119]

[118] Königsberger lectured in the same term *Elements of the Quantum Theory.*

[119] [Bio10], p. 130.

1.4.2 Disciples

Within the institute, Mie predominantly appeared as an experimental physicist. During his activity as a professor in Greifswald, Halle and Freiburg he did not supervise purely theoretical theses.[120] In this sense, Mie, concerning his theoretical work, did not form a school in the way Sommerfeld did. Nevertheless, he promoted several young physicists in their career systematically. As an example, August Julius Herweg[121] came from W. Wien (1864–1928) (Würzburg) to Greifswald. After Herweg's habilitation Mie requested the title professor on his behalf and tried to obtain him a teaching appointment for spectroscopy.[122] This request for the teaching appointment was declined by the minister for religious and educational affairs. Herweg followed Mie to Halle. In Halle however, a teaching appointment for "Research Methods of latest Physics" was then obtained for Herweg. In 1923 Herweg accepted an offer as a regular extraordinary professor for high-frequency engineering, outlines of physics and photography to the Technical University Hannover. The vacant assistantship was occupied by Wilhelm Kast,[123] who later also followed Mie, as assistant, to Freiburg and in 1937 became full professor and director of the Physical Institute in Halle.

1.4.3 Textbooks and Popular Scientific Brochures

Gustav Mie wrote a series of popular scientific brochures and papers on actual problems of physics for the general public. The brochures "Ions and Electrons" [Mie14] and "Molecules, Atoms, World Ether" [Mie18] are two examples. Mie also published a popular description of Einstein's theory in 1921 [Mie50]. His final treatise "Foundations of Mechanics" dates back to 1950 [Mie86]. His book entitled, "Textbook on Electricity and Magnetism. An Experimental Physics of the World Ether for Physicists, Chemists an Electrotechnicians". [Mie29] is highly

[120] See: Jahresverzeichnis der an den deutschen Universitäten erschienenen Schriften (1887–1913), Jahresverzeichnis der an den deutschen Universitäten und Technischen Hochschulen erschienenen Schriften (19-14-1925), Jahresverzeichnis der an den deutschen Universitäten und Hochschulen erschienenen Schriften (1926–1936). Purly theoretical Ph.D thesis supervised by Mie could not be found.

[121] August Julius Herweg (1879–1936), studies in Würzburg, scholar of W. Wien, graduation Würzburg 1905, habilitation Greifswald 1907, 1913 professor, 1914–1918 army service, 1921 extraordinary professor in Halle, from 1923, extraordinary professor at TH Hannover.

[122] UAG, Acta of the Phil.Fak. of the Königl. Univ. Greifswald I 324 (Dekanat Mie), J.Nr. 1245.

[123] Wilhelm Kast (1898–1980), studies in Halle, 1922 graduation, 1924 Freiburg, 1927 habilitation, 1933 appointment as extraordinary Professor, 1937 full professor and director of institute in Halle, 1947 Bayer AG Leverkusen, 1954 honorary professor University Cologne, 1957 honorary professor university Freiburg.

regarded, was printed in three editions 1910, 1941 and 1948, Mie was 80-years old when he finished working on the final edition of his textbook.

Mie's book was the first to contain a consequent classification of the field quantities in quantities of intensity and quantities of quality. He also developed an electromagnetic system of units, based on Volt, Coulomb, Centimeter and Second, which is no longer relevant. The first edition starts from the physical phenomenon rather than the mathematical approach.

Mie's textbook, dedicated to Engelbert Arnold and Otto Lehmann in memory of his Karlsruhe years of apprenticeship, appeared in the publishing house of Ferdinand Enke, Stuttgart. Mie kept connections with this publisher for over 40 years. The exchange of letters between Enke and Mie highlights the relationship, which was not entirely frictionless at times, between a scientist and the publisher, who also had economic interests.[124] Enke visited Mie in Halle at the beginning of 1924, to discuss details of a second edition after the success of the first edition. They reached an agreement and settled on the deadline of April 1925. This deadline was exceeded by more than 15 years in the end. Mie wrote in a letter to Enke on 29 November 1925:

> I started to rewrite my book to get a shorter version after your visit to Halle, but I cannot do this without completely changing the character of the book. Therefore I finally decided to follow the second possible way, namely to adapt it to a truly complete textbook which includes all elements of the mathematical theory. ...I assume, that I will need nearly three years to get it done. It will be a totally new book.[125]

Schäfer quoted in his recension of the second edition regarding the relevance of Mie's textbook as follows:

> The first edition of Mie's "Textbook of Electricity and Magnetism" appeared in 1910. This was the first experimental textbook which took the introduction and realisation of the field concept seriously into account, introducing the reader in the mindscape of Faraday and Maxwell following a thoroughly thought out plan. Mie did not go back to ideas and concepts of the theory of remote action. To the expert it is unquestionable that this book has had a profound impact to the textbook literature after 1910, even if the authors did not mention their source explicitly.[126]

The following remark of Schäfer about the style in Mie's textbook seems to be characteristic of Mie's mode of practice: "The diction is sober, objective, clear, and empathic. The formulation of the important theorems is presented in almost pedantic accuracy to avoid any misunderstanding about the meaning of the material: The learner cannot expect a better leader".[127]

---

[124] vergl. UAF—Nachlass Mie, Bestand E12/30 Verlagsverhandlungen 1912–1946.

[125] ibid.

[126] Mie, Gustav: Lehrbuch der Elektrizität und des Magnetismus, Stuttgart 1941, Besprechung durch Cl. Schaefer, Die Naturwissenschaften H. 17/18, S. 267f (1942).

[127] ibid., S. 268.

1.5 Scientific Community and Society

Gustav Mie was strongly affected by the events of his times, in particular in his work as a university teacher. His time in Greifswald coincides with World War I. In these awkward times, he also officiated as rector of Greifswald University. The time in Halle was accompanied by the revolution of 1918–1919, the Kapp Putsch of 1920, the Central-German rebellion in March 1921 and the effects of hyper-inflation in 1922–1923. During his last years at Freiburg University, Mie felt the seizure of power of National Socialists and their increasing influence on the University. Mie's retirement was marked by the experience of the horrors of World War II, as well as the beginning of the democratic reconstruction of Western Germany. It is beyond the scope of this text to give a complete analysis of the development of the academic, political and ideological opinions of Mie. In addition, among other things, a detailed evaluation of the diaries would be necessary.[128]

Gustav Mie was fully integrated into the network of theoretical physicists at that time. He came into contact with Arnold Sommerfeld over his work on the parallel wire system [Mie9]. At almost the same time, Mie began an exchange of letters with W. Wien in 1900.[129] Soon, Mie belonged to the community of prominent German theoretical physicists, who met for skiing holidays, often in Mittenwald. Mie was friends with W. Wien, A. Sommerfeld, M. von Laue and P. Debye. The appreciation of his work was also expressed in the fact that his colleagues recommended him for full professorships several times. He was suggested for the post of the full professor in Frankfurt/Main by Sommerfeld along with Laue and Debye in 1913, Wien suggested Mie for a full professor in Tübingen as the second choice after Stark in 1916.

Mie's extensive scientific exchange of letters as well as the numerous letters of appraisal in appointment procedures[130] show that Mie was strongly integrated into the physics community of that time. S.L. Wolff places Mie in a network of reactionary physicists during the time of the Weimar Republic [107, 108]. Gustav Mie's basic political orientation can be regarded as national conservative.[131]

Gustav Mie's mainly conservative and somewhat nationalistic point of view can be deduced, for example, from his publication "Werner Siemens as a Physicist" [Mie40] from 1916. At the end of the essay he characterised Siemens as a person. Mie wrote:

> If we consider Siemens as a person, as we can see him from the memoirs of his life, one can conclude that he possess to a large degree the typical characteristics of a German. It is

[128] UAF, inventory E012/68-E012/72 (copies), Nachlass Gustav Mie.

[129] UAF, E012/48, Briefwechsel Mie-Wien, see also Archives Deutsches Museum Munich, Nachlass Wilhelm Wien, Briefwechsel.

[130] see UAF: E012/33, E012/34, Nachlass Gustav Mie, Berufungen.

[131] Diary entries to current events prove Mie's political point of view, UAF, inventory E012/68-E012/72.

definitely impossible to conceive this man to have grown up anywhere else but in Germany. Also the peculiar connection of the desire for clear, scientific and objective insight with a strong compulsion to influence the environment in an active and creative way has to be counted as the typical properties which, in the end, makes the difference between the German technology in general and that of other peoples. This unification of energy and spirit of research in Germans, which allowed Siemens, with ingenious peculiarity, to pace through his life as a hero, faced by the largest difficulties, which will hopefully also contribute such that our native country will overcome the actual ordeal.[132]

World War I, involving all industrial nations, also gave rise to industrial conflicts, in which scientists were professionally mobilised [109]. At the beginning of the war, scientists tried to justify the points of view of the respective belligerent parties with invocations and corresponding replies. This development is characterised by the term "War of the Spirit".[133] After an article by Sir J.J. Thompson in The Times on the 1 August 1914 on the 4 October 1914 the appeal "An die Kulturwelt! Ein Aufruf" followed as a reply.

In this time also the *Aufforderung*, initiated by Wilhelm Wien appeared, polemising against the "Ausländerei" and in particular the strong English influence. Finally it led to a politically motivated regimentation of the citations of scientific work. Mie was one of the 16 signatories of the *Aufforderung*, 700 copies of which were sent to all Universities in Germany and Austria.

To evaluate Mie's attitude at the time of the National Socialism, one must consider that he had already reached his retirement age in 1933 [110]. In the conflict centred around "German Physics" the article "German and Jewish Physics" by W. Menzel was published in the "Völkischer Beobachter" on 29 January 1936. Heisenberg answered in the same journal on the 28 February followed by another reply by Stark. Thereupon, Heisenberg prepared a memorandum to the Reichserziehungsminister Rust, which pointed out the difficulties of physics, and in particular the importance of theoretical education and the cooperation between theoretical and experimental physics. The memorandum was signed by 75 experimental and theoretical professors in the summer of 1936. Being asked by Werner Heisenberg to sign the memorandum on 11 May 1936, Mie replied, that he would do it with "great pleasure".[134] In addition to the examples specified by Heisenberg, Mie pointed out another example of fruitful collaboration between theory and experiment:

A more impressive example, albeit from older times, ...is the discovery of the electromagnetic waves propagating freely in air by Heinrich Hertz. It is safe to state that without Maxwell's theory no one would ever have got the idea of performing such experiments. H. Hertz considered them as an explicit verification of Maxwell's theory. As I know from a trustworthy source in Karlsruhe, any practical applications were far from his

[132] [Mie40], p. 776.

[133] After the title of the book "Krieg der Geister" from Hermann Kellermann, Dresden, 1915. In this book corresponding calls and signatory lists are printed.

[134] UAF, inventory E012/41, exchange of letters Mie-Heisenberg, 1a, letter Heisenberg at Mie of 11 1936, 1b, letter Mie at Heisenberg 15 May 1936, 2b text of the memorandum Leipzig, 30 May 1936.

considerations. What was more amazing is to see the world being reshaped by the electromagnetic waves.[135]

Besides physics and mathematics, philosophy (interest in Kant's philosophy) and theology played an important role in Mie's life. His talk at the beginning of his rectorate in Greifswald in 1916, entitled "Law of Nature and Spirit" [Mie41], dealing with questions of causality, received, to Mie's surprise, much attention amongst the members of the theological faculty. After 1930, he published a series of writings of religious and ideological content [Mie76, Mie78, Mie82, Mie84, Mie87]. In Freiburg, Mie, along with the philosopher Jonas Cohn (1869–1947) and the zoologist Spemann (1869–1941) founded a scientific society called "Pentathlon", which met monthly for lectures and discussions. The society ceased to exist in 1933, when Cohn was forced to leave Germany.

During the November pogroms on the night from 10 to 11 November 1938, amongst many others, the synagogue of Freiburg was set on fire by the SS. These events led to the establishment of the "Freiburger Kreis", a resistance group, to which Mie and his wife belonged [111–113]. The Group organised monthly discussions, in private dwellings, of persons involved until September 1944.

1.6 Concluding Remarks

Single aspects of Mie's scientific work have been illuminated in the past, often in the context of different contributions to the history of physics, especially larger historical studies about field theory and the theory of relativity [33, 98, 114]. Parts of his extensive correspondence can be found in the edited correspondence of Arnold Sommerfeld [115] as well as in the "Collected Papers of Albert Einstein" [116]. Gustav Mie's archives[136] are, to a large extent, unexplored.

In the 1970s an initiative existed to publish parts of scientific legacy, in particular the correspondence of Gustav Mie.[137] This project was driven in particular by Helmut Hönl[138] in collaboration with Herbert Fröhlich[139] and Ernst Plötze. The Mie edition should have been published with the support of the Leopoldina. The correspondence between Helmut Hönl and Heinz Bethge (1919–2001), president of the Leopoldina in Halle at that time, documents this attempt. In the

135 ibid., 1b.

136 UAF, E0012, Nachlass Gustav Mie.

137 see UAF: C136, materials for the edition Gustav Mie, worked on by Rolf Steinmann 2001; E14, Nachlass Helmut Hönl, worked on by David Geiger, 2003.

138 Helmut Hönl (1903–1981), Studies in Heidelberg, Göttingen and Munich, 1926 graduation with A. Sommerfeld, 1933 habilitation in Stuttgart, 1940 extraordinary Professor in Erlangen, 1943 professor in Freiburg, 1971 emeritus.

139 Herbert Fröhlich (1905–1991), graduation 1930 with Sommerfeld, private lecturer in Freiburg, as Jew forced to leave Germany, 1935 University of Bristol, 1948 University of Liverpool.

end, the project could not be realised. A detailed investigation of the life and scientific work of Gustav Mie is thus still pending.

Acknowledgments I appreciate the support of the staff members of the university archives of the Universities of Greifswald, Halle, Heidelberg and Freiburg, the archives of Karlsruhe Institute of Technology as well as the archives of the National Academy of the Sciences Leopoldina during my search. My special gratitude applies to Vincent Loke, Thomas Wriedt and Christian Matyssek for the critical reading of the manuscript and Thomas Wriedt for suggestions and references.

Publications of Gustav Mie

[Mie1] G. Mie, Zum Fundamentalsatz über die Existenz von Integralen partieller Differentialgleichungen, Dissertationes Philosoph. Heidelbergensis 1892, Dresden, Druck von B.G. Teubner, 1892

[Mie2] G. Mie, Fundamentalsatz über die Existenz von Integralen partieller Differentialgleichungen, Zeitschrift für Mathematik 37, 151–170 (1892)

[Mie3] G. Mie, Fundamentalsatz über die Existenz von Integralen partieller Differentialgleichungen, Zeitschrift für Mathematik 37, 193–211 (1892)

[Mie4] G. Mie Die Röntgen'sche Strahlung, S. V-IX, Sonderdruck aus: Beilage zur Nr. 7 der Zeitschrift des Vereins für Finanzassistenten, 1896

[Mie5] G. Mie, Beweis der Integrirbarkeit gewöhnlicher Differentialgleichungen nach Peano, Mathemat. Annalen 43, 553–568 (1893)

[Mie6] G. Mie, Entwurf einer allgemeinen Theorie der Energieübertragung, Sitzungsber. d. K. Akad. d. Wiss. Wien, Math.-nat. Klasse 107 Abt. IIa, 1113–1180 (1898) (Habilitationsschrift)

[Mie7] G. Mie, Über mögliche Ätherbewegungen, Ann. der Physik 68, 129–134 (1899)

[Mie8] G. Mie und E. Arnold, Über den Kurzschluß der Spulen und die Commutation des Stromes eines Gleichstromankers, Elektrotechn. Zeitschrift 20, 97–101, 136–138, 150–152 (1899)

[Mie9] G. Mie, Elektrische Wellen an zwei parallelen Drähten, Ann. der Physik 2, 201–249 (1900)

[Mie10] G. Mie, Die mechanische Erklärbarkeit der Naturerscheinungen: Maxwell-Helmholtz-Hertz, Nat. Ver. Verh. (Karlsruhe) 13, 402–420 (1900)

[Mie11] G. Mie, Ein Beispiel für Poyntings Theorem, Zeitschrift für physikal. Chemie 34, 522–526 (1900)

[Mie12] G. Mie, Über die Bewegung eines als flüssig angenommenen Äthers, Physikal. Zeitschrift 2, 319–325 (1901)

[Mie13] G. Mie, Zur kinetischen Gastheorie der einatomigen Körper, Ann. der Physik 11, 657–697 (1903)

[Mie14] G. Mie, Die neueren Forschungen über Ionen und Elektronen, Sonderausgabe aus der Sammlung elektrotechnischer Vorträge Bd. 4, Enke Stuttgart, 1903, 40 S. (2. Aufl. 1906)

[Mie15] G. Mie, Der elektrische Strom in ionisierter Luft in einem ebenen Kondensator, Ann. der Physik 13, 857–889 (1904)

[Mie16] G. Mie, Über eine Methode, das spezifische Gewicht sehr verdünnter Lösungen zu bestimmen, Festschrift Ludwig Boltzmann gewidmet zum 60. Geburtstag, Ambrosisus Barth,Leipzig 1904, 326–332

[Mie17] G. Mie, Das Problem der Wärmeleitung in einem verseilten elektrischen Kabel, Neu-Vorpommern u. Rügen, Nat.-Ver. Mitteil. 36, 155–188 (1904)

[Mie18] G. Mie, Moleküle, Atome, Weltäther, Verlag von B.G. Teubner, Leipzig, 1. Auflage 1904, *Aus Natur und Geisteswelt* Sammlung wissenschaftlich-gemeinverständlicher Darstellungen 58. Bändchen (2. Aufl. 1907, 3. Aufl. 1911, 4. Aufl. 1919)

[Mie19] G. Mie, Das Problem der Wärmeleitung in einem verseilten elektrischen Kabel, Elektrotechn. Zeitschrift 137, 137–143 (1905)
[Mie20] G. Mie, Über die Kurzschlußstromkurve eines Gleichstromankers, Zeitschrift Math. Phys. 53, 37–60 (1906)
[Mie21] G. Mie, Experimentelle Darstellung elektrischer Kraftlinien, Zeitschrift für d. phys. u. chem. Unterricht 19, 154–156 (1906)
[Mie22] G. Mie, Erwiderung auf Riebesells Abhandlung: Kommutation des Stromes in Gleichstromgeneratoren, Zeitschrift Math. Phys. 55, 143–146 (1907)
[Mie23] G. Mie, Die optischen Eigenschaften kolloidaler Goldlösungen, Zeitschrift für Chemie und Industrie der Kolloide 2, H. 5, 129–133 (1907)
[Mie24] G. Mie, Die optischen Eigenschaften kolloidaler Goldlösungen, Berichte der dt. phys. Ges. 5, 492–500 (1907)
[Mie25] G. Mie, Beiträge zur Optik trüber Medien, speziell kolloidaler Metallösungen, Ann. der Physik 25, 377–445 (1908)
[Mie26] G. Mie, Sättigungsstrom und Stromkurve einer schlecht leitenden Flüssigkeit, Ann. der Physik 25, 597–614 (1908)
[Mie27] G. Mie, Eine bequeme Methode zur Erzeugung ganz schwach gedämpfter elektrischer Schwingungen von kleiner Wellenlänge, Verh. der dt. phys. Ges. 12, 860–866 (1910)
[Mie28] G. Mie, Hydratisierung und Molekularwärme der Ionen in sehr verdünnten wäss rigen Lösungen, Ann. der Physik 33, 381–399 (1910)
[Mie29] G. Mie, Lehrbuch der Elektrizität und des Magnetismus. Eine Experimentalphysik des Weltäthers für Physiker, Chemiker und Elektrotechniker, 736 S., Stuttgart 1910, 2. vollständig umgearbeitete Aufl. 1941, 3. Aufl. 1948, span. Ausgabe Barcelona 1944
[Mie30] G. Mie, Antwort auf die Bemerkung des Hrn. Seibt zu der Arbeit des Hrn. K. Settnik: Entstehung von sehr wenig gedämpften Wellen usw., Ann. der Physik 36, 207–208 (1911)
[Mie31] G. Mie, Die Materie, 32 S., Verlag von Ferdinand Enke, Stuttgart 1912 (Vortrag gehalten am 27. Januar 1912 (Kaisers Geburtstag) in der Aula der Universität Greifswald)
[Mie32] G. Mie, Grundlagen einer Theorie der Materie I, Ann. der Physik 37, 511–534 (1912)
[Mie33] G. Mie, Grundlagen einer Theorie der Materie II, Ann. der Physik 39, 1–40 (1912)
[Mie34] G. Mie, Bemerkungen zu der Arbeit des Herrn W. Sorkau über Turbulenzreibung, Physikal. Zeitschrift 14, 93–95 (1913)
[Mie35] G. Mie, Bemerkungen zu der Arbeit des Herren Gotthelf Leimbach: Energieaufnahme elektrischer Sender von kleiner Wellenlänge, Physikal. Zeitschrift 14, 723–725 (1913)
[Mie36] G. Mie, Grundlagen einer Theorie der Materie III, Ann. der Physik 40, 1–66 (1913)
[Mie37] G. Mie, Bemerkungen zu Einsteins Gravitationstheorie I, Physikal. Zeitschrift 15, 115–122 (1914)
[Mie38] G. Mie, Bemerkungen zu Einsteins Gravitationstheorie II, Physikal. Zeitschrift 15, 169–176 (1914)
[Mie39] G. Mie, Prizip von der Relativität des Gravitationspotentials, Festschrift für J. Elster und H. Geitel (1915), S. 251–268
[Mie40] G. Mie, Werner Siemens als Physiker, Die Naturwissenschaften 4, 771–775 (1916)
[Mie41] G. Mie, Naturgesetz und Geist, Deutsche Revue 41, 150–163 (1916)
[Mie42] Bilder aus Greifswalds Vergangenheit, Gewidmet von der Universität Greifswald ihren Kriegsstudenten (Vorwort G. Mie), Verlag Brucken & Co. Greifswald, 1917
[Mie43] G. Mie, Die Einsteinsche Gravitationstheorie und das Problem der Materie, Physikal. Zeitschrift 18, 551–555 (1917)
[Mie44] G. Mie, Die Einsteinsche Gravitationstheorie und das Problem der Materie, Physikal. Zeitschrift 18, 574–580 (1917)
[Mie45] G. Mie, Die Einsteinsche Gravitationstheorie und das Problem der Materie, Physikal. Zeitschrift 18, 596–602 (1917)
[Mie46] G. Mie, Einführung eines vernunftgemäßen Koordinatensystems in die Einsteinsche Gravitationstheorie und das Gravitationsfeld einer schweren Kugel, Ann. der Physik 62, 46–74 (1920)
[Mie47] G. Mie, Die Gesetzmäßigkeit des Naturgeschehens, Bertelsmann, Gütersloh, 1920

[Mie48] G. Mie, Das elektrische Feld eines um ein Gravitationszentrum rotierenden geladenen Partikelchens, Physikal. Zeitschrift 21, 651–658 (1920)
[Mie49] G. Mie, Über die Abklingung der Lichtemission eines Atoms, Ann. der Physik 66, 237–260 (1921)
[Mie50] Die Einsteinsche Gravitationstheorie. Versuch einer allgemeinverständlichen Darstellung der Theorie, 4+69 S., Leipzig, Hirzel 1921 (2. verbesserte Aufl. 1923)
[Mie51] Die Relativitätstheorie von Einstein. Zeitschrift des Vereins Deutscher Ingenieure 65, 26. Februar, S. 236 (1921)
[Mie52] Die Einsteinsche Gravitationstheorie, Deutsche Rundschau, 187, 171–184, 310–342 (1921)
[Mie53] G. Mie und J. Herweg, Die Zahl der von sehr schwachen Röntgenstrahlen ausgelösten Elektronen, Ann. der Physik 68, 120–126 (1922)
[Mie54] G. Mie, Träge und schwere Masse, Ann. der Physik 69, 1–51 (1922)
[Mie55] G. Mie, Thèorie Einsteinienne de la Gravitation. Essai de vulgarisat. de la théorie, Hermann, Paris 1922
[Mie56] G. Mie, Echte optische Resonanz bei Röntgenstrahlen, Zeitschrift für Physik 15, 56–57 (1923)
[Mie57] G. Mie, Echte optische Resonanz bei Röntgenstrahlen, Zeitschrift für Physik 18, 105–108 (1923)
[Mie58] G. Mie, Das elektrische Feld eines schweren elektrisch geladenen Kügelchens, das um ein Gravitationszentrum kreist, Ann. der Physik 70, 489–558 (1923)
[Mie59] G. Mie, Abklingzeit und Verweilzeit angeregter Atome, Ann. der Physik 73, 195–208 (1924)
[Mie60] G. Mie, Theorie der Bremsstrahlung und der Comptonschen Streustrahlung (Vortrag auf dem III. Dt. Physikertag, Danzig, 10.-16.9.1925) Physikal. Zeitschrift 26, 665–669 (1925)
[Mie61] G. Mie, Bremsstrahlung und Comptonsche Streustrahlung, Zeitschrift für Physik 33, 33–41 (1925)
[Mie62] G. Mie, Theorie der Bremsstrahlung und der Comptonschen Streustrahlung, Physikal. Zeitschrift, 26, 665–669 (1925)
[Mie63] G. Mie, Das Problem der Materie und die Relativitätstheorie, Scientia (Mailand) 37, 149–156 (1925)
[Mie64] G. Mie, Das Problem der Materie und die Relativitätstheorie, Scientia (Mailand) 37, 225–234 (1925)
[Mie65] G. Mie, Das Problem der Materie (Öffentliche Antrittsrede, gehalten am 26. Januar1925), Speyer & Kärner, Freiburg i. Br., 1925
[Mie66] G. Mie, Die Grundlagen der Quantentheorie (Vortrag in der Freiburger Naturforschenden Gesellschaft), 38 S., Speyer & Kärner, Freiburg i. Br., 1925
[Mie67] G. Mie, Ein Linienspektrum bei Wellenlängen von mehreren Dezimetern (nach Messungen von E. Frankenberger), Physikal. Zeitschrift 27, 792–795 (1926)
[Mie68] G. Mie, H. Staudinger, R. Signer, H. Johner und J. Hengstenberg, Der polymere Formaldehyd, ein Modell der Zellulose, Zeitschrift für physikal. Chemie 126, 425–448 (1927)
[Mie69] G. Mie und J. Hengstenberg, Zur Arbeit: E. Ott, Röntgenometrische Untersuchungen an hochpolymeren organischen Substanzen, Helvet. chim. Acta 11, 1052 (1928)
[Mie70] G. Mie, Probleme der Quantenelektrik, Ann. der Physik 85, 711–729 (1928)
[Mie71] G. Mie, Einleitung in die Physik, in: A. Eucken, O. Lummer, E. Waetzmann (Hrsg.) Müller-Pouillets Lehrbuch der Physik, 11. Auflage, Erster Band, Erster Teil, 1–70, Friedr.Vieweg und Sohn, Braunschweig 1929
[Mie72] G. Mie, Quantentheorie und Elektrodynamik, Verh. Dt. Phys. Ges. 12, 46 (1931)
[Mie73] G. Mie und E. Frankenberger, Über Präzisionsmessungen von elektrischen Brechungsexponenten nach der 2. Drudeschen Methode, Festschrift zur 10-j. Tätigkeit des magnetischenLabors. Moskau (1931) S. 13–20
[Mie74] G. Mie, Elektrodynamik, in: W. Wien und F. Harms (Hrsg.) Handbuch der Experimentalphysik Bd. 11, Teil 1, XII u. 502 S., Akad. Verlagsges. Leipzig 1932

[Mie75] G. Mie, Elektron, Bd. 2 S. 500–512, Materie, Bd. 6 S. 786–790, Weltäther, Bd.10 S. 649–655, in: R. Dittler, G. Joos, E. Korschelt, G. Link, F. Oltmanns, K. Schaum (Hrsg.) Handbuch der Naturwissenschaften, 2. Auflage, Jena, Gustav Fischer 1932–1935
[Mie76] G. Mie, Naturwissenschaft und Theologie, Akad. Verlagsgesellschaft, Leipzig 1932, 40S., Sonderdruck aus der Zeitschrift *Die christliche Welt*, Nr.22 und 23 (Leopold Klotz Verlag, Gotha)
[Mie77] G. Mie, Geometrie der Spinoren, Ann. der Physik 17, 465–500 (1933)
[Mie78] G. Mie, Die geistige Struktur der Physik, Studien des apologetischen Seminars Bd. 38, 36 S., Bertelsmann, Gütersloh 1934
[Mie79] G. Mie, Zur Erinnerung an Ch.A. Coulomb und A.M. Ampère, Die Naturwissenschaften 24, 369 (1936)
[Mie80] G. Mie, Julius Herweg †, Zeitschrift für techn. Physik 17, 321–323 (1936)
[Mie81] G. Mie, Raum und Zeit in der Physik, Festskrift til Professor Dr. Anathon Aall (1937)
[Mie82] G. Mie, Die Denkweise der Physik und ihr Einfluß auf die geistige Stellung des heutigen Menschen, 37 S., Stuttgart 1937
[Mie83] G. Mie, Der Trägheitsfaktor, Zeitschrift für d. phys. u. chem. Unterricht 51, 90 (1938)
[Mie84] G. Mie, Die göttliche Ordnung in der Natur, 3 Aufsätze, 31 S., Tübingen 1946
[Mie85] G. Mie, Aus meinem Leben, Zeitenwende 19, 733–743 (1948)
[Mie86] G. Mie, Die Grundlagen der Mechanik, VI u. 80 S., Ferdinand Enke Verlag, Stuttgart 1950
[Mie87] G. Mie, Gott in der Natur, in: Wunder der Gnade Gottes in unserem Leben - Gesammelte Zeugnisse, Verlagsbuchhandlung Bethel, Hamburg 1952, S. 94–97

Biographic and Bibliographic Communications about GustavMie

[Bio1] Poggendorffs biographisch-literarisches Handwörterbuch 4 (1904) S. 1009; 5 (1926) S. 852; 6,3 (1938) S. 1728; 7a,3 (1958) S. 304
[Bio2] Reichshandbuch der Deutschen Gesellschaft 2 Wirtschaftsverlag, Berlin 1931, S.1252 (m. Bildn.)
[Bio3] O. Eberhard, Zeugnisse deutscher Frömmigkeit, Leipzig 1938, S. 224–226
[Bio4] um 70. Geburtstag, anonym, Annalen der Physik (5) 33 (1938) S. 185 (m. Bildn. u. Faks.)
[Bio5] F. Emde, Gustav Mie 80 Jahre, Physikalische Blätter, 4, 349–50 (1948)
[Bio6] K.H. Wagner Archiv der elektrischen Übertragung 3 (1949) S. 112
[Bio7] Anonym, Naturwissenschaftliche Rundschau 2 (1949) S. 42–43
[Bio8] H. Hönl, Gustav Mie 85 Jahre, Physikalische Blätter, 9, 508–510 (1953)
[Bio9] Anonym, Naturwissenschaftliche Rundschau 6 (1953) S. 523
[Bio10] W. Kast, Gustav Mie †, Physikalische Blätter, 13, 129–131 (1957)
[Bio11] Anonym, Naturwissenschaftliche Rundschau 10 (1957) S. 161
[Bio12] Anonym, Pharmazeutische Zeitung 102 (1957) S. 201
[Bio13] H. Hönl, Gustav Mie zum Gedächtnis, Gedenkrede auf dem 22. Deutschen Physikertag in Heidelberg 1957, ...
[Bio14] Gerhard Neumann, Naturwissenschaftler und Christ - Gustav Mie, in: Der Sonntag, Gemeideblatt der Ev.-Luth. Landeskirche Sachsens 20 (1965) Nr. 47 S. 4
[Bio15] H. Hönl, Intensitäts- und Quantitätsgrößen, In memoriam Gustav Mie zu seinem hundertsten Geburtstag, Physikalische Blätter, 24, 498–502 (1968)
[Bio16] H. Spehl, Mie, Gustav Adolf Feodor Wilhelm Ludwig, in: B. Ottnad (Hrsg.) Badische Biographien, Neue Folge Bd.III, W. Kohlhammer Verlag Stuttgart, 1990
[Bio17] P. Lilienfeld, Gustav Mie: The Person, Appl. Optics 30, 4696–98 (1991).
[Bio18] H. Rechenberg, Mie, Gustav in: Neue Deutsche Biographie XVII München, 1994

[Bio19] Berühmter unbekannter Rostocker: Gustav Mie-in der Heimat vergessen, an der Uni Freiburg und in Nachschlagwerken geehrt, Mecklenburg-Magazin: Regionalbeilage der Schweriner Volkszeitung, Band 44, 22 (2005)
[Bio20] U. Kreibig, Optische Eigenschaften von Nanopartikeln: 100 Jahre Mie-Theorie, Phys. in unserer Zeit 39 H. 6 (2008), S. 281–287
[Bio21] M.I. Mishchenko und L.D. Travis, Gustav Mie and the evolving discipline of electromagnetic scattering by particles, Bull. Am. Met. Soc., 89, 1853–1861 (2008)
[Bio22] M.I. Mishchenko, Gustav Mie and the fundamental concept of electromagnetic scattering by particles: A perspective, J. Quant. Spec. and Rad. Trans. 110, 1210–1222 (2009)

References

1. D. Cahan, Hist. Stud. Phy. Sci. **15**, 1 (1985)
2. C. Jungnickel, R. McCormmach, *Intellectual Mastery of Nature–Theoretical Physics from Ohm to Einstein,* Vol. 2 The Now Mighty Theoretical Physics 187–1925 (University of Chicago Press, Chicago and London, 1986).
3. G. Willgeroth, *Die Mecklenburg-Schwerinischen Pfarren seit dem Dreißigjährigen Kriege* (Im Selbstverlag des Verfassers, Wismar, 1924)
4. S. Grimme (ed.), *Scharnebeck- gestern und heute* (Verein füer Heimatkunde im Raum Scharnebeck e.V., 2002).
5. W. Neumann, *Die Große Stadtschule zu Rostock in 3 1/2 Jahrhunderten* (Rostock, 1930).
6. G. Schott, Wissenschaftliche Zeitschrift der Universität Rostock **18**, 981 (1969)
7. W. Sarich, Beiträge zur Geschichte der Wilhelm-Pieck-Universität Rostock **13**, 49 (1989)
8. K. Uhle, Wissenschaftliche Zeitschrift der Wilhelm-Pieck-Universität Rostock **34**, 12 (1985)
9. R. Mahnke, Beiträge zur Geschichte der Universität Rostock **17**, 92 (1991)
10. Rektor der Universität Rostock (ed.), *Mögen viele Lehrmeinungen um die eine Wahrheit ringen-575 Jahre Universität Rostock* (Konrad Reich Verlag, Rostock, 1994).
11. G. Becherer, Wiss. Zeitschr. der Univ. Rostock **16**, 825 (1967)
12. L. Königsberger, *Mein Leben* (Carl Winters Universitätsbuchhandlung, Heidelberg, 1919)
13. G. Kern, *Die Entwicklung des Fachs Mathematik an der Universität Heidelberg 1835–1914* (Universität Heidelberg, Staatsexamensarbeit, 1992)
14. O.H. Erdmannsdörffer, Sitzungsberichte der Heidelberger Akademie der Wissenschaften: Math.-naturwiss. Klasse (1936). 3. Abhandlung.
15. E. Wülfing, Sitzungsberichte der Heidelberger Akademie der Wissenschaften: Math.-naturwiss. Klasse (1914). 8. Abhandlung.
16. W. Lorey, Das Studium der Mathematik an den Deutschen Universitäten seit Anfang des 19. Jahrhunderts (B.G. Teubner, Leipzig und Berlin, 1916).
17. O. Lehmann, *Geschichte des Physikalischen Instituts der Technischen Hochschule Karlsruhe (G.* Braunsche Hofbuchdruckerei und Verlag, Karlsruhe i. B, 1911)
18. J. Fricks, *Physikalische Technik oder Anleitung zu Experimentalvorträgen sowie zur Selbstherstellung einfacher Demonstrationsapparate*, (Vieweg, Braunschweig, 1904–1909).
19. P. Knoll, H. Kelker, *Otto Lehmann, Erforscher der Flüssigen Kristalle* (Ettlingen und Frankfurt, 1988)
20. H. Wessel (ed.), *Energie-Information-Innovation: 100 Jahre Verband Deutscher Elektrotechniker*, Geschichte der Elektrotechnik 12 (VDE Verlag GmbH Berlin, 1993).
21. D. Mlynski, Fridericiana **52**, 24 (1996)
22. M. Meyer, Fridericiana **52**, 19 (1996)
23. D. Baird, R. Hughes, A. Nordmann, *Heinrich Hertz: Classical Physicist, Modern Philosopher* (Kluwer Academic Publishers, Dordrecht/Boston/London, 1998)
24. A. Beu, G. Sokoll, *Greifswalder Tafelrunde* (Schelzky & Jeep, Greifswald, 1996)
25. K. Fesser (ed.), *Festschrift aus Anlaß des Jubiläums "150 Jahre Physik in Greifswald"* (Universität Greifswald, Greifswald, Institut für Physik, 1998)

26. D. Alvermann, K.H. Spiess, *Universität und Gesellschaft–Festschrift zur 550-Jahrfeier der Universität Greifswald*, Band I: Die Geschichte der Fakultäten im 19. Und 20. Jahrhundert (Hinstorff, Rostock, 2006).
27. Festschrift zur 500-Jahrfeier der Universität Greifswald, Band II (Universität Greifswald, Greifswald, 1956).
28. F. Schubel, *Universität Greifswald* (Wolfgang Weidlich, Frankfurt am Main, 1960)
29. A. Kleinert, *Johannes Stark: Erinnerungen eines Deutschen Naturforschers* (Bionomica-Verlag, Mannheim, 1987)
30. W. Hergert, Martin-Luther-Universität Wissenschaftliche Beiträge 1990/33 (O32) (1990).
31. W. Hergert, scientia halensis (3), 37 (1995).
32. C. Stephan, Die Naturforschende Gesellschaft in Halle: eine lokale Gesellschaft und das Engagement ihrer Mitglieder insbesondere auf den Gebieten der Physik und Chemie. Ph.D. Thesis, Universität Marburg, Marburg (2006).
33. G. Kohl, Relativität in der Schwebe: Die Rolle von Gustav Mie. No. 209 in Preprint Series (Max Planck Institut for the History of Science, Berlin, 2002).
34. W. Hergert, Scientia halensis (3), 13 (2005).
35. R.K. Merton, *The Sociology of Science: Theoretical and Empirical Investigations* (The University of Chicago Press, Chicago and London, 1973)
36. S. Stigler, Trans. New York Acad. Sci. **39**, 147 (1980)
37. J. Jackson, Am. J. Phys. **76**, 704 (2008)
38. G. Stigler, Transactions of the New York Academy of Sciences **39**, 143 (1980)
39. D. Ballentyne, D. Lovett, *A Dictionary of Named Effects and Laws in Chemistry, Physics and Mathematics* (Chapmann and Hall, Londan, 1980)
40. M. Jammer, *Der Begriff der Masse in der Physik* (Wissenschaftliche Buchgesellschaft, Darmstadt, 1981)
41. D. Plappert, Praxis der Naturwissenschaften-Physik in der Schule **57**, 22 (2008)
42. H. Geiger, K. Scheel (eds.), *Handbuch der Physik, Band X: Thermische Eigenschaften der Stoffe* (Springer, Heidelberg, 1926)
43. E. Grüneisen, Annalen der Physik **39**, 257 (1912)
44. R. Furth, Proc. R. Soc. Lond. **183**, 87 (1944)
45. J. Jones, Proc. R. Soc London A **106**, 441 (1924)
46. J. Jones, proc. R. Soc. London A **106**, 463 (1924)
47. J. Jones, Proc. R. Soc. London A **106**, 709 (1924)
48. S. Erkoç, R. Sever, Phys. Rev. D **33**, 588 (1986)
49. S. Erkoç, R. Sever, Phys. Rev. D **30**, 2117 (1984)
50. E. Kasap, B. Gönül, M. Simsek, Chem. Phys. Lett. **172**, 499 (1990)
51. R. Sever, C. Tezcan, M. Bucurgat, Ö. Yesiltas, J. Math, Chem. 43, 749 (2007).
52. R.S.S. Ikhdair, J. Mol. Struc. (Theochem) **855**, 13 (2008)
53. S.Ö.A. Erkoç, E. Iltan, Chem. Phys. Lett. **135**, 582 (1987).
54. H. Ledbetter, M. Lei, R. Rao, Physica B **159**, 265 (1989)
55. K. Jayachandran, C. Menon, J. Phys. Chem. Sol. **60**, 267 (1999)
56. T. Barakat, O. Al-Dossary, A. Alharbi, Int. J. Nanosci. **6**, 461 (2007)
57. S. Flügge (ed.), *Handbuch der Physik, Astrophysik III, Das Sonnensystem* (Springer, Berlin, 1959)
58. E. Lecher, Annalen der Physik **41**, 850 (1890)
59. A. Sommerfeld, Ann. der Physik und Chemie **67**, 233 (1899)
60. A. Sommerfeld, Nuovo Cimento **9**, 151 (1943)
61. D. Hondros, Über elektromagnetische Drahtwellen. Ph.D. Thesis, Ludwig-Maximilians-Universität München (1909).
62. D. Hondros, Annalen der Physik **335**, 905 (1909)
63. K. Jäger, *Die Entwicklung der Starkstromtechnik in Deutschland*, Teil 2: Von 1890 bis 1920 (VDE-verlag Berlin, 1991).
64. M. Cardona, W. Marx, Physik Journal **3**(11), 27 (2004)
65. W. Marx, Phys. in unserer Zeit **38**(1), 34 (2007)

66. K. Degen, *Beiträge zur Kenntnis kolloidaler Metalllösungen* (Julius Abel, Greifswald, 1903)
67. E. Lischner, *Über die elliptische Polarisation des Lichtes bei der Reflexion an Lösungen anomal dispergierender Substanzen* (Greifswald, 1903).
68. E. Lischner, Ann. der Physik **317**, 964 (1903)
69. H. Siedentopf, R. Zsigmondy, Ann. der Physik Journal **10**, 1 (1903)
70. R. Zsigmondy, *Zur Erkenntnis der Kolloide* (Verlag von Gustav Fischer, Jena, 1905)
71. M. Planck, Ann. d. Physik **306**, 69 (1900)
72. M. Planck, Ann. der Physik **311**, 818 (1901)
73. J.M. Garnett, Trans. Roy. Soc. CCIII A-370, 385 (1904).
74. W. Steubing, *Über die optischen Eigenschaften kolloidaler Goldlösungen* (Johann Ambrosius Barth, Leipzig, 1908)
75. M. Kerker, Appl. Optics **30**, 4699 (1991)
76. N. Logan, J. Opt. Soc. Am. **52**, 342 (1962)
77. N. Logan, Proceedings of the IEEE **53**, 773 (1965)
78. A. Clebsch, J. für Mathematik **61**, 195 (1863)
79. L. Rayleigh, Phil. Mag. **41**, 447 (1871)
80. L. Rayleigh, Proc. Math. Soc. (London) **4**, 253 (1872)
81. L. Rayleigh, Phil. Mag. **12**, 81 (1881)
82. H. Lamb, Proc. Math. Soc. (London) **13**, 51 (1881)
83. L. Lorenz, Vidensk. Selsk. Skrifter **6**, 1 (1890)
84. P. Debye, Ann. d. Physik **30**, 57 (1909)
85. J. Thomson, *Recent Researches in Electricity and Magnetism* (University Press, Oxford, 1893)
86. G.W. Walker, Quart. J. Math. **30**, 204 (1899)
87. G. Walker, Quart. J. Math. **31**, 36 (1900)
88. J.W. Nicholson, Proc. Math. Soc. (London) **9**, 67 (1910)
89. H. Bateman, *The Mathematical Analysis of Electrical and Optical Wave Motion on the Basis of Maxwell's Equations* (University Press, Cambridge, 1915)
90. A.J. Proudman, A. Doodson, G. Kennedy, Phil. Trans. Roy. Soc. **217**, 279 (1917)
91. T.J. Bromwich, Phil. Trans. Roy. Soc. **220**, 189 (1920)
92. G.N. Watson, Phil. Trans. Roy. Soc. **220**, 189 (1920)
93. L. Lorenz, *Oeuvres Sciéntifiques de L. Lorenz* (Librairie Lehmann, Copenhagen, 1896)
94. H. Kragh, Applied Optics **30**, 4688 (1991)
95. M. Pihl, Der Physiker L.V. Lorenz-Eine kritische Untersuchung (Einar Munksgaard, Kopenhagen, 1939).
96. L. Lorenz, Wied. Ann. der Phys. **11**, 70 (1880)
97. P. Debye, Zeitschrift für Physik **9**, 775 (1908)
98. V. Vizgin, *Unified Field Theories* (Birkhäuser Verlag, Basel, Boston, Berlin, 1994)
99. J. Renn (ed.), *The Genesis of General Relativity*, vol. 4 (Gravitation in the Twilight of Classical Physics, The Promise of Mathematics (Springer Verlag, Dordrecht, 2007)
100. J. Renn (ed.), *The Genesis of General Relativity* Vol. 3: Gravitation in the Twilight of Classical Physics: Between Mechanics, Field Theory and Astronomy (Springer, Dordrecht, 2007).
101. L. Corry, Stud. Hist. Phil. Mod. Phys. **30**, 159 (1999)
102. M. Born, L. Infeld, Proc. R. Soc. A **144**, 425 (1934)
103. H. Weyl, *Raum - Zeit - Materie*, 5th edn. (Springer, Dordrecht, 1923)
104. E. Scholz (ed.), *Hermann Weyl's Raum-Zeit-Materie and a General Introduction to His Scinetific Work* (Birkhäuser Publishing Ltd, Switzerland, 2001)
105. J. Illy, in Einstein Studies Vol. 3, Studies in the History of General Relativity, ed. by J. Eisenstaedt, A. Knox (Birkhäuser Publishing Ltd, Switzerland 1992), pp. 244–259.
106. H. Weyl, Jahresbericht der Dt. Mathematiker-Vereinigung **31**, 51 (1922)
107. S. Wolff, Ber. Wissenschaftsgesch. **31**, 372 (2008)
108. S. Wolff, Acta Historica Leopoldina **48**, 41 (2007)
109. R. MacLeod, J. War Culture Studies **2**, 37 (2009)
110. W. Heisenberg, *Deutsche und Jüdische Physik* (Piper Verlag, München, 1992)

111. N. Goldschmidt, Historisch-politische Mitteilungen: Archiv für christlich-demokratische Politik **4**, 1 (1997)
112. U. Kluge, Freiburger Universitätsblätter **102**, 19 (1988)
113. D. Rübsam, H. Schadek (eds.), *Der Freiburger Kreis–Widerstand und Nachkriegsplanung 1933–1945* (Stadtarchiv Freiburg, Freiburg, 1989)
114. K. Hentschel, *Interpretationen und Fehlinterpretation der speziellen und allgemeinen Relativitätstheorie durch Zeitgenossen Albert Einsteins* (Birkhäuser Publishing Ltd, Switzerland, 1990)
115. M. Eckert, K. Märker (eds.), *Arnold Sommerfeld Wissenschaftlicher Briefwechsel* (Deutsches Museum Verlag für Geschichte der Naturwissenschaften und der Technik, Berlin Diepholz München, 2000)
116. A. Einstein, *The collected papers of Albert Einstein* (Princton University Press, UK, 1987)

Chapter 2
Mie Theory: A Review

Thomas Wriedt

Abstract In optical particle characterisation and aerosol science today light scattering simulations are regarded as an indispensable tool to develop new particle characterisation techniques or in solving inverse light scattering problems. Mie scattering and related computational methods have evolved rapidly during the past decade such that scattering computations for spherical scatterers a few order of magnitudes larger, than the incident wavelength can be easily performed. This significant progress has resulted from rapid advances in computational algorithms developed in this field and from improved computer hardware. In this chapter the history and a review of the recent progress of Mie scattering and Mie-related light scattering theories and available computational programs is presented. We will focus on Mie scattering theories but as there is much overlap to related scattering theories they will also be mentioned where appropriate. Short outlines of the various methods are given. This review is of course biased by my interest in optical particle characterisation and my daily reading.

2.1 Introduction

When Gustav Mie wrote his classic paper on light scattering by dielectric absorbing spherical particles in 1908 he was interested in explaining the colourful effects connected with colloidal Gold solutions. Nowadays, the interest in Mie's theory is much broader. Interests range from areas in physics problems involving interstellar dust, near-field optics and plasmonics to engineering subjects like optical particle characterisation. Mie theory is still being applied in many areas because scattering particles or objects are often homogeneous isotropic spheres or can be approximated in such a way that Mie's theory is applicable.

T. Wriedt (✉)
Institut für Werkstofftechnik, Badgasteiner Str., 328359 Bremen, Germany
e-mail: thw@iwt.uni-bremen.de

W. Hergert and T. Wriedt (eds.), *The Mie Theory*, Springer Series in Optical Sciences 169, DOI: 10.1007/978-3-642-28738-1_2,

On the 100th anniversary of Mie's theory a number of review papers, special papers and conference proceedings were published [1–10] commemorating his 1908th paper.

In this chapter I would like to review the history and the state of the art in Mie scattering as it developed over the previous decades. I will give a short description of the various extensions to Mie scattering available and provide information about computation programs making reference to review papers where available. For more profound reviews the interested reader is referred to the cited review articles.

2.2 Nonspherical Particles

In this review I am concerned with spherical scattering particles. Nevertheless, to guide the reader interested in the broader aspects of light scattering I will also briefly consider nonspherical particles in this section.

A number of light scattering theories also suitable for nonspherical particles have been developed, all having their pros and cons. Extensive overviews of available theories have been published by Wriedt [11], Kahnert [12] and more recently by Veronis and Fan [13]. A review of this subject with emphasis on colloid science has been written by Niklasson and Vargas [14]. A paper by Zhao et al. [15] provides an overview of the methods that are currently being used to study the electromagnetics of silver and gold nanoparticles. Of historic interest might be the critical review by Bouwkamp [16] who presents the progress in classical diffraction theory up to 1954. In this paper both scalar and electromagnetic problems are discussed. This report may also serve as an introduction to general diffraction theory. Interesting reviews on related subjects of nanooptics and metamaterials have recently been published by Myroshnychenko et al. [17] and by Veselago et al. [18].

2.3 History of Mie's Theory

In 1908, Gustav Mie published his famous paper on simulation of the colour effects connected with colloidal Gold particles [19]. In this paper he gave a first outline of how to compute light scattering by small spherical particles using Maxwell's electromagnetic theory. With his first computations he managed to explain the colour of gold colloids changing with diameter of the Gold spheres, which was later interpreted in terms of surface plasmon resonances. First computations of scattering diagrams for larger spheres of diameters up to 3.2λ are presented by *Richard Gans* (1880–1954) [20] and *Hans Blumer* [21] in 1925. The early study of scattering by a sphere is traced by *Nelson A. Logan* [22], who comments on Blumer's results, that he missed the regular undulations in the scattering diagram because of numerical mistakes. This paper is interesting reading for all interested in the history of light scattering.

Electromagnetic scattering by a homogeneous, isotropic sphere is commonly referred to as Mie theory, although *Gustav Mie* (1868–1957) was not the first to formulate this electromagnetic scattering problem. Before him *Alfred Clebsch* (1833–72), solving the elastic point source scattering problem of a perfectly rigid sphere using potential functions [23], and *Ludvig Lorenz* (1829–91) [24, 25] contributed to this problem. In 1909, *Peter Debye* (1884–1966) considered the related problem of radiation pressure on a spherical particle [26] utilising two scalar potential functions like Mie. Therefore, plane wave scattering by a homogeneous isotropic sphere is also referred to as Lorenz-Mie theory [27], or even Lorenz-Mie-Debye theory [28]. The incorrect name Lorentz-Mie theory is also quite commonly used (e.g. in Burlak [29]). *Milton Kerker* (1920) provides an extensive postscript on the history of scattering by a sphere in his book [30].

In 1890, Danish physics *Ludvig Lorenz* published essentially the same calculation on the scattering of radiation by spheres [24]. This paper was not much noticed at the time because it was written in Danish and it remained hardly known even after it had been translated to French [25]. Also, Lorenz did not connect his derivation with the Maxwells theory of electromagnetics [31, 32]. For information about Ludvig Lorenz, see the article by *Helge Kragh* [33] who outlines his career and his contributions to optical theory.

Cardona and Marx [34] commented that Mie's 1908 paper was almost ignored until about 1945 but its importance rose with increasing interest in colloids starting from the 1950s such that hundred years after publication the paper is still much cited [35] with currently 291 citations a year according to the *Web of Science*. *Google* currently gives 4,064 citations and the *Web of Science* 4,958. The paper is called Dornröschen (Sleeping Beauty) by the researchers at Information Retrieval Services of Max Planck Society [36] because of its low recognition considering the number of citations during the first years after its publication. But in the 1930s, contemporary scientists acknowledged the importance of his contribution. In a special issue devoted to Mie's 70th birthday *Ilse Fränz-Gotthold* and *Max von Laue* [37] dedicate their paper to "*Gustav Mie* whom physics owes the mathematical treatment of diffraction by a sphere".

Apparently one of the first English language versions of Mie theory was published by *Harry Bateman* (1882–1946) [38]. Mie's original paper was translated into the English language as late as 1976 by the Royal Aircraft Establishment in the UK [39] and 2 years later by Sandia Laboratories in US [40]. Recently, a Spanish translation of the original paper became available [41] and Chinese, Hebrew and Italian translations are on the way [42]. For information about typographical errors in the original paper please refer to the Mie Translation Project [42]. Those typographical errors have been corrected in the Spanish version of the paper [41].

Early citations of Mie's theory make reference to the classical textbook on electromagnetic theory by *Julius Adams Stratton* (1901–1994) [43]. In outlining the theory, Stratton made use of so-called vector spherical wave functions (VSWF) as first introduced by *William Webster Hansen* (1909–1949) [44]. Later references were made to the classic optics book by *Max Born* (1882–1970) and *Emil Wolf* (1922) [45] who used the Debye potential in their derivation. *Hendrik Christoffel van de Hulst* (1918–2000) [46] and Bohren and Hufman [47] follow the Stratton approach.

An up-to-date version of the derivation of Mie's theories has been published in Appendix H of the recent book by Le Ru and Etchegin [48] and in a book chapter by Enguehard [49]. Both provide an outline on how to solve the electromagnetic scattering problem in the case of a spherical particle using Mie's theory which give the most useful mathematical expressions of Mie theory and its derivatives and also information about implementation.

2.4 Mie Algorithms

Understandably, prior to the development of electronic computers in thc middle of the last century there were not many papers written on computing scattering problems using Mie's theory since the computational labour involved in evaluating functions such as Riccati-Bessel functions was quite extreme.

Even with the rise of the computer it took some time before stable algorithms were developed. Gradually, several generally reliable and stable scattering programs were published. Early well-known algorithms were published by Giese [50] and Dave [51]. In a report Cantrell [52] reviews the different numerical methods for the accurate calculation of spherical Bessel functions as needed in Mie's theory. Deirmendjann [53] presents computational results of scattering parameters of polydispersions in a number of tables, produced by application of the Mie theory. According to Cantrell [52], Deirmendjian et al. [54] appear to have been the first to calculate the logarithmic derivative of the Riccati-Bessel function to evaluate the Mie coefficients. Dave [51] improved upon the approach of Deirmendjian et al. by applying downward recursion to the correct calculation of logarithmic derivative of the Riccati-Bessel function. The IBM report by Dave from 1968 [51] was still sent out on request in the 1990s.

Apparently at that time there was also independent research in this field such that one can find papers not often cited such as Metz and Dettmar [55]. With multiple databases and easy Internet search available today, the existence of other early and at that time available Fortran program can be confirmed such as for example the program by Maguire [56].

Nowadays, a number of efficient algorithms and Fortran programs are available. A major step was the program *MIEV0* written by Wiscombe [57, 58], which is based on Lentz's continued-fraction method for the calculation of spherical Bessel functions [59]. The program is well-tested and widely used. The Supermidi program by Gréhan and Gouesbet [60] is also based on Lentz's algorithm. The authors give numerical results over a wide range of size parameters and refractive indices. The advantage of Lentz's method is that errors do not accumulate as can occur with the use of faster recurrence relation methods [61].

It has been demonstrated that Mie's theory can now be successfully applied up to size parameters of 10.000 [62–64]. Recently, Gogoi et al. [65] presented a new efficient and reliable computer program written in the C programming language to compute the scattering matrix elements.

2.5 Spheres in an Absorbing Medium

Originally, the Mie theory was restricted to a nonabsorbing ambient medium although some real-world applications require that the absorbing surrounding media be accounted for in the theory. This is a topic in colloid science and computer graphics.

Apparently one of the first derivation of the Lorenz-Mie theory for a scattering sphere immersed in an absorbing host medium is presented by Mundy et al. [66] and by Bohren and Gilra [67]. Similar extension to the Mie theory was studied by Quinten and Rostalski [68], Sudiarta [69] and Frisvad et al. [70]. It was found that the absorption of the ambient medium can alter the scattering efficiency and the scattering pattern of a sphere. The effect on the absorption efficiency is much less. To extend such theories for light scattering by coated sphere immersed in an absorbing medium is a small step [71].

2.6 Coated Spheres

As it is easy to consider spherical scatterers there are many extensions of the Mie theory covering different aspects. A theory for a coated dielectric sphere was first published by Aden and Kerker [72]. The Fortran code BART by Arturo Quirantes [73] is based on the Aden-Kerker theory to calculate light-scattering properties for coated spherical particles. In the program polydispersity is included for either core, coating or entire particle.

A basic Fortan (BHCOAT) code is printed in the appendix of the book by Bohren and Huffman [47]. An advanced algorithm is given by Toon and Ackerman [74]. An algorithm for a sphere having two coatings has been presented by Kaiser and Schweiger [75]. These theories have also been extended to spherical particles consisting of multiple layers by Li Kai [76]. Another algorithm for plane wave and shaped beam scattering by a multilayered sphere has been published by Wu et al. [77]. Even in recent times, improved algorithms have been published on this subject [78].

2.7 Distorted Spheres

To compute scattering by a slightly distorted sphere Martin [79, 80] developed a first-order perturbation theory. He derives formulae bearing close resemblance to the zero-order ones encountered in Mie scattering theory, expressing the perturbation applied to the surface of the sphere in angular functions identical to those used in the spherical harmonic expansion. The approximation is that the particle be smooth and not deviate far from sphericality.

2.8 Magnetic Spheres

With growing interest in magnetic nanostructures there is also interest in Mie scattering of magnetic spheres. A Mie theory allowing for magnetic media where the media properties includes a nonzero permeability μ has been developed early by Kerker et al. [81, 82]

A Fortran code for spherical scatterers with both a complex permittivity and a complex permeability is listed in a report by Milham [83]. Tarento et al. [84] considered Mie scattering of magnetic spheres extending the classical Mie scattering approach to a media where the dielectric constant is no more a real number but a tensor with a gyrotropic form. A usable code for scattering by a sphere having a different magnetic permeability than the surrounding is included in the Matlab program by *Christian Mätzler* [85].

2.9 Chiral and Anisotropic Spheres

During the previous decades, much attention has been focused on the light scattering interaction between chiral and anisotropic particles and electromagnetic waves, as a result of numerous applications in the electromagnetic scattering and antenna theory. The historical background, and a general description, of the subject of electromagnetic chirality and its applications can be found in the books by Lakhtakia and Varadan [86] and Weiglhofer and Lakhtakia [87]. A isotropic chiral medium or an optical active medium is rotationally symmetric but not mirror symmetric. Only circularly polarised plane waves can propagate in the chiral medium without a change in their state of polarisation.

A scattering sphere can also be chiral. The extension of Mie's theory to an optically active sphere was published by Bohren [88] and later extended to a sphere with a chiral shell [89]. Bohren devised a transformation to decompose the problem such that he could consider two independent modes of propagation in the chiral medium, namely left- and right-circularly polarised waves. In the limit of no chirality the solution is identical to the Mie solution.

Theories and programs for such types of scatterers have been published by Bohren in his thesis for a chiral sphere [90].

Using the Bohren decomposition Hinders and Rhodes [91] investigate the problem of scattering by chiral spheres embedded in a chiral host medium.

As colour pigments and crystals are often anisotropic there is some interest to extend Mie's theory to such kind of scattering particles. Stout et al. [92] established a vector spherical harmonic expansion of the electromagnetic field propagating inside an arbitrary anisotropic medium to solve this problem. A similar problem of a uniaxial anisotropic sphere was considered by Geng et al. [93] to find the coefficients of the scattered field. Droplets of liquid crystals may be considered as spherical particles with radial anisotropy. This scattering problem is discussed by Qiu et al. [94] using an extension of Mie theory.

2.10 Scattering by a Short Pulse

The history of the investigation of time domain scattering by a sphere, given an incident short pulse, can be traced back at least to Kennaugh [95]. The interest was to investigate short-pulse radar systems for target discrimination. The incident pulse is expanded in a Fourier series. For each wavelength in the Fourier series Mie scattering is computed.

Ito et al. [96] analyse the transient responses of a perfectly conducting sphere. For numerical calculation a Laplace transform algorithm is used. This method was later extended to a dielectric sphere [97] to analyse the multiple reflections within a sphere. Bech et al. [98] provided an extension by way of the Fourier-Lorenz-Mie-Theory (FLMT) [99], which permits the scattering of a laser beam and the separation of individual scattered light orders using the Debye series expansion to compute scattering by a femtosecond Laser pulse to determine the particle diameter from the time differences between the scattered light orders of the Debye series. The Debye series expansion is analogous to a multiple internal reflection treatment such that it is similar to geometrical optics where each term has a clear physical interpretation such as diffraction, reflection, refraction and higher order refractions.

2.11 Nanosized Spheres

With a nanosized nobel metal particle of size lower than about 20 nm, various modifications, extensions and corrections to Mie's original theory are needed to take into account that "sharp" boundary conditions do not hold in the nanoscale [100]. In his recent survey paper [101] Kreibig lists among others the following supplementary models to the Mie theory, i.e. non plane-wave incident field, non step-like boundary condition and a particle size-dependent dielectric function. Applying these extensions help to explain measured absorption spectra of Ag nanoparticles and plasmon polaritons.

2.12 Gaussian Beam Scattering

The traditional Lorenz-Mie theory describes the scattering by a spherical homogeneous dielectric particles illuminated by an incident plane wave. However, in many applications such as optical particle sizers, confocal microscopes, optical trapping and optical manipulation scattering by a focused laser beam is quite often of interest. In his recent book Quinten [102] refers to Möglich [103] who in 1933 derived expression for the expansion coefficients of the field in the focal point of a lens system to compute scattering in the Rayleigh approximation.

There are different concepts available to handle the problem of laser beam scattering. With a point matching approach you just need an analytical description of the incident field at the matching points on the particle surface. This method which is also suitable for a spherical scatterer has been implemented in the Multiple Multipole Program (MMP) by Evers et al. [104]. But in this review we are concerned with the Mie theory and within this theory various beam expansion methods have been developed.

A laser beam with Gaussian intensity distribution can be expanded into spherical vector wave functions or into a spectrum of plane waves [105]. The Generalised Lorenz Mie Theory (GLMT) developed by *Gérard Gouesbet* and co-workers is based on the first approach computing beam shape coefficients. It has recently been reviewed by Gouesbet [106]. Plane wave expansion is used by Albrecht et al. [99] and it is integrated into the Null-Field Methods with Discrete Sources (NFM-DS) developed by Doicu and co-workers [107]. Doicu et al. [105] give quantitative comparison of the localised approximation method and the plane wave spectrum method for the beam shape coefficients of an off-axis Gaussian beam. Usually the plane wave expansion method is considered computationally inefficient compared to the GLMT method [108].

Since its foundation at the end of the 1970s *Gérard Gouesbet* and his co-worker *Gérard Gréhand* developed the Generalised Lorenz-Mie Theory (GLMT). Initially the theory only included a single spherical scatterer but the theory was later extended to aggregates of spheres [109], coated spheres [110], a sphere with an eccentrically located spherical inclusion [111] and pulsed Laser scattering [112]. Based on the Davis formulation of the Gaussian beam [113] Gouesbet and co-workers developed a theory to expand the beam into vector spherical wave functions which is compatible with the Lorenz-Mie Theory. In the first implementation the expansion was restricted to on-axis spheres. This theory was later extended to an arbitrarily positioned sphere. Parallel to this development the localised approximation was investigated to reduce the computational demand in computing the beam shape coefficients. Since 1998 the Livre GMTL, fully outlining the GLMT theory and providing printouts of Fortran programs, is available via the Internet [114]. This year Gouesbet and Gréhan published a revised version with the Springer publisher [106].

Moore et al. [115] developed an alternate generalisation of the Lorenz-Mie theory wherein the incident fields are complex focus fields which are nonparaxial generalisations of Gaussian beams. This approach results in an easily calculable closed form for the coefficients in the multipolar expansion of the incident field that results from a beam passing through a high numerical aperture lens system.

2.13 Near Fields

With the recent rise of plasmonics there seems to be an increasing interest in computing the near field or the internal field by scattering particles. There is especially interest in morphological resonances and in plasmon resonances. There are not many programs available which focus on near field computation. Most programs consider the far field approximation of the radiating spherical vector wave functions to compute scattering patterns in the far field. The T-Matrix programs on the disk accompanying the book by Barber and Hill [116] allow for the simulation of the internal and external near field intensity distribution by a scattering sphere.

Near field and internal field computations of a spherical particle in a Gaussian laser beam can be done using the Windows program GLMT Champ Internes by *Loic Mees* [117].

To compute near fields by a number of scattering spheres you can use the extension of *Yu-lin Xu's GMM* program by *Moritz Ringer*. He extended GMM while working on his PhD thesis [118]. His Fortran program GMM-Field allows for the computation of the near field. GMM-Dip gives the near field with a dipole as the incident field. Apparently *Giovanni Pellegrini* did a similar extension of the GMM program in this PhD thesis [119]. A paper on parallising multiple scattering and near field computation by coated spheres was recently published by Boyde et al. [120].

The LightScatPro Matlab program by *Sylvain Lecler* used for his thesis [121] is also suitable for near field computation for light scattering by a number of spheres.

2.14 Longitudal Modes

Basically, in the Mie theory, transverse wave modes are dominant. However, there is the question about what the theory would look like if there were also longitudinal modes created inside the scattering sphere.

Ruppin [122] extended Mie theory to the case of small metal spheres whose material sustains the propagation of longitudinal waves. This theory is also outlined in the book by Borghese et al. [123]. Ruppin demonstrated that a slight shift of the surface plasmon peak towards higher frequencies would occur. According to Quinten in his recent book [102], all effects connected with longitudal modes have not yet been identified experimentally. Travis and Guck [124] recently revisited the Mie theory and also include longitudinal vector spherical wavefunctions in the expansion of the internal field. The authors argue that in the optical wavelength range the longitudinal wavenumber is almost purely imaginary such that for particles larger than about 2 nm the longitudinal plasma modes are evanescent and the normal Mie approximation is acceptable.

2.15 Aggregates of Spheres

Scattering by an aggregate or cluster of spheres will be considered in this section. There is a recent extensive review on this topic by Okada [125] giving a timeline of the development in the field since the 1960s. Multiple scattering by a number of spheres dates at least as far back as Trinks [126], who used multipole expansions combined with the translation addition theorem for spherical vector wave functions to express the scattered fields of all other spheres at the origin of a sphere.

Levine et al. [127] extended the work of Trinks by considering two Rayleigh particles, arbitrarily positioned in relation to the incident plane wave. Lenine et al. also made use of Debye scalar potentials as did Trinks. In the following years various rigorous methods extending the Mie theory were developed. Bruning and Lo [128, 129] presented the first solution for two spheres using recursive relations for the translation coefficients.

Later, Fuller et al. [130] considered the two-sphere problem using the order-of-scattering approach, a chain of spheres [131] and an arbitrary cluster of spheres [132].

Next, two well-known rigorous methods were developed to simulate the light scattering by an aggregate of spheres, Mackowski [133] and Mackowski and Mishchenko [134] developed the multiple scattering T-matrix algorithm and the generalised multisphere Mie-solution (GMM) method developed by Xu [135, 136]. Today, there are efficient Fortran programs made available by Mackowski [134] and Yu-lin Xu [135].

These multiple scattering programs are commonly used to study the scattering and absorption properties of soot aggregates [137].

The theory has recently been extended to clusters of rotationally symmetric particles [138] and arbitrary shaped particles [6] using the T-Matrix method.

2.16 Parallelisation

Another current issue is the parallelisation of executing programs. Various methods such as Open Multiple-Processing (OpenMP) and Message Passing Interface (MPI) are available. Within the T-Matrix methods the NFM-DS has recently been parallelised [139]. With the T-Matrix method the computational demand lies mostly in the need to compute the surface integrals. This can of course be easily done by dividing the surface of the scatterer into different sections and then computing the integral of this sections using a separate parallel thread. Mackowski et al. [140] recently implemented parallelisation of a multiple scattering program. A paper on parallizing multiple scattering by coated spheres was published by Boyde et al. [120].

2.17 Further Topics

Not all extensions of the Mie theory can be extensively covered in this chapter. Therefore, in this section some interesting topics not treated above will be briefly mentioned.

Travis and Guck revised the Mie theory [124] using a modern T-matrix formalism and discuss inclusion of longitudinal components of the dielectric permittivity tensor.

Tagviashvili [141] considered the classical Mie theory for electromagnetic radiation scattering by homogeneous spherical particles in the dielectric constant (ε) $\rightarrow 0$ limit and TE field resonances in the visible spectrum are demonstrated.

With Mie scattering morphological resonances show up. This can be used to determine the diameter of a spherical droplet to a high precision. Ward et al. [142] used a white light LED for illumination to study the broadband Mie backscattering from optically levitated aerosol droplets to observe the morphological Mie resonances simultaneously across a spectral range from 480 to 700 nm. Correlating the measured resonances to the mode order and mode number using Mie theory the droplet size can be determined with an accuracy of ± 2 nm.

As the scattering of a plane electromagnetic wave by a dielectric sphere is considered a canonical problem, Mie's theory is still widely used as a standard reference to validate methods intended for more complex scattering problems [143, 144]. For example, Khoury et al. [145] compare COMSOL's finite element method (FEM) algorithm with the Mie theory for solving the electromagnetic fields in the vicinity of a silica–silver core–shell nanoparticle. It is demonstrated that the COMSOL FEM algorithm also generates accurate solutions of the near field. Takano and Liou [146] developed a geometrical optics ray-tracing program to compute scattering for concentrically stratified spheres and in the validation section show that it produces the same general scattering features as the "exact" Lorenz–Mie theory for a size parameter of 600.

2.18 Further Reading

There are various topics not considered in this review. For these topics I suggest some further reading. For the basics on the Mie theory readers may refer to the classical books by Stratton [43], van de Hulst [46] (available as a budget Dover edition), Born and Wolf [45], Kerker [30] and Bohren and Huffman [47]. For a more recent and extensive discussion on the topic refer to Grandy [27]. Morphological resonances in Mie scattering are extensively covered in the book by Davis and Schweiger [28].

Scattering particles are commonly positioned on a plane interface and particle–surface scattering interaction will have to be considered in the corresponding extension of Mie theory. This is for example a problem when simulation particle surface scanners detect particle conterminants on wafers. For these who are interested in such kinds of problems there are methods based on the T-matrix method taking account of the plane surface. Such methods are fully outlined in the books by Doicu et al. [147] and Borghese et al. [123]. There is a section on supported nanoparticles in the recent book by Quinten [102]. For those interested in plasmonics and the optics of nanosized noble metalparticles, Quinten's book gives an overview of analytical and numerical models for the optical response of nanoparticles and nanoparticle systems. Nanooptics is the topic of the book by Novotny and Hecht [148] who also care for the topic of particles on surfaces in an extensive way.

2.19 Available Programs

Those looking for computational programs to compute Mie scattering or to solve related scattering problems there are various web sites available which provide information about these topics or give links to available computer programs. First of all there is the portal *ScattPort* [149] which is intended to be a Light Scattering Information Portal for the light scattering community. The basic concept of this information portal includes features such as database of available computer programs and up-to-date information related to the subject of light scattering, e.g. conference announcements, available jobs, new books, etc. Many of the listed programs are based on Mie theory or are extensions of Mie theory. Further information about the history and concept of the portal is presented by Hellmers and Wriedt [150].

SCATTERLIB by Flatau [151] is another web site with an emphasis to list open access computational programs suitable to compute light scattering by particles. In the list Mie theory has been implemented in various traditional programming languages originating in Fortran, but are also available in Pascal, C++ and Java. It has also been programmed in computer algebra systems such as Maple and Mathematica. Additionally, it is available in numerical mathematical systems such as IDL and Matlab [149] and 3 years ago the Mie theory has even been programmed on a Java enabled Mobile Phone using a Matlab clone [152].

2.20 Sample Simulation Results

To graphically demonstrate the differences of various kinds of extensions of the Mie's theory we next plot some scattering diagrams. All programs used are available from *ScattPort* [149]. As basic parameters for this comparison, we choose a particle diameter of $d = 700\,\text{nm}$ and an incident wavelength of $\lambda = 500\,\text{nm}$. Figure 2.1 shows the scattering pattern of a dielectric sphere with a refractive index of $n = 1.5$ calculated using the Mie theory. Both p- and s-polarisation are plotted. The next figures demonstrate the variation in the scattering pattern with change in the properties of the scattering sphere. Even a small amount of absorption gives a change in the scattering pattern as seen from Fig. 2.2. Also, a coating with a shell having a slightly lower refractive index gives another scattering diagram (Fig. 2.3). The next

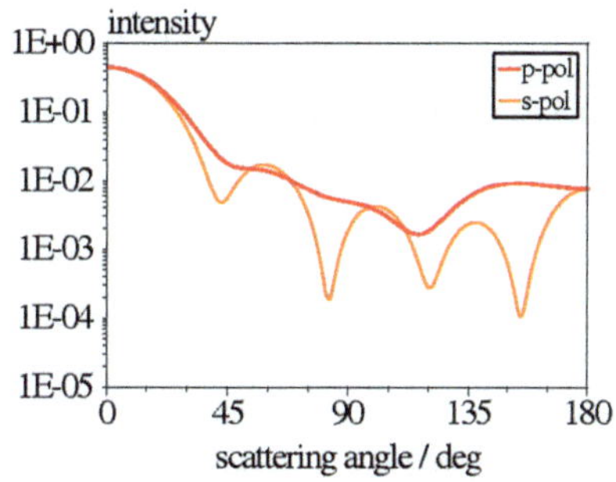

Fig. 2.1 Scattering pattern of a homogeneous nonabsorbing sphere $d = 700\,\text{nm}$, $\lambda = 500\,\text{nm}$, $n = 1.5$

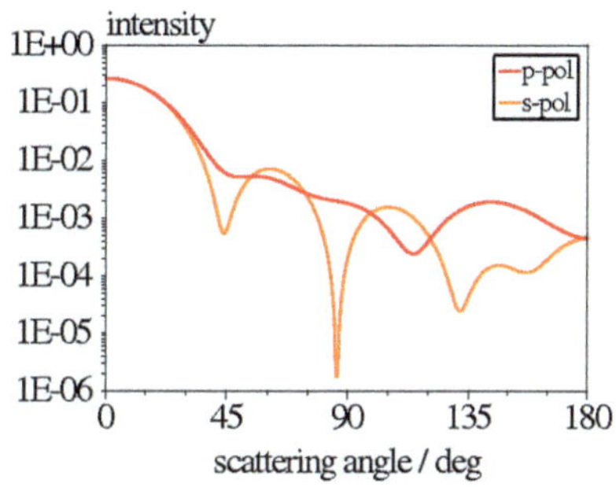

Fig. 2.2 Scattering pattern of a homogeneous absorbing sphere $d = 700\,\text{nm}$, $\lambda = 500\,\text{nm}$, $n = 1.5 + \text{i}\,0.1$

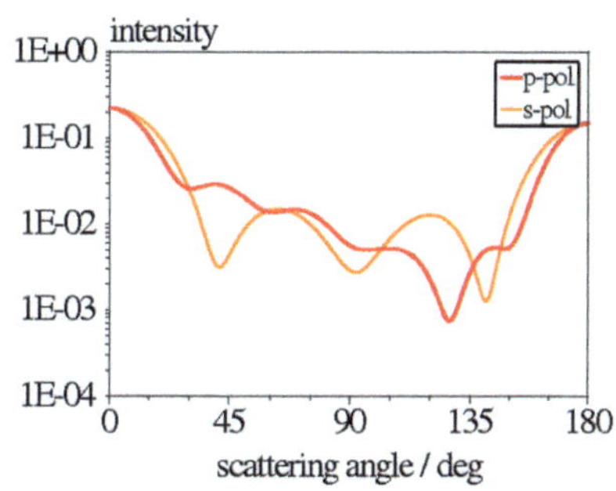

Fig. 2.3 Scattering pattern of a homogeneous coated sphere $d1 = 700\,\text{nm}$, $d2 = 500\,\text{nm}$, $\lambda = 500\,\text{nm}$, $n1 = 1.5$, $n2 = 1.33$

scattering pattern (Fig. 2.4) is for a sphere possessing a refractive index gradient gradient with $n = 1.33$ at the centre of the sphere and $n = 1.5$ at the boundary of the sphere. This gradient is approximated by 50 steps. With a chiral or optically active sphere there are also great differences in the scattering pattern (Fig. 2.5).

If the sphere is positioned in a highly focused laser beam with Gaussian intensity distribution pronounced differences to the first scattering diagram show up (Fig. 2.6). A slightly perturbed sphere gives also only slight perturbation in the scattering diagram (Fig. 2.7). With an unisotropic particle having three different refractive indices in the three cartesian coordinates the scattering pattern is no longer rotationally symmetric (Fig. 2.8).

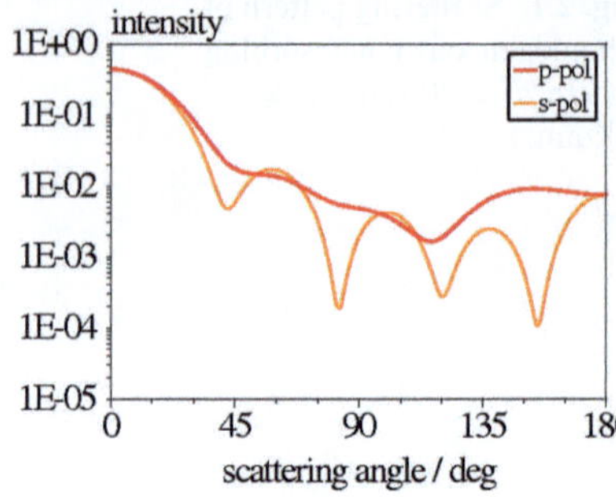

Fig. 2.4 Scattering pattern of a sphere possessing a refractive index gradient $d = 700\,\text{nm}$, $\lambda = 500\,\text{nm}$, gradient $n = 1.33$–1.5

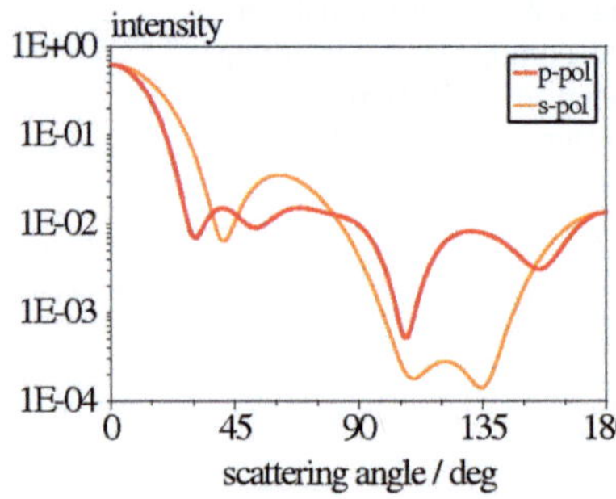

Fig. 2.5 Scattering pattern of a homogeneous chiral sphere $d = 700\,\text{nm}$. $\lambda = 500\,\text{nm}$, $n = 1.5$, kb=0.1

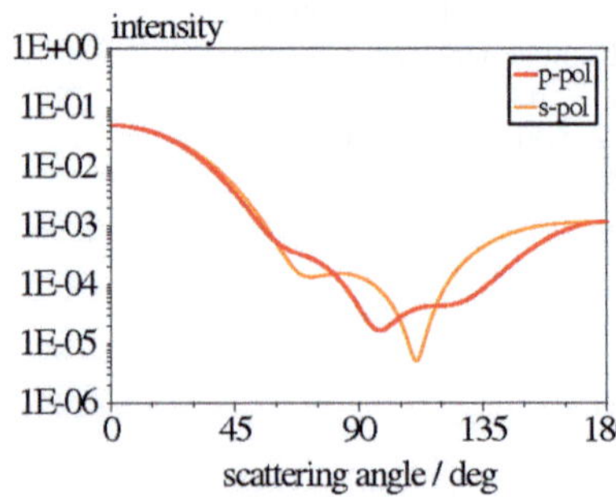

Fig. 2.6 Scattering pattern of a homogeneous sphere in a Gaussian laser beam $d = 700\,\text{nm}$, $\lambda = 500\,\text{nm}$, beam waist diameter = 500 nm, $n = 1.5$

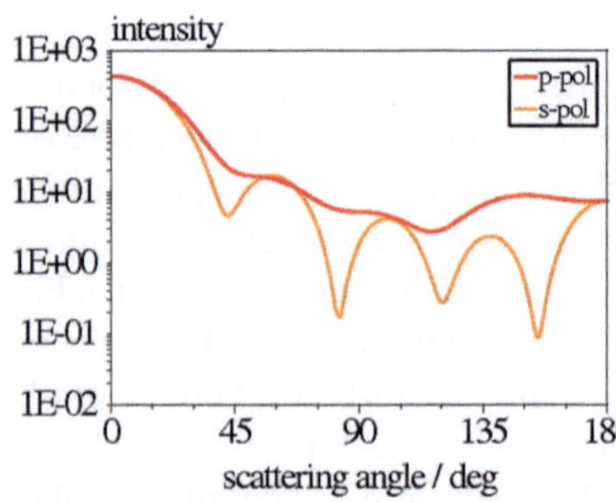

Fig. 2.7 Scattering pattern of a homogeneous slighly rough sphere beam $d = 700\,\text{nm}$, $\lambda = 500\,\text{nm}$, epsilon (perturbation) = 0.2, $n = 1.5$

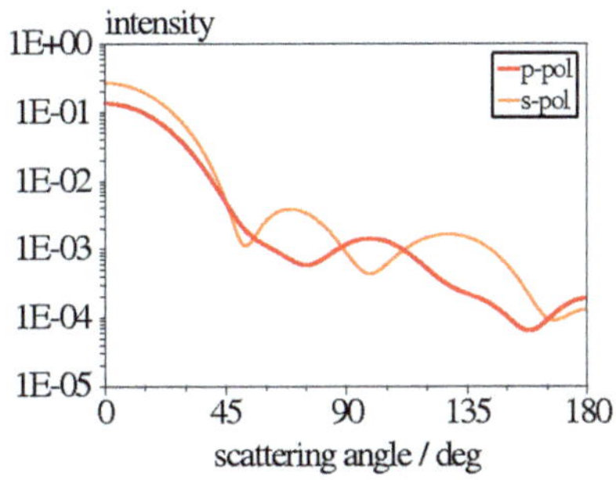

Fig. 2.8 Scattering pattern of an unisotropic sphere beam $d = 700\,\text{nm}$, $\lambda = 500\,\text{nm}$, $n_x = 1.2$, $n_y = 1.3$, $n_z = 1.5$

2.21 Conclusion

What is the current state of Mie's theory? It seems to be as lively as ever. It is continuously used for code validation. Wherever particles can be considered spherical, Mie scattering is applicable. It is applied everyday in diffraction-based instruments to characterise particles. To this day, new approaches in computing Mie scattering are developed and the methods are continuously extended to cover related scattering problems such as coated spheres.

It is not only used for educating students and the validation of more advanced theories but also it is the basis of radiative transfer, Lidar and optical particle characterisation. Even today there is still progress in programming and new programs which are based on Mie theory or which extend Mie theory in some respect show up every year.

In this chapter an overview of the progress in developing light scattering programs suitable for Mie-related scattering problems has been given. A short description of each technique has been presented and suitable references have been provided such that the reader can find more detailed information about the various methods and sources of computer programs.

The state of the art in numerical light scattering modelling is progressing rapidly especially with fast advancing research fields such as nanophotonics and near-field optics. With almost all concepts there was much progress in the recent years.

The recent advances in computer hardware and the development of fast algorithms with reduced computational demands and memory requirements have made the exact numerical solution of the problem of scattering from large scattering particles highly feasible. Today, scattering by spherical particles of realistic sizes can be computed in an efficient way. Especially, the number of open source programs and development projects is increasing continuously.

Acknowledgments I acknowledge the support of this work by Deutsche Forschungsgemeinschaft (DFG). I like to thank Jannis Saalfeld and Vincent Loke for language editing.

References

1. M.I. Mishchenko, L.D. Travis, Bull. Am. Meteorol. Soc. **89**(12), 1853 (2008)
2. M.I. Mishchenko, J. Quant. Spectrosc. Radiat. Transf. **110**(14–16), 1210 (2009)
3. U. Kreibig, Physik in unserer Zeit **39**(6), 281 (2008)
4. W. Hergert, T. Wriedt (eds.), *Mie Theory 1908–2008* (Universität Bremen, Bremen, 2008)
5. T. Stübinger, U. Köhler, W. Witt, U. Köhler, W. Witt, in *Particulate Systems Analysis 2008* (Stratford-upon-Avon, UK, 2008)
6. T. Wriedt, in *Mie Theory 1908–2008*, ed. by W. Hergert, T. Wriedt (Universität Bremen. Bremen, 2008), pp. 17–21
7. H. Horvath, J. Quant. Spectrosc. Radiat. Transf. **110**(11), 783 (2009)
8. H. Horvath, J. Quant. Spectrosc. Radiat. Transf. **110**(11), 787 (2009)
9. V. Garbin, G. Volpe, E. Ferrari, M. Versluis, D. Cojoc, D. Petrov, New J. Phys. **11**(1), 013046 (2009)
10. M. Kolwas, Comput. Methods Sci. Technol. **16** Special Issue (2), 108 (2010)
11. T. Wriedt, Part. & Part. Syst. Charact. **15**(2), 67 (1998)
12. F.M. Kahnert, J. Quant. Spectrosc. Radiat. Transf. **79–80**, 775 (2003)
13. G. Veronis, S. Fan, *in Surface Plasmon Nanophotonics* (Springer, Dortrecht, 2007), pp. 169–182
14. G.A. Niklasson, W.E. Vargas, *Encyclopedia of Surface and Colloid Science*, 2nd edn. pp. 3346–3358 (2006)
15. J. Zhao, A.O. Pinchuk, J.M. McMahon, S. Li, L.K. Ausman, A.L. Atkinson, G.C. Schatz, Acc. Chem. Res. **41**(12), 1710 (2008)
16. C.J. Bouwkamp, Rep. Prog. Phys. **17**(1), 35 (1954)
17. V. Myroshnychenko, J. Rodríguez-Fernández, I. Pastoriza-Santos, A.M. Funston, C. Novo, P. Mulvaney, L.M. Liz-Marzán, F.J. F. Javier García de Abajo, Chem. Soc. Rev. (2008)
18. V. Veselago, L. Braginsky, V. Shklover, C. Hafner, J. Comput. Theor. Nanosci. **3**, 189 (2006)
19. G. Mie, Annalen der Physik **330**(3), 377 (1908)
20. R. Gans, Annalen der Physik **381**(1), 29 (1925)
21. H. Blumer, Zeitschrift für Physik **32**, 119 (1925)
22. N.A. Logan, Proc. IEEE pp. 773–785 (1965)
23. A. Clebsch, Journal für Mathematik, Band **61**, Heft 3, 195 (1863)
24. L. Lorenz, Det Kongelige Danske Videnskabernes Selskabs Skrifter **6**. Raekke, 6. Bind 1, 1 (1890)
25. L. Lorenz, Oeuvres scientifiques de L. Lorenz. Revues et annotées par H. Valentiner (Libraire Lehmann & Stage, Copenhague, 1898), chap. Sur la lumière réfléchie et réfractée par une sphère (surface) transparente., pp. 403–529
26. P. Debye, Annalen der Physik, Vierte Folge, Band **30**. No. 1, 57 (1909)
27. W.T. Grandy, *Scattering of waves from large spheres* (Cambridge Univ. Press, Cambridge, 2000)
28. E. Davis, G. Schweiger, *The Airborne Microparticle* (Springer, Berlin, Heidelberg, 2002)
29. G. Burlak, *The Classical and Quantum Dynamics of the Multispherical Nanostructures* (Imperial College Press, London, 2004)
30. M. Kerker, *The Scattering of Light, and Other Electromagnetic Radiation: and other electromagnetic radiation* (Academic Press, New York, London, 1969)
31. P. Pesic, Sky in a bottle (MIT Press, Cambridge, 2005)
32. O. Keller, Prog. Opt. **43**, 257 (2002)
33. H. Kragh, Appl. Opt. **30**(33), 4688 (1991)
34. M. Cardona, W. Marx, Physik Journal **11**, 27 (2004)
35. W. Marx, Phys. Unserer Zeit, pp. 34–39 (2007)
36. B.F. Bowman, www.biochem.mpg.de/iv/AgFN_Bibliometrie.pdf. accessed 15. Aug. 2008. (2008)
37. I. Fränz-Gotthold, M. von Laue, Annalen der Physik **425**(3), 249 (1938)

38. H. Bateman, Cambridge University Press, 1915. http://www.archive.org/download/mathematicalanal00baterich/mathematicalanal00baterich.pdf (1915)
39. G. Mie, Contributions to the optics of turbid media, particularly of colloidal metal solutions. Tech. Rep. Library Translation 1873, RAE-Lit-Trans-1873, Royal Aircraft Establishment (1976). http://diogenes.iwt.uni-bremen.de/vt/laser/papers/RAE-LT1873-1976-Mie-1908-translation.pdf
40. G. Mie, Contributions on the optics of turbid media, particularly colloidal metal solutions – Translation. Technical report. SAND78-6018. National Translation Center, Chicago, ILL, Translation 79–21946, Sandia Laboratories, Albuquerque, New Mexico (1978)
41. G. Mie, Consideraciones sobre la óptica de los medios turbios, especialmente soluciones coloidales. Traducción: Arturo Quirantes Sierra. Technical report http://www.ugr.es/aquiran/ciencia/mie/mie1908_spanish.pdf, Universidad de Granada (2007)
42. A. Quirantes, http://www.ugr.es/aquiran/mie.htm. (2007)
43. J.A. Stratton, *Electromagnetic theory* (McGraw-Hill, New York, 1941)
44. W.W. Hansen, Phys. Rev. **47**, 139 (1935)
45. M. Born, E. Wolf, *Principles of Optics*, 7th edn. (Cambridge University Press, 1999)
46. H.C. van de Hulst, *Light Scattering by Small Particles* (Dover Publications, New York, 1981)
47. C.F. Bohren, D.R. Huffman, *Absorption and Scattering of Light by Small Particles* (Wiley-VCH, Berlin, 1983)
48. E.C.L. Ru, P.G. *Etchegoin, Prinziples of surface-enhanced Raman spectroscopy and related plasmonic effects* (Elsevier, Amsterdam, 2009)
49. S.V. (ed.), *Thermal Nanosystems and Nanomaterials* (Springer, Heidelberg, 2009) Mie Theory and the Discrete Dipole Approximation, Franck Enguehard, 2009
50. R.H. Giese, Elektron. Rechenanl. **3**, 240 (1961)
51. J. Dave, Subroutines for computing the parameters of the electromagnetic radiation scattered by a sphere, Rep. No. 320–3237. Technical report, IBM Scientific Center, Palo Alto, Calif. (1968)
52. C.D. Cantrell, Numerical methods for the accurate calculation of spherical Bessel functions and the location of Mie resonances. Tech. Rep. http://www.utdallas.edu/cantrell/ee6481/lectures/bessres1.pdf
53. D. Deirmendjann, *The Electromagnetic Scattering on Spherical Polydispersion* (Elsevier, New York, 1969)
54. D. Deirmendjann, R. Clasen, W. Viezee, J. Opt. Soc. Am. **51**(6), 620 (1961)
55. H.J. Metz, H.K. Dettmar, Kolloid-Zeitschrift und Zeitschrift für Polymere **192**(1–2), 107 (1963)
56. B. Maguire, J. Aerosol Sci. **2**(4), 417 (1971)
57. W. Wiscombe, Mie scattering calculations: Advances in technique and fast, vector-speed computer codes. Technical report, Technical Note NCAR/TN-140+STR, National Center for, Atmospheric Research (1979)
58. W.J. Wiscombe, Appl. Opt. **19**(9), 1505 (1980)
59. W.J. Lentz, Appl. Opt. **15**(3), 668 (1976)
60. G. Grehan, G. Gouesbet, Appl. Opt. **18**(20), 3489 (1979)
61. A.R. Jones, J. Phys. D: Appl. Phys. **16**(3), L49 (1983)
62. G.G. Siu, L. Cheng, J. Opt. Soc. Am. B **19**(8), 1922 (2002)
63. H. Du, Appl. Opt. **43**(9), 1951 (2004)
64. J. Shen, PIERS Online **1**, 691 (2005)
65. A. Gogoi, A. Choudhury, G. Ahmed, J. Mod. Opt. **57**, 2192 (2010)
66. W.C. Mundy, J.A. Roux, A.M. Smith, J. Opt. Soc. Am. **64**(12), 1593 (1974)
67. C.F. Bohren, D.P. Gilra, J. Colloid Interface Sci. **72**(2), 215 (1979)
68. M. Quinten, J. Rostalski, Part. & Parti. Syst. Charact. **13**(2), 89 (1996)
69. I.W. Sudiarta, P. Chylek, J. Quant. Spectrosc. Radiat. Transf. **70**(4–6), 709 (2001)
70. J.R. Frisvad, N.J. Christensen, H.W. Jensen, ACM Trans. Graph. **26** (2007)
71. W. Sun, N.G. Loeb, Q. Fu, J. Quant. Spectrosc. Radiat. Transf. **83**(3–4), 483 (2004)
72. A.L. Aden, M. Kerker, J. Appl. Phys. **22**(10), 1242 (1951)

73. A.Q. Sierra, http://www.ugr.es/aquiran/codigos.htm (2007)
74. O.B. Toon, T.P. Ackerman, Appl. Opt. **20**(20), 3657 (1981)
75. T. Kaiser, G. Schweiger, Comput. Phys. **7**(6), 682 (1993)
76. L. Kai, P. Massoli, Appl. Opt. **33**(3), 501 (1994)
77. Z.S. Wu, L.X. Guo, K.F. Ren, G. Gouesbet, G. Gréhan, Appl. Opt. **36**(21), 5188 (1997)
78. L. Liu, H. Wang, B. Yu, Y. Xua, J. Shen, China Part. **5**, 230 (2007)
79. R.J. Martin, J. Mod. Opt. **40**(12), 2467 (1993)
80. R. Martin, J. Mod. Opt. **42**(1), 157 (1995)
81. M. Kerker, C.L. Giles, D.S.Y. Wang, J. Opt. Soc. Am. **72**, 1826 (1982)
82. M. Kerker, D.S. Wang, C.L. Giles, J. Opt. Soc. Am. **73**(6), 765 (1983)
83. M.E. Milham, Electromagnetic scattering by magnetic spheres: theory and algorithms. Technical report ERDEC-TR-207, ADA289798, Edgewood Research Development and Engineering Center (1994)
84. R.J. Tarento, K.H. Bennemann, P. Joyes, J. Van de Walle, Phys. Rev. E **69**(2), 026606 (2004)
85. C. Mätzler, Matlab functions for mie scattering and absorption. Technical report, Research Report No. 2002–08, Institute of Applied Physics, University of Bern (2002)
86. A. Lakhtakia, V.K. Varadan, V.V. Varadan, *Time-Harmonic Electromagnetic Fields in Chiral Media* (Springer, Berlin, 1989)
87. W.S. Weiglhofer, *A* (Introduction to complex mediums for optics and electromagnetics (SPIE Press, Lakhtakia, 2003)
88. C.F. Bohren, Chem. Phys. Lett. **29**, 458 (1974)
89. C.F. Bohren, J. Chem. Phys. **62**(4), 1566 (1975)
90. C.F. Bohren, Light scattering by optically active particles (University of Arizona, Tucson 1975)
91. M. Hinders, B. Rhodes, Il Nuovo Cimento D **14**, 575 (1992). 10.1007/BF02462344
92. B. Stout, M. Neviére, E. Popov, J. Opt. Soc. Am. A **23**(5), 1111 (2006)
93. Y.L. Geng, X.B. Wu, L.W. Li, B.R. Guan, Phys. Rev. E **70**(5), 056609 (2004)
94. C.W. Qiu, B. Luk'yanchuk, J. Opt. Soc. Am. A **25**(7), 1623 (2008)
95. E. Kennaugh, Proc. Inst. Radio Eng. **49**, 380 (1961)
96. A. Itoh, T. Hosono, IEICE Trans. Electron. E **75**C(1), 107 (1992)
97. A. Itoh, T. Hosono, Electron. Commun. Japan (Part II: Electronics) **78**(11), 10 (1995)
98. H. Bech, A. Leder, Optik—International Journal for Light and Electron Optics **117**(1), 40 (2006)
99. H.E. Albrecht, H. Bech, N. Damaschke, M. Feleke, Optik **100**, 118 (1995)
100. U. Kreibig, M. Vollmer, *Opt. properties met. clust.* (Springer, Berlin, 1995)
101. U. Kreibig, Appl. Phys. B: Lasers Opt. **93**(1), 79 (2008)
102. M. Quinten, *Optical Properties of Nanoparticle Systems, Mie and Beyond* (Wiley-VCH, Berlin, 2011)
103. F. Möglich, Annalen der Physik **409**(8), 825 (1933)
104. T. Evers, H. Dahl, T. Wriedt, Electron. Lett. **32**(15), 1356 (1996)
105. A. Doicu, T. Wriedt, Opt. Commun. **136**(1–2), 114 (1997)
106. G. Gouesbet, *G* (Gréhan, Generalized Lorenz-Mie Theories (Springer-Verlag Berlin Heidelberg, 2011)
107. A. Doicu, T. Wriedt, Y. Eremin, *Light Scattering by Systems of Particles, Null-Field Method with Discrete Sources: Theory and Programs*, vol. 124 (Springer, Berlin; New York, 2006)
108. J.A. Lock, Appl. Opt. **34**(3), 559 (1995)
109. G. Gouesbet, G. Grehan, J. Opt. Pure Appl. Opt. **1**(6), 706 (1999)
110. L. Mees, G. Gouesbet, G. Grehan, Opt. Commun. **282**(21), 4189 (2009)
111. J.J. Wang, G. Gouesbet, Y.P. Han, G. Grehan, J. Opt. Soc. Am. Opt. Image Sci. Vis. **28**(1), 24 (2011)
112. L. Mees, G. Gouesbet, G. Grehan, Appl. Opt. **40**(15), 2546 (2001)
113. L.W. Davis, Phys. Rev. A **19**(3), 1177 (1979)
114. G. Gouesbet, G. Grehan, B. Maheu, K.F. Ren, Electromagnetic scattering of shaped beams (Generalized Lorenz-Mie Theory). Technical report, LESP, CORIA, INSA de Rouen (1998). http://ren.perso.neuf.fr/ThesisHDRBooks/Livre_glmt.pdf

115. N.J. Moore, M.A. Alonso, Opt. Express **16**(8), 5926 (2008)
116. P.W. Barber, S.C. Hill, *Light scattering by particles: Computational methods* (World Scientific Publishing, Singapore, 1990)
117. L. Mees. GLMT Champ Internes (2008). http://www.scattport.org/index.php/programs-menu/mie-type-codes-menu/130-glmt-champ-internes
118. M. Ringler, Plasmonische Nahfeldresonatoren aus zwei biokonjugierten Goldnanopartikeln. (2008)
119. G. Pellegrini, Modeling the optical properties of nanocluster-based functional plasmonic materials. Ph.D. thesis, Tesi di dottorato, Università degli Studi di Padova, Padova (2008)
120. L. Boyde, K.J. Chalut, J. Guck, Phys. Rev. E **83**(2), 026701 (2011)
121. S. Lecler, Etude de la diffusion de la lumière par des particules sub-microniques. Ph.D. thesis, Thèse. l'Université Louis Pasteur, Strasbourg (2005)
122. R. Ruppin, Phys. Rev. B **11**, 2871 (1975)
123. F. Borghese, P. Denti, R. Saija, *Scattering from model nonspherical particles: theory and applications to environmental physics* (Springer- Berlin, New York, 2007)
124. K. Travis, J. Guck, Biophys. Rev. Lett. **2**, 179 (2006)
125. Y. Okada, in *Light Scattering Reviews*, ed. by A.A. Kokhanovsky, 5th edn. Springer Praxis Books (Springer Berlin, New York, 2010) pp. 3–35
126. W. Trinks, Annalen der Physik **414**(6), 561 (1935)
127. S. Levine, G.O. Olaofe, J. Colloid Interface Sci. **27**(3), 442 (1968)
128. J. Bruning, Y. Lo, IEEE Trans. Antennas Propag. **19**(3), 378 (1971)
129. J. Bruning, Y. Lo, IEEE Trans. Antennas Propag. **19**(3), 391 (1971)
130. K.A. Fuller, G.W. Kattawar, R.T. Wang, Appl. Opt. **25**(15), 2521 (1986)
131. K.A. Fuller, G.W. Kattawar, Opt. Lett. **13**(2), 90 (1988)
132. K.A. Fuller, G.W. Kattawar, Opt. Lett. **13**(12), 1063 (1988)
133. D.W. Mackowski, J. Opt. Soc. Am. A **11**(11), 2851 (1994)
134. D.W. Mackowski, M.I. Mishchenko, J. Opt. Soc. Am. A **13**(11), 2266 (1996)
135. Y. lin Xu, Appl. Opt. **34**(21), 4573 (1995)
136. Y.l. Xu, Appl. Opt. **36**(36), 9496 (1997)
137. L. Liu, M.I. Mishchenko, W.P. Arnott, J. Quant. Spectrosc. Radiat. Transf. **109**(15), 2656 (2008)
138. Y.l. Xu, J. Quant. Spectrosc. Radiat. Transf. **89**(1–4), 385 (2004)
139. V. Schmidt, T. Wriedt, J. Quant. Spectrosc. Radiat. Transf. **110**(14–16), 1392 (2009)
140. D. Mackowski, M. Mishchenko, J. Quant. Spectrosc. Radiat. Transf. **112**, 2182 (2011)
141. M. Tagviashvili, Phys. Rev. A **81**, 045802 (2010)
142. A.D. Ward, M. Zhang, O. Hunt, Opt. Express **16**(21), 16390 (2008)
143. T. Grosges, A. Vial, D. Barchiesi, Opt. Express **13**(21), 8483 (2005)
144. R.J. Zhu, J. Wang, G.F. Jin, Optik—Int. J. Light Electr. Opt. **116**(9), 419 (2005)
145. C.G. Khoury, S.J. Norton, T. Vo-Dinh, Nanotechnology **21**(31), 315203 (2010)
146. Y. Takano, K.N. Liou, Appl. Opt. **49**(20), 3990 (2010)
147. A. Doicu, R. Schuh, T. Wriedt, in *Light Scattering Reviews*, 3 edn. by A.A. Kokhanovsky (Springer-Berlin, Germany, 2008), pp. 109–130
148. L. Novotny, B. Hecht, *Principles of Nano-Optics* (Cambridge Univ. Press, Cambridge, 2006)
149. T. Wriedt, http://www.scattport.org
150. J. Hellmers, T. Wriedt, J. Quant. Spectrosc. Radiat. Transf. **110**(14–16), 1511 (2009)
151. P.J. Flatau, http://code.google.com/p/scatterlib/
152. T. Wriedt, J. Quant. Spectrosc. Radiat. Transf. **109**(8), 1543 (2008)

115. T.J. Moore, M.A. Alonso, Opt. Express 16(8), 5926 (2008)
116. P.W. Barber, S.C. Hill, Light Scattering by Particles: Computational Methods (World Scientific Publishing, Singapore, 1990)
117. [illegible] (2013), http://www.scatport.org/index.php/programs[illegible]
118. M. Kerker, [illegible] (1969)
119. [illegible]
120. [illegible], J. Quant. Spectr. Rad. Transf. 112, [illegible] (2011)
121. [illegible] la lumière par des particules sub-micronique. [illegible] (2005)
122. [illegible] (1972)
123. [illegible]
124. [illegible]
125. [illegible]
126. [illegible]
127. [illegible]
128. [illegible]
129. [illegible]
130. [illegible]
131. [illegible]
132. [illegible]
133. [illegible]
134. [illegible]
135. [illegible]
136. [illegible]
137. [illegible]
138. [illegible]
139. [illegible]
140. [illegible]
141. [illegible]
142. [illegible]
143. [illegible]
144. [illegible] Springer, Berlin, Germany, 2004) pp. 105–130
145. [illegible]
146. [illegible], http://www.scattport.org
147. [illegible]
148. [illegible]
149. [illegible] 109, 83 (2008)

Chapter 3
From Theories by Lorenz and Mie to Ontological Underdetermination of Theories by Experiments

Gerard Gouesbet

Abstract In 1908, relying on Maxwell's electromagnetism, Gustav Mie published a theory describing the interaction between an electromagnetic plane wave and a sphere characterised by its diameter and its complex refractive index. A theory treating the same problem has however been published by Lorenz about 20 years before. Furthermore, this theory did not rely on Maxwell's electromagnetism, but did rely on a mechanical theory of ether. Yet, although relying on two deeply different ontologies, both theories generate the same experimental predictions. Experts in epistemology would say that they are empirically equivalent. The existence of conflicting theories being empirically equivalent may be discussed in the framework of a thesis called Quine's thesis of underdetermination of theories by experiments to which this chapter is devoted.

3.1 Introduction

In light scattering theory, the most famous theory is likely to be the one published by Gustav Mie in 1908 [1]. This theory describes the quasi-elastic interaction between an electromagnetic plane wave and a homogeneous sphere defined by its (arbitrary) diameter and its (arbitrary) complex refractive index. It allows one to calculate scattered fields outside the sphere, internal fields, phase relations and various cross sections, including for radiation pressure forces. The theory is built using Maxwell's electromagnetism.

The year 2008 was the year of the 100th anniversary of Mie's paper which has been commemorated in several conferences (i) GAeF conference on "Light scattering: Mie and More", 3 and 4 July 2008, in Karlsruhe, Germany [2] (ii) 11th conference on electromagnetic and light scattering, 7–12 September 2008, in Hatfield, UK [3]

G. Gouesbet (✉)
LESP, CORIA, UMR CNRS, Rouen University,
6614 Rouen, France
e-mail: gerard.gouesbet@coria.fr

W. Hergert and T. Wriedt (eds.), *The Mie Theory*, Springer Series in Optical Sciences 169, DOI: 10.1007/978-3-642-28738-1_3,

(iii) "Mie theory 1908–2008: Present developments and interdisciplinary aspects of light scattering", 15–17 September 2008, University of Halle-Wittenberg, Germany (present book) and (iv) International Radiation Symposium IRS 2008, 3–8 August, Foz do Iguaçu, Brazil [4].

Nevertheless, actually, about 20 years before Mie's paper, another theory, treating the same problem, had been produced by Lorenz [5, 6]. Hence, rather than speaking of Mie's theory, I always preferred to give the name of Lorenz–Mie to the theory under discussion, namely Mie's theory. Several historical papers are available concerning the formulation of Lorenz, and its relationship with that of Mie [7–10]. According to these references, Mie's and Lorenz' theories lead to the same experimental predictions, that is to say, they are empirically equivalent. The work of Lorenz has been overlooked, certainly in part because it has been written in Danish but, most importantly, certainly as another reason, it happens that the theoretical framework used by Lorenz was not that of Maxwell, but another one, of a mechanical nature, relying on the existence of ether. Therefore, although the two theories are empirically equivalent, they rely on two different visions of the world, i.e. two different ontologies. In other words, we are facing two conflicting empirically equivalent theories. When I heard of this story, I found it weird. It took me much time before being able to discuss this issue in a satisfactory enough context. This context pertains to epistemology and is described by what can be called the Quine's thesis of underdetermination of theories by experiments (Quine's underdetermination in short), to which this paper is essentially devoted, i.e. I am taking the pretext of theories by Lorenz and Mie to take the reader to a tour and detour in the landscape of epistemology.

Quine's underdetermination is related to the issue of realism in science. Many scientists have opted for a strong (sometimes also called: naive) realistic interpretation of science. Such a realistic vision has been challenged by many philosophical doctrines, since a long time, such as by various kinds of idealisms, obviously including the transcendental idealism of Kant. More recently, the realistic interpretation has been also challenged by quantum mechanics. Quine's underdetermination thesis constitutes another source of troubles for those who would like to share a realistic interpretation of science. This thesis may receive various formulations. One of them states that, for any one theory formulation, there is another one that is empirically equivalent to it but logically incompatible with it, and cannot be rendered logically equivalent to it by any reconstrual of predicates [11]. This is a rather abstract enunciation but, for this chapter, it is sufficient to remark that the previous statement is in agreement with the following one: we may face several theories which are empirically equivalent but conflicting. There are different ways for theories to conflict. One of them is indeed at the stage when we are facing two logically incompatible theories. However, we shall consider another possible way to encounter conflicting theories, namely theories may conflict when they exhibit different ontologies (furnitures of the world). Such a situation defines what I call the ontological underdetermination. Ontological underdetermination is easier to grasp than Quine's underdetermination. In particular, while Quine's underdetermination thesis has not been satisfactorily exemplified (in the viewpoint of physicists, to the best of my present knowledge), I may provide quite decent examples for ontological underdetermination.

This chapter is organised as follows. After a discussion concerning the concept of underdetermination of theories by experiments (Sect. 3.2), I shall discuss, in Sect. 3.3, three examples, or case studies, relevant to physical sciences (i) one pertaining to classical mechanics: Newton's formulation of classical mechanics versus Hamilton-Jacobi's formulation (ii) causal interpretations of quantum mechanics (mainly: pilot wave theory) versus Copenhagen interpretation and, last but not least, at least for a scatterist (iii) Mie's theory versus Lorenz' theory. In Sect. 3.4, discriminations between undecidable ontologically under-determined theories are later achieved using ampliative arguments, that is to say arguments allowing one to decide between undecidables. Section 3.5 is the conclusion.

3.2 Quine's and Ontological Underdeterminations, with Precursors

Before any attempt to systematise the concepts of underdetermination and associated undecidability, it may be useful to provide a flavour of them by invoking the history of sciences. As far as I have been able to find, the first relevant historical notice is to be found in Osiander (laconically mentioned by Harré [12], but without referring to Quine). Osiander has been in charge of printing the famous book by Copernicus (*De Revolutionibus Orbium Coelestium*). In a rather contested famous preface [13], [14], he wrote that "Copernicus hypotheses did not need to be true or even probable. It would be sufficient if they *saved the appearances*". Answering a letter from Copernicus, wondering whether he should publish his book, Osiander replied that "as far as he is concerned, he always believed that hypotheses are not articles of faith but starting points for computations, so that, even if they are false, it does not matter, if they simply exactly represent the phenomena". In another letter, he wrote that "Aristotelians and theologians will be easily pacified if they are told that several hypotheses can be used to explain the same apparent movements; and that the present hypotheses are not proposed because they are actually true, but because they are the most convenient to compute the apparent combined motions."

Later on, in the same context (geocentrism versus heliocentrism), but regarding the famous Galileo Galilei's affair, the Pope Urbain VIII stated that a hypothesis which gives results does not necessarily represent the reality, for there may be several ways to explain how the Lord produces phenomena. This statement from Urbain VIII is reflected in a statement from Simplicio [15] concerning the ebb and flow of tides: ... "God ... can give to water the alternated motion that we observe ... by using various ways, unthinkable to our mind." Let us however mention that this statement might have been motivated by a political and strategic desire to pacify the catholic hierarchy (although, this is a statement from Simplicio, not from Salviati).

Later, we have to mention Descartes, quoted by both Duhem [16] and Van Fraassen [17]. In the third part of his principles of philosophy [18], Descartes indeed stated that what he wrote should be taken as a hypothesis, possibly far remote from the truth. But, even in this case, Descartes believed he would have done much if only all the things that can be deduced from the causes he exhibited would entirely agree with the experiments. Furthermore, such a false hypothesis, agreeing with experiments, would be as useful to the life as a correct one, because it can be used in the same way as the correct one to produce the effects that we want to produce.

There are actually many places in the principles of philosophy by Descartes where such considerations are put forward. Let us simply mention another one, taken from the fourth part of the principles [19], in which Descartes wrote that, "... in the same way than an industrious watchmaker can produce two watches which indicate the same time, without any difference which can be perceived for the external eye ... it is certain that God possesses an infinity of various means, allowing him to make that all the things of this world look as they now look, without however any possibility for the human mind to know, among all the means it could use, which one he actually used." And Descartes reiterated the idea that, in some sense, it does not matter: ..and I would believe having done enough if the causes I explained are such that all the effects they can produce are like the ones that we see in the world, without informing myself whether it is by them or by others that they are produced. I even believe that it is equivalently useful in life to know the causes we have imagined or to know the true ones.

Just after, Descartes, referring to Aristotle, also wrote that regarding the things which our senses do not perceive, it is sufficient to explain what they could be, though perhaps they are not as we describe them. For Van Fraassen [17], Descartes here *voiced a thoroughly empirical sentiment. Under a given interpretation, a physical theory describes how things are - one way they might indeed be - but the story is not unique, and for an empiricist it need not be unique*. The idea that we may have several stories is indeed in the spirit of Quine's underdetermination.

In most cases, the history of Quine's underdetermination, as has been reported up to now, heavily relies on the invocation of the Supreme Being, namely God. The process from a theological understanding of Quine's underdetermination to its secular interpretation is already achieved with Pascal. In a letter to the Révérend Père Noël [20], he wrote: "In the same way than a same cause may produce several different effects, a same effect can be produced by several different causes. Therefore, when one discusses, with our human abilities, of the motion and of the stability of the Earth, all phenomena of movements and retrogradations of the planets may perfectly be derived from the hypotheses of Ptolemy, of Tycho, of Copernicus and of many others, although from all of them only one can be correct".

With Duhem, and his famous book on the structure of physical theories [16], we arrive at one of the two thinkers who gave his name to a Duhem-Quine thesis (holism), not to be confused however with Quine's underdetermination [11], although the name of Duhem can also be (and has been) associated with Quine's underdetermination.

Duhem expressed the issue under discussion, in general terms, by comparing, and opposing, the French and the English styles of making physics. As he stated, for geometricians, like Laplace or Ampère (on the French side, but forgetting Descartes, a theological precursor of Duhem), it would be absurd to give to a law of physics two different theoretical explanations, and to express the idea that two such different explanations can be simultaneously valid. But, in contrast, for scientists building models, like Thomson or Maxwell (on the English side), there is no contradiction in the fact that a given law could be apprehended by two different models.

Lifting the issue to still more general terms, beyond a plain comparison between different national ways of thinking and working, he raised the question to know whether it is allowed to symbolise several distinct groups of experimental laws, or even a single group of laws, by using several theories, each one relying on hypotheses which cannot be reconciled with the hypotheses underlying the other ones. For Duhem, the answer is that, "*if we restrain ourselves to the use of pure logical reasons* (emphasised because Duhem himself emphasised), then one cannot prevent a physicist to use several irreconcilable theories to represent various sets of laws, or even a single set of laws." And, he concluded, one cannot condemn incoherence in the physical theory. This is a point of view which agrees with the opinion of Poincaré, quoted by Duhem, according to which two contradictory theories, if we do not combine them, and if we do not search with them to do things thoroughly, can both be useful tools for research.

Among other discussions from Duhem, not always convincing, there are discussions of physical laws or of theoretical facts that have nothing to do with formally and consistently structured theories such as relativity or quantum mechanics (that Duhem could not know) or Newtonian mechanics and Maxwell's electromagnetism (that Duhem did know).

Up to now, with all these precursors, we possess some general formulations of Quine's underdetermination without however any concrete example, nor demonstration, that could usually satisfy a physicist. Could possibly Quine himself be more convincing? The reader might refer to "Word and Object", regarded as the most important book from Quine [21] or to the "Pursuit of Truth" [22], his last book. The French reader may also find a synthetic account of the work of Quine in Ref.[23]. But a specific and specialised report is available in a famous paper by Quine [11] to which I shall better refer.

At the beginning of his paper, Quine stated what could be the essence of the underdetermination thesis as follows: *If all observable events can be accounted for in one comprehensive scientific theory- one system of the world, to echo Duhem's echo of Newton- then we may expect that they can all be accounted for equally in another, conflicting system of the world.* Such an expectation may rely on the examination of *how scientists work. For they do not rest with mere inductive generalizations of their observations: mere extrapolation to observable events from similar observed events. Scientists invent hypotheses that talk of things beyond the reach of observation. These hypotheses are related to observation only by a kind of one-way implication, namely, the events we observe are what a belief in the hypotheses would have led us to expect. These observable consequences of the hypotheses do not, conversely, imply the hypotheses.* In the words of Quine, this defines a doctrine, a doctrine saying that *natural science is empirically under-determined,* a doctrine from which we can say that it *is plausible insofar as it is intelligible, but it is less readily intelligible that it may seem.* The doctrine is afterward exposed in general terms (i.e. again without any convincing example) which alone do not allow one to be clearly convinced, something which is implicitly acknowledged by Quine when, instead of using the word "truthfulness", he is content of using the word "plausibility". Redefining more carefully the thesis, Quine arrived to the following formulation: "*underdetermination says that for any one theory formulation there is another that is empirically equivalent to it but logically incompatible with it, and cannot be rendered logically equivalent to it by any reconstrual of predicates.*" Eventually, after more discussions, he landed on the following form, that "*our system of the world is bound to have empirically equivalent alternatives which, if we were to discover them, we would see no way of reconciling by reconstrual of predicates.*" Whether the thesis, under this *vague and modest* form, is true or not, this remains for Quine *an open question*. However Quine, as he said, believed to it. A logician might roar in face of such an incongruous object: a (scientific? philosophical? rational?) thesis which cannot be demonstrated but must be a matter of faith. At this stage, we may certainly agree with Harré [12] when he spoke of the *myth of the underdetermination of theories by data.*

One of the most difficult things to correctly state Quine's underdetermination thesis is to precisely define what is a theory and/or to define to which kinds of theories the thesis applies. Consider for instance the old theory of Bohr correctly explaining the energy levels of the hydrogen atom. The energy levels predicted by Bohr are in perfect agreement with the energy levels predicted by Schrödinger from his equation. Should we consider that we have here a simple example of the application of Quine's underdetermination thesis? Certainly not, and this for at least one reason, namely that all predictions of Bohr's model do not agree with all predictions of Schrödinger's wave mechanics. Therefore, we have here two theories (let us accept the name of theory) which are not experimentally equivalent in all aspects.

Quine himself was aware of this difficulty. On the one hand, one of the formulations of the thesis, the one I provided at the beginning of this section, refers to *all observable events*. If we interpret this expression as concerning what it means, i.e. all observable events, whether they have been already actually observed, in the past, or are to be potentially observed, in the future, then the underdetermination thesis should be applied to the whole of science, however, not yet completed. The validity of the thesis when applied to the whole of science is certainly something overwhelmingly difficult to convincingly establish. On the other hand, the thesis is of no value for weak theories, i.e. *theories that imply no rich store of observation conditionals,* cases where the thesis can be trivially and therefore uninterestingly satisfied. So, at the best of our present common understanding, examples of Quine's underdetermination thesis should be searched for theories which are rich enough without being, because unrealistic, the whole of science.

Another complementary point of view may be gained if we attach a domain of validity (or of application) to each theory, from the simple ones (fire burns, water is humid) to the most achieved one, the *ultimate theory* which would generate the whole of science, encompassing all domains of validity, if any. If we affirm, by a *fiat*, that we are just interested with energy levels of the hydrogen atom, therefore defining in this way a domain of validity, then we could take the agreement between Bohr's model and the results from Schrödinger's equation as a decent example of Quine's underdetermination. This, however, is something that we are very reluctant to accept, because the levels of the hydrogen atom define a very weak domain of application. Therefore, the problem is not only to define exactly what can be named a theory, but also what are admissible domains of validity (or of application). There is indeed a quasi-continuum of theoretical constructions, from a simple experimental law to a consistent theory like relativity, and there is also a quasi-continuum of domains of validity, from a certain domain, like the quantum mechanical domain, to subdomains, and sub-subdomains, like the one concerning only the predictions of energy levels of the hydrogen atom.

A pertinent analysis of Quine's anthropology is available from Laugier-Rabaté [24]. According to this author, it appears that Quine did not provide convincing examples for his thesis of the underdetermination of the radical translation (another famous thesis from Quine, not discussed in this paper). And, as other authors like Duhem, the lack of convincing examples is also characterising the thesis of the underdetermination of theories by experiments. Quine himself would certainly agree with this statement if we refer to the mood expressed in his Erkentniss-paper [11].

Another author who is often quoted as relevant to the issue of Quine's underdetermination is Van Fraassen, in particular in reference to his book "Laws and Symmetries" [25]. Van Fraassen distinguished between theoretical equivalence and empirical equivalence. For him, if two statements are logically (theoretically) equivalent, then they are saying the same thing. However, if we consider a theory as a kind of statement, say a meta statement, we have at least one example to demonstrate that the affirmation of Van Fraassen is erroneous. This example concerns the Newton's and Hamilton-Jacobi's formulations of classical mechanics (soon to be discussed). As we shall see, both theories may be viewed as logically, theoretically, equivalent by the simple fact that they are mathematically equivalent. But they are not saying the same thing. Contrarily, they are proposing two drastically different visions of the world, one with local trajectories, the other with local trajectories embedded into a non-local field.

Also, Van Fraassen believed (as everyone he said!) that there exist numerous theories, may be not yet formulated, but which are in agreement with the experimental facts which are up to now known, and these theories are at least as explanatory as the best theories that we possess nowadays. However, he added, the big majority of these theories must be erroneous because they may disagree, in many different manners, with experimental facts not yet known. He concluded that our best theory is very unlikely to be correct. I would like to express this idea with a probabilistic flavour: if there is only one truth, we are sure to be in error.

Elsewhere, Van Fraassen [17] discussed the lack of unicity asserted by Quine's underdetermination in association with the existence of different interpretations of quantum mechanics. As he wrote, "*why then be interested in interpretation at all? If we are not interested in the metaphysical question of what the world is really like, what need is there to look into these issues? Well, we should still be interested in the question of how the world could be the way quantum mechanics–in its metaphysical vagueness but empirical audacity–says it is. That is the real question of understanding. To understand a scientific theory, we need to see how the world could be the way that the theory says it is. An interpretation tells us that.*" Later on, he added:

"the answer is not unique, because the question 'How could the world be the way the theory says it is?' is not the sort of question to call for a unique answer."

The issue of interpretation is also considered by Cushing [26]: "*Very loosely, the formalism refers to the equations and calculational rules that prove empirically adequate (i.e. getting the numbers right) and the interpretation refers to the accompanying representation the theory gives us about the physical universe (i.e. the picture story that goes with the equations of what our theory 'really' tells us about the world). Since a (successful) formalism does not uniquely determine its interpretation, there may be two radically different interpretations (and ontologies) corresponding equally well to one adequate formalism. This can be taken as an instantiation of the Duhem-Quine thesis of undetermination of theory by an empirical base. Even if one wants to restrict (and, arguably, that would be a mistake) the Duhem-Quine thesis to different formalisms each handling equally well a given body of empirical information, there nevertheless remains the interesting and important point of opposing ontologies equally well supported by a common empirical base.*"

The correct understanding of this Cushing's quotation requires us to tell more on the terminologies we may use. We have extensively discussed Quine's thesis of underdetermination. As underlined by Quine himself [11], this thesis "*is not to be confused with holism. It is holism that has rightly been called the Duhem thesis and also, rather generously, the Duhem-Quine thesis. It says that scientific statements are not separately vulnerable to adverse observations, because it is only jointly as a theory that they imply their observable consequences.*" In contradiction with this quotation from Quine, but in agreement with other authors, Cushing used the terminology "Duhem-Quine" to designate Quine's underdetermination. Sometimes, the expression of Duhem-Quine theorem (a philosophical theorem not demonstrated!) may be found to express an enunciation of Quine's underdetermination thesis.

Now, both van Fraassen and Cushing, driving us towards the issues of interpretations and ontologies, allow one to put forward another point of view on the interpretation of Quine's underdetermination thesis, a point of view concerning ontologies associated with theories and interpretations of theories. Let us enunciate a Duhem-Quine theorem under the following form: we may have several conflicting theories which are empirically equivalent. Now, such an enunciation is not complete if we do not specify the meaning of the word "conflicting". The meaning could be as given by Quine, for example: two theories are conflicting when they are logically incompatible, and cannot be rendered logically equivalent by any reconstrual of predicates. From now on, we use another definition: two theories are conflicting when they exhibit different ontologies. The thesis of ontological underdetermination then states that we may have empirically equivalent theories which do not exhibit the same ontologies. It remains to define what is an ontology: the ontology of a theory tells us which kinds of entities are populating the world. It describes the *furniture of the world* as implied by the interpretation of the theory. For instance, the furniture of classical physics is made out from localised objects and fields (possibly taking the form of waves).

We are now going to provide explicit and well-defined examples of ontological underdetermination, relying on theories which are rich enough, although not describing the whole of science, namely classical mechanics, quantum mechanics and electromagnetism.

3.3 Case studies

The possibility of an underdetermination of theories by experiments is usually much disliked by physicists, and the vagueness which remains attached to this concept after the somewhat fuzzy efforts of philosophers does not help. Squires [27] expressed this point with a bit of humour or even irony by saying that "*the statement that two theories, both of which fit the data, are equally good can be seen to be unreasonable if we note that a theory in which the sun always turns into cream cheese as it disappears over the horizon, and turns back again later, gives a perfectly adequate account of my observation.*" Hence, to be convinced, the physicist demands convincing examples, concerning rich enough theories.

3.3.1 From Classical Mechanics

This section introduces the first example of ontological underdetermination with a rich enough domain of validity which is that of (nonrelativistic) classical mechanics. We restrict ourselves to the mechanics of matter points without any loss of generality; however, insofar as an extended body may be viewed as a collection of matter points (in Newtonian framework).

It is known that classical mechanics can be declined under four different formulations, which are empirically equivalent. These are Newton's, Lagrange's, Hamilton's and Hamilton-Jacobi's formulations. We only need in this chapter to discuss Newton's and Hamilton-Jacobi's formulations. Discussing Newton's formulation is fast. It is sufficient (and necessary) to recall that, using the basic law telling us that force is equal to mass multiplied by acceleration, and integrating with initial conditions, we can build trajectories of matter points. Hamilton-Jacobi's formulation is however less familiar, see for instance Louis de Broglie [28], Blotkhintsev [29], Landau and Lifchitz [30] and Holland [31]. For several reasons, in particular to save room, I am not going to provide here a thorough exposition of this formulation, I mean using mathematical expressions. The essence may be conveyed in plain words. The typical main quantity of the Hamilton-Jacobi's formulation is an extended field S having the dimension of an action, satisfying a certain partial derivative equation, for which a general solution is to be found. The trajectories of the Newton's formulation are orthogonal to iso-values surfaces of S. For the first time, in the history of science, the motion of a localised object (a matter point of classical mechanics) was associated with the existence of an extended field.

Therefore, we now indeed possess two formulations of classical mechanics, the Newton's and Hamilton-Jacobi's formulations, which are empirically equivalent, but which are in contradiction insofar as they do not say the same thing on the world, i.e. they do not exhibit the same ontologies. Indeed, Newton's formulation only deals with localised objects following trajectories, while Hamilton-Jacobi's formulation deals with localised objects following trajectories dressed by an unobservable, non-local, action field filling the space. Because they are conflicting, these two formulations may be counted as two different theories. They have to be taken as forming an example of ontological underdetermination, for rich enough theories, except if we just shrug our shoulders and rashly sweep S under the carpet.

A possible escape could be as follows. Since we are used to observe trajectories of macroscopic objects in our everyday experience (macroscopic objects, not matter points however), we are keen to believe that the field S of the Hamilton-Jacobi's formulation does not have any physical reality, but is simply an intermediary tool for computations. After all, this field does not pertain to our sense data. Then, we would have found a way to discriminate between Newton's and Hamilton-Jacobi's formulations (theories) in favour of Newton's formulation, by denying S to be an ontological entity. As we shall see, this point of view, although highly reasonable, will reveal itself to be erroneous.

3.3.2 From Quantum Mechanics

For the second case study, we are going to compare two different kinds of interpretations (actually two different theories) of quantum mechanics, the usual one, the one in textbooks, which is conventional, standard, orthodox and the causal theories, the heretical and unorthodox ones, mainly developed by Louis de Broglie and David Bohm. As for the previous case, we shall avoid mathematical formulae to express the essence in plain words.

The usual quantum mechanics, on the one hand, exhibits an intrinsic undeterminacy, associated with the Von Neumann's projection postulate governing any quantum measurement process. The causal theories, on the other hand, restore the determinism in quantum mechanics by invoking a subquantum level containing Newton-like trajectories of hidden particles. These causal theories are examples of what is called "hidden variables theories".

By definition, "causal theories" specifically refers to two theories initiated by Louis de Broglie (double solution and pilot wave) intending to restore determinism in quantum mechanics, hence producing a picture of the world much different from the one provided by quantum mechanics. A good enough understanding of the issue may be gained by reporting on a brief history of causal theories. Reporting on such a history is a fairly simple exercise because we possess many invaluable testimonies, in particular from the two main protagonists of the enterprise, namely Louis de Broglie and Bohm, e.g. [28, 32–39] or [40], for a selection. Complementary historical analyses and reports are also available from Vigier [41], Jammer [42, 43], Pinch [44], Wheeler and Zurek [45], Holland [31] or Cushing [46], among others. A more general point of view, not only on causal theories, but also on larger classes of hidden variables theories, is also discussed by Freistadt [47], Belinfante [48], Pipkin [49] or d'Espagnat [50].

It happened that, for many reasons, Louis de Broglie was dissatisfied with quantum mechanics, and then committed himself to the development of alternative theories. Indeed, he certainly has been one of the most severe and most eloquent opponents to the Copenhagen orthodoxy. He took many opportunities to claim his displeasure, and even his reluctance, in particular with respect to the following items: the impossibility to represent the motion of a quantum object in ordinary space and time and, in contrast, the use of an abstract configuration space; more generally, the impossibility of reconciling quantum mechanics with the clear ideas of classical physics; or the fictitious and subjective character of the wave function (its probabilistic interpretation, or his lack of objectivity when it is interpreted as representing only a state of knowledge). Indeed, for Louis de Broglie whose mind was educated with classical physics, the ideas of the new physics, out of space, out of time and out of causality, generated a head-on conceptual collision. Also, Louis de Broglie complained about what has been called elsewhere the complacency to quantum mechanics, for instance in [37] where he expressed his increasing conviction that the reasonings underlying the present interpretation of quantum mechanics are not as decisive as they look, and contain many small cracks. He later also expressed his feeling that most of the present theoreticians, capitulating in face of exaggerated abstract tendencies, renounced too easily to the building of an intelligible picture of quantum physical phenomena.

Very likely, the most important clash in Louis de Broglie's mind arose from the conflict between the concepts of particles and waves. This may be inferred from the very writings of Louis de Broglie, in particular when he stated [34] that the duality between waves and particles constituted the big tragedy of contemporary physics. Quantum mechanics succeeded in a magnificent way to dissolve the duality between waves and particles by a synthesis. Quantum objects are not classical waves and they are not classical particles: not wave *nor* particle, but something else. For Louis de Broglie, however, the price to pay was too heavy. In building hidden-variables theories, he made the choice of another option: wave *and* particle [42]. The duality became a complementarity, not however in the sense of Bohr. As soon as Louis de Broglie conceived his first ideas on wave mechanics, there was no doubt for him that it was necessary to find a new theory, synthesising the concepts of waves and particles, and retaining as far as possible their traditional aspects. Concerning particles, although quantum mechanics rejects the very concept of trajectory, Louis de Broglie wanted to preserve it. For him, the two points of view (rejecting and preserving) were not deeply conflicting because [35], from the fact that quantum measuring processes do not allow us to simultaneously assign a position and a state of motion to a corpuscle, it does not necessarily follow that, *in reality* , the corpuscle does not have any position nor velocity. Concerning waves, for Louis de Broglie, they had to be preserved too. In particular, Louis de Broglie was impressed by the fact that the theory of atoms introduced integers to deal with the stationary motion of electrons. But, in classical physics, such occurrences of integers only arose in interference phenomena, and in oscillatory normal modes. Therefore, for Louis de Broglie, corpuscles and waves were to become, in some way, two aspects of the same thing. As he expressed during the Solvay Congress of 1927 [51], the existence of elementary corpuscles for matter and radiation being admitted as a matter of fact, these corpuscles are supposed to be endowed with a periodicity. In such a vision, a matter point is no more conceived as a static entity only concerning a tiny part of space, but as the centre of a periodic phenomenon spreading around it. This idea that a periodic phenomenon must be associated with a corpuscle, has been an obsession to which Louis de Broglie dedicated most of his life.

The way to achieve the synthesis was to consider the corpuscle as a kind of permanent singularity embedded in an extended wavy phenomenon. This has led Louis de Broglie to a theory known as the double solution. The name of the theory comes from the idea (about 1925–1927) that the equation of motion of wave mechanics (Schrödinger's equation of course) is satisfied simultaneously by two coupled equations. One solution, denoted u, represents a singularity, describing really the corpuscle associated with the wave while the other solution, with a continuously variable amplitude, describes only the statistical behaviour of the motion of a cloud of particles. The first singular solution exhibits very high values of amplitudes, even infinite in an early version, around a moving point of space. We are then facing a local moving singularity which can be identified with a corpuscle. For Louis de Broglie, although the motion of this corpuscle should admittedly satisfy new dynamical laws, it should develop itself along a trajectory, with a well-defined velocity at each point of it. The second solution formally identifies with the Ψ of quantum mechanics, but with a different interpretation. For Louis de Broglie, the Ψ of the double solution, as the Ψ of quantum mechanics, does not exactly describe the physical reality. It leads indeed to a statistical theory, representing some kind of average, when the actual trajectories followed by the corpuscles are totally unknown, with the exhibition of two layers, the second layer, the deepest one, being of a classical nature and therefore deterministic. Then, at the level of the first layer , Ψ can be viewed as fictitious, statistical, subjective, without however any epistemological damage.

The quantum mechanical indeterminacy then results from our lack of knowledge concerning the initial conditions required to build the trajectories of the corpuscles in the cloud. Therefore, determinism is restored in a classical way, with indeterminism as the consequence of a lack of knowledge, instead of being of an intrinsic nature. We obtain a hidden variables theory with hidden variables associated with quantities describing the hidden trajectories. This is a brief summary of the first causal theory of Louis de Broglie.

To gain a better understanding of the theory of the double solution, and more important, of where it may come from, let us return to the Hamilton-Jacobi's formulation of classical mechanics. Let us recall that, in Hamilton-Jacobi's formulation of classical mechanics, we are facing trajectories which are orthogonal to iso-value surfaces of the action S. Actually, we may consider a cloud of particles, generating a compact set of trajectories, embedded in and orthogonal to a field phenomenon extended in space. This is a classical structure that Louis de Broglie attempted to generalise to a quantum mechanical structure. The motivation for this attempt was that Louis de Broglie believed that it was wrong to abandon the notion of trajectories, followed by classical particles. The Hamilton-Jacobi structure then provided a natural already ready framework for an efficient generalisation because, now, in quantum mechanics as in classical mechanics, we have trajectories embedded in a field. In the corresponding structure for quantum mechanics, quantum trajectories are now orthogonal to a field formed by the phase of Ψ. It may be demonstrated that, in the classical limit, this phase of Ψ identifies with the field S (within a multiplicative constant).

Next, Brussels, in 1927, has been the very place where occurred a living confrontation between eminent experts of the time, the very place where the Copenhagen interpretation has become the dominating one [42, 51, 52]. It also has been a turning point for the history of the newly born causal theories, the very place where Louis de Broglie has been defeated. This happened as follows. In the spring of 1927, Lorentz asked Louis de Broglie to prepare a report on wave mechanics for the forthcoming Solvay Congress to be held in october. Louis de Broglie, at this time, was fighting with the many mathematical difficulties he encountered when trying to make his double solution satisfactory enough. He then found it too hazardous to expose his not yet ready theory, and decided to fall back on a simpler half-hearted point of view, named the pilot wave theory, the second causal theory of Louis de Broglie, that he himself called a degenerated version of the double solution.

In the pilot wave theory, Louis de Broglie held fast on to the existence of corpuscles and also admitted the validity of the wave equation of wave mechanics, even if only of a statistical nature. He then placed the corpuscle inside the wave and assumed that its motion is driven by the propagation of the wave. The driving process is described by a guidance formula which made trajectories orthogonal to the isophases of Ψ (again our now familiar Hamilton-Jacobi's picture). Therefore, the corpuscle is viewed as an entity which is piloted by the wave Ψ, hence the name given to the theory. Something is lost however with respect to the double solution. The corpuscle is guided by the wave but it is no more incorporated in a tight way inside the wave. Also, the wave-particle duality is acknowledged as a matter of fact, but there is no more any understanding of its nature. However, the intuitively appealing notion of a localised particle with a deterministic motion along its trajectory is preserved [53].

During the Solvay Congress, discussions were sharp: Bohr, Heisenberg, Born defended the Copenhagen probabilistic interpretation. In particular, Bohr re-expressed his idea of complementarity (presented previously in a Como lecture), the fact that underlying quantum events are outside of space and time, and that the space/time description and the causal description are complementary [54]. Schrödinger attempted to put forward its interpretation of a corpuscle viewed as a wave packet, i.e. he did not believe in the existence of corpuscles as basic entities. He expressed his displeasure in the face of what would become the orthodox interpretation. For him, it seems that there is an indication that, when our knowledge has progressed enough, everything will eventually become understandable again in a 3D space. But he did not support Louis de Broglie [55]. Lorentz rejected the probabilistic approach and insisted on the fact that physics should rely on clear notions in the framework of space and time. Pauli attacked the pilot wave, stating that he did not believe that the representation of Louis de Broglie could be developed in a satisfactory way. Einstein, deeply hostile to a genuine probabilistic interpretation, vigorously attacked it, against Bohr and did not win. But more important and most disappointing for Louis de Broglie is that Einstein, although in principle in favour of a deterministic interpretation, did not support him. For details, we may return to the printed discussion following the presentation of Louis de Broglie [53] which is not very enthusiastic, and also to the final discussion following the bulk of the Congress [51].

As a whole, besides the laconic indifference of Einstein, the pilot wave theory of Louis de Broglie was severely criticised, possibly in a much more acerbic way than the one we could draw by simply studying the accounts of the Congress [51]. You know, during a congress, there is not only a conference room for public debates but also coffee breaks, corridors and all sorts of informal gossips, particularly if, as was the case, all participants shared the same hotel. Indeed, according to a testimony from Heisenberg [56], the most animated discussions occurred during meals at the hotel. This is the place where Einstein put forward, several times, his famous objection that God does not play dices, to which Bohr could only reply that it is not to us to prescribe how God should rule the world [56]. It might however be significant that the name of Louis de Broglie is not mentioned at all by Heisenberg in this book [56] where he reported so many details on what has been going on during the Solvay Congress of 1927. This was however not a complete boycott of Louis de Broglie. Indeed, Heisenberg let him express extensively in one of his books [57], although not on the issue of hidden variables. A dramatic feature is that Pauli made an objection that Louis de Broglie could not successively strike down. This objection was that the pilot wave theory (or model) could not be applied in a coherent way to the case of a two-body scattering process, specifically to the Fermi quantised rotator. Louis de Broglie had become a lonesome warrior.

Returning from Brussels to Paris, thinking more about the pilot wave and on what happened during the Solvay Congress (no support from Einstein, at least publicly, Pauli's objection, generally little support to him even from the parts of Schrödinger or Lorentz ...), Louis de Broglie felt poignantly alone (Bohm's testimony in [44]). It is not a big step forward to believe that he must have felt depressed too. He thought that Pauli's objection could possibly be refuted, but more important, he found new objections by himself (such objections are discussed in [28]). The most significant one is that, in the pilot wave model, the objective corpuscle was guided by a wave Ψ which, insofar as it was reflecting our state of knowledge, was of a subjective nature. How something subjective could possibly guide something objective? Another objection, again associated with the status of Ψ, concerns the many-body problem because, for this case, it is even more difficult to consider the pilot wave theory as giving a genuine physical picture of phenomena, due to the fictitious and abstract character of the propagation of a wave in the configuration space. Also, there was little comfort with the double solution. It did not suffer from the above objections but was buried in mathematical difficulties that, actually, Louis de Broglie never succeeded to solve. For Louis de Broglie however, if the causal theories he started to develop were one day or another day able to grow again from their ashes, it would be from the double solution approach. Such a phoenix process would certainly be impossible from the pilot wave model (he erroneously thought), because this degenerated version was absolutely indefensible.

Facing such a situation, Louis de Broglie felt so discouraged that he gave up the causal theories, only a few years after he introduced them. He then capitulated, defeated by the probabilistic interpretation, and even accepted to teach it, although, we guess, reluctantly. Later on, Born will take notice of the acceptance by Louis de Broglie of his defeat, and of his acknowledgment that the probabilistic interpretation (the one of Born) is the only adequate one [58]. In 1932, five years after Brussels, 1927, von Neumann, in his celebrated book [59], produced a proof of the impossibility of hidden variables which, seemingly, closed the issue forever. Apparently, this was the final knock.

Ironically enough, it was not the double solution but, in contradiction with Louis de Broglie's expectation, the pilot wave theory which made the legendary Arabian bird born anew from ashes. In 1952, Bohm published two papers on a pilot wave approach [38, 60] in which he pointed out the fallacy of von Neumann's proof of impossibility, provided a rebuttal to Pauli's objection and proposed a consistent solution to the measurement problem, among other issues. From an epistemological and also methodological point of view, one of the most interesting features of Bohm's approach is the way he dealt with one of Louis de Broglie objections, namely the one concerning the subjectivity of Ψ and the damning question to know how such a subjective entity could pilot an objective corpuscle. For Bohm, the answer is incredibly simple. By a kind of *fiat*, he just decreed from the beginning that Ψ is an objective wave. This is however made consistent by the fact that Bohm provided an objective solution to the measurement problem.

I shall not investigate deeply whether Bohm was aware or not of Louis de Broglie's work when he started to develop his theory, although we have the following testimony from Bohm [39]: "*then, in 1951 ... the author* (that is to say Bohm) *began to seek such a model, and indeed shortly thereafter, he found a simple causal explanation of the quantum mechanics which, as he later learned, had already been proposed by de Broglie in 1927.*" But, in any case, it is well attested that he was aware of it in 1951 because, one year before having his seminal papers published in 1952 in the Physical Review, he sent preprints to Louis de Broglie. According to the latter, Bohm fully picked up the pilot wave theory as it was exposed at the Solvay Congress of 1927, including certain developments such as the introduction of a quantum potential. As a matter of fact, he also acknowledged that Bohm was not aware of this work produced 25 years before, when he engaged himself on causal theories. He also acknowledged that Bohm achieved a few more steps, in particular regarding Pauli's objection and the measurement problem [34]. The official reaction of Louis de Broglie has then been to publish a C.R.A.S. to recall his indisputable priority in the topic, and also the difficulties which made it repudiating the pilot wave after 1927 [33]. Objections against the pilot wave are also summarised by Louis de Broglie in Refs. [28] and [34].

According to a testimony of Bohm to Pinch [44], after having received the 1951 preprints, Louis de Broglie *wrote back saying that he had already done this sort of thing.* Upon looking at the preprints, Pauli manifested a similar reaction, saying that "*it was old stuff, dealt with long ago.*" See Pauli [61] reporting on de Broglie-Bohm story. But Louis de Broglie, in his reply to Bohm, also explicitly discussed the objections against the pilot wave that made him give it up, and Bohm found that he could answer them. The rebuttals are included in Appendix B of the second 1952 paper of Bohm [60] , saying: "*all the objections of de Broglie and Pauli could have been met if only de Broglie had carried his ideas to their logical conclusion.*" Later on, in many places, Bohm referred to Louis de Broglie as the primary father of the filiation. For instance, in Bohm and Hiley [40], he wrote: "*thus de Broglie proposed very early what is, in essence, the germ of our approach,*" and he considers his theory as "'*an extension' of de Broglie which answered all the objections.*" But, as a whole, the primary reception of Bohm's ideas by Louis de Broglie was rather negative. In the words of Belinfante [48], "*when Bohm in his 1951 theory revived de Broglie's 1927 theory of the pilot wave without attaching to it a theory of the double solution, de Broglie objected against this by arguments that might have been presented by a pupil of the Copenhagen school.*"

Although Louis de Broglie's reactions to Bohm's work were not, at the beginning, very enthusiastic (to say the least), he changed his mind rather fast, likely in part due to the relevant answers of Bohm to his objections. According to Pinch [44], "*these answers were in part enough to convince de Broglie who then took up his original interpretation again.*" In the words of Cushing [46], "*David Bohm's 1952 paper was to have a profound reconversion effect on de Broglie.*" This was a new start for Louis de Broglie who felt encouraged to take up his old ideas again, and to develop them in various ways, with however his previous recurrent belief that the double solution was on the right track. As testified in 1956 [35], Louis de Broglie has been wondering since 1951 (the year when he received Bohm's preprints) whether his first idea, the double solution, after all, was not the correct one. In fact, already in 1952 [32], he published a work in which he examined how the objective and causal theory of the double solution could be taken up again.

Improvements and modifications are then introduced by Louis de Broglie. May be the most innovative one concerns a new hypothesis telling that the equation of propagation of the singularity wave u should be nonlinear, in contrast with the one for Ψ which remains linear, although both equations are nearly everywhere identical. Such a late focus on nonlinear equations was motivated by some developments due to Einstein, in general relativity and unified field theory, in which besides the gravitational field there also exist superimposed gravity singularities having to satisfy nonlinear equations. We are here facing a strong similarity between the original ideas of Louis de Broglie's double solution and latest researches of Einstein, a similarity pointed out by Vigier, and that Louis de Broglie tentatively exploited [35, 39, 41].

It is easy to pass from the double solution to the pilot wave. Essentially, what we have to do is to forget anything concerning the second solution, that is to say concerning the u-waves and their singularities, and to preserve the first solution Ψ and the guidance formula driving the underlying hidden objective particles. Then, these particles are piloted by the wave Ψ, hence the name of the theory. We are still left with two versions, the pilot wave theory of Louis de Broglie, and the pilot wave theory of Bohm. They essentially differ by the epistemological status of Ψ, subjective for Louis de Broglie and objective for Bohm, and by the fact that Bohm developed the theory much further, up to its logical conclusion, as he wrote. In particular, objections which have been addressed to the version by Louis de Broglie have been successfully enough answered by Bohm. From now on, when referring to the pilot wave, we meant the 1952 version of Bohm [38, 60], which is enough for further discussions.

In short, avoiding again, as in the previous case study, any mathematical development, let us summarise by telling the reader that Bohm accepted the validity of Schrödinger's equation to describe the wave Ψ. However, Ψ is now viewed as an objective field from the very beginning, exactly such as an electromagnetic field satisfying Maxwell's equations. We do not know from first principles why there should be an electromagnetic field, nor do we know *what is* an electromagnetic field. Therefore, why should we not adopt the same complacency for Ψ as the one we have for **E** and **H**, or for their unification in an electromagnetic tensor H_{ij} ? In the words of Bohm: "*in the last analysis, there is, of course, no reason why a particle should not be acted on by a Ψ-field, a gravitational field, a set of meson fields, and perhaps by still other fields that have not yet been discovered.*"

Elaborating on Schrödinger's equation, Bohm demonstrated that the usual quantum mechanics can be recast in another framework exhibiting Newton-like hidden trajectories, forming a deterministic sublevel below the quantum level. In connection with Hamilton-Jacobi's formulation which inspired this approach, and to which it is formally closely related, the velocity of the particles is orthogonal to the iso-phase surfaces of Ψ which, therefore, indeed plays the role of a pilot wave.

Now, a most important fact is that the pilot wave is empirically equivalent to quantum mechanics. Following Bohm, "*all the results of the usual interpretation are obtained from our interpretation if we make the following three special assumptions which are mutually consistent*:

(1) That the Ψ-field satisfies Schrödinger's equation.

(2) That the particle momentum is restricted to $p_j = \partial S/\partial x_j$.

(3) That we do not predict or control the precise location of the particle, but have, in practice, a statistical ensemble with probability density $P = |\Psi|^2$. The use of statistics is, however, not inherent in the conceptual structure, but merely a consequence of our ignorance of the precise initial conditions of the particle."

Therefore, we again have, in this example, two empirically equivalent theories with conflicting ontologies: the orthodox quantum mechanics exhibiting an intrinsic indeterminacy, and the heretical pilot wave relying on the deterministic motion of hidden trajectories. In the pilot wave framework, indeterminacy is no more of an intrinsic nature. It is simply the consequence of our lack of knowledge concerning hidden trajectories. Probabilities are no more intrinsic, but now receive a classical interpretation, like when tossing a coin, where our inability to predict the outcome of the toss merely reflects our ignorance of initial conditions (and our inability to carry out computations which are much too complicated and CPU-demanding).

3.4 An Escape to Underdetermination: Ampliative Arguments

3.4.1 Undecidability

The diagnosis of an underdetermination by experiments (whether ontological or not) is the same as a diagnosis of undecidability. Let us comment on this issue by focusing on the undecidability between usual quantum mechanics and pilot wave.

It has been argued that the pilot wave is simply quantum mechanics recast in another language and, if this were true, the issue of undecidability could be of little significance. We could say: just use the language you have learnt, except if you are inquisitive enough to feel the desire to learn another language. But learning another language means entering another world, facing a quite different way of thinking. And, if this is true for natural languages, it is still true, or even more, when we compare quantum mechanics and pilot wave. These two empirically equivalent, ontologically conflicting theories, offer different visions and interpretations of the world.

This looks like a severe blow against realism, in some sense, that is to say against the very idea that science could pretend to really tell us something on the world around us. This might drive us towards cultural relativism. But what our case study does imply is not an absolute relativism. It precisely tells us the following: in some cases, we may be facing some undecidable statements because they pertain to undecidable and even contradictory theories, which have however to be simultaneously accepted, insofar as they are empirically equivalent. It does not tell us that undecidability is the ultimate fate in *all cases,* nor *in all aspects*. For instance, both quantum mechanics and pilot wave agree on the nonlocality of microscopic phenomena (in agreement with experiments), i.e. some of the aspects of the veiled reality could possibly be unveiled [62].

The most significant undecidable issue concerning quantum mechanics and pilot wave is that of determinism, actually an issue which boosted the search for hidden variables. Indeed, if we cannot decide between quantum mechanics and pilot wave, we cannot decide between the intrinsic indeterminacy of quantum mechanics and the determinism postulated, and later constructed, by causal theories. Both Einstein (an opponent to quantum mechanics) and Born (a defender of quantum mechanics) even agreed on the undecidability between determinism and indeterminism. In a letter to Einstein, Born wrote: "*You are absolutely right that an assertion about the possible future acceptance or rejection of determinism cannot be logically justified. For there can always be an interpretation which lies one layer deeper than the one we know (as your example of the kinetic theory as against the macroscopic theory shows)*" [63]. Bohm used a similar argument and explicitly extended it to defend the undecidability under question. For him, starting from an deterministic (indeterministic) level of description, we can always imagine and construct a sub-level which would be indeterministic (deterministic), and this *ad infinitum.* For instance, determinism at a certain upper level may be the result of indeterministic, or possibly stochastic, processes at the next lower level, e.g. macroscopic determinism resulting from averages over quantum indeterminist processes or atomic classical stochastic processes, at a microscopic level. And indeterminism at a certain upper level may be the result of deterministic processes at the next lower level, e.g. quantum indeterminacy underlied by deterministic hidden variables. In the words of Bohm himself [39], "... *one sees that the possibility of treating causal laws as statistical approximation to laws of chance is balanced by a corresponding possibility of treating laws of probability as statistical approximations to the effects of causal laws... The assumption that any particular kind of fluctuations are arbitrary and lawless relative to all possible contexts, like the similar assumption that there exists an absolute and final determinate law, is therefore evidently not capable of being based on any experimental or theoretical developments arising out of specific scientific problems, but it is instead a purely philosophical assumption.*"

Interestingly enough, Bohm and Hiley [40] took advantage of the issue of undecidability to advocate a peaceful relationship between Bohm's interpretation and quantum mechanics. For them, and also for us at the present time, "*there does not seem to be any valid reason ... to decide finally what would be the accepted interpretation,*" and they asked: "*is there a valid reason why we need to make such a decision at all?*" So, they went on, "*would it not be better to keep all options open and to consider the meaning of the interpretations on its own merits, as well as in comparison with others? This implies that there should be a kind of dialogue between different interpretations rather than a struggle to establish the primacy of any of them.*" If we observe that these quotations are contained in the last book of Bohm (with Hiley), published one year after his death, we can view the above statements as forming a kind of final testamentary auto appraisal. But, is it true that we cannot have any argument to decide? This is the issue of ampliative arguments.

3.4.2 Ampliative Arguments

We can actually, at least in some cases, decide between undecidable physical theories. When this is possible, it is achieved by using what is called ampliative arguments. Harré [12], and Van Frassen quoting and rewording Harré [25] defined what an acceptable theory is. Elaborating on these authors, I came to the following proposal. A theory is satisfactory (acceptable, admissible...) if (1) it agrees with experimental facts, (2) it is logically consistent and (3) it satisfies other demands. These other demands, let us call them ampliative arguments. The word "ampliative" means that these arguments enlarge our possibilities of choosing between several theories, the enlargement being made with respect to the underdetermination by experiments. In other words, ampliative arguments allow us to decide between undecidables. Therefore, we may now build a new list of criteria as follows. A theory is satisfactory if (1) it agrees with experimental facts, (2) it is logically consistent and (3) it satisfies ampliative arguments.

It is interesting to remark that Einstein had something to tell us concerning Quine's underdetermination (without referring to it) and ampliative arguments (without using this terminology). Indeed, in an address he delivered in 1918 before the Physical Society of Berlin on the occasion of Planck 60th birthday, he expressed himself as follows [26]: "*there is no logical paths in these laws; only intuition, resting on sympathetic understanding of experience, can reach them. In this methodological uncertainty, one might suppose that there were any number of possible systems of theoretical physics all equally well justified; and this opinion is no doubt correct, theoretically. But the development of physics has shown that at any given moment, out of all conceivable constructions, a single one has always proved itself decidedly superior to all the rest. Nobody who has really gone deeply into the matter will deny that in practice the world of phenomena uniquely determines the theoretical system, in spite of the fact that there is no logical bridge between phenomena and their theoretical principles.*"

In this chapter, I do not intend to take the risk of listing what could be acceptable ampliative arguments (this might be postponed to future epistemological researches). I shall be content in using one of them, convincing enough, at least for many individuals, as an example, and as a tool to proceed further. This example relies on a non-singularity principle which, if accepted, implies the falseness of Newton's formulation of classical mechanics (and, more generally, of classical mechanics as a whole), the rational necessity of wave mechanics, and the inadequacy of the pilot wave theory. We shall use another ampliative argument to deal with the case study from electromagnetism.

3.4.3 The Mechanical Rainbow

Similarly as for the optical rainbow, there exists a mechanical rainbow. To introduce it, let us consider classical mechanics. A subtopic of classical mechanics is classical scattering, e.g. [64, 65]. Following Nussenzveig [65], let us discuss the scattering of a nonrelativistic particle of mass m by a central potential $V(r)$. The trajectory of the incoming particle is deflected by an angle $\overline{\theta}$ called the deflection angle which depends on the impact parameter b. For a repulsive interaction, $\overline{\theta}$ ranges from 0 to π but, for an attractive interaction, it can take arbitrary large values (in modulus). There is a simple relationship between the deflection angle $\overline{\theta}$ and the more usual scattering angle θ, namely: $\overline{\theta} + 2n\pi = \pm\theta, n = 0, 1, 2, ...$ It is then established that the differential cross-section in the direction θ contains terms of the form $\left|d\theta/db_j\right|^{-1}$, where the index j arises from the fact that, in general, there exist several impact parameters b_j that lead to a same scattering angle θ. From this result, we see that we have a divergence whenever $d\theta/db = 0$, for at least one j, and for some angle $\theta_R = \theta$, that is to say when the deflection function, say $\theta(b)$, goes through an extremum. We then have a stationary trajectory and the singularity is a rainbow singularity occurring at the rainbow angle θ_R. Other singularities involved in the phenomenon are glory, forward scattering, orbiting, but we do not need to discuss them here. I then invoke a non-singularity principle telling us that such singularities are inadmissible. From this, I conclude, without referring to any experiment (i.e. *a priori*), that classical mechanics, which therefore contained the germ of its own destruction, is an approximation to a more general theory. This more general theory must be a wave theory, say a wave mechanics, in order to smooth out the rainbow singularity, and any other singularity, as well. We therefore also conclude to the rational necessity of a wave mechanics.

3.4.4 On the Nonsingularity Principle

I have used above a nonsingularity principle that I may call a special non-singularity principle. This special principle might be an example of a more general nonsingularity principle that I am not going, even tentatively, to formulate in this paper. Let us however recall that there has been a long and venerable history concerning the concept of infinity, both in mathematics and physics, even starting from the presocratic age. And, nowadays, most physicists would certainly agree with the special nonsingularity principle used above and with its consequences. Indeed, physicists are most usually much reluctant to accept the actual existence of local infinite quantities in the world they study. Whenever a physicist encounters a local infinity in physics, he instinctively immediately recognises that something is going deeply wrong, or that he has been dealing with convenient idealisations which cannot be actual, and possibly that he has to think more. In other words, the non-singularity principle is indeed something, at least implicitly, anchored in the mind of physicists, intimely associated with their intuition. This does not imply that this intuition is correct, and does not mean that it should be shared by everyone. Also, in any case, we possibly have to accept the idea that any statement is revisable, even logical statements [66–68]. Nevertheless, those who accept the special nonsingularity principle possess a decent ampliative instrument to proceed further.

3.4.5 Newtonian Trajectories of Matter Points do not Exist

We are all used to the concept of trajectory, so easy to extract from classical mechanics, in deep agreement with our intuition, and with our sense data: trajectories of cars, of balls, or even of planets. There is no apparent difficulty to consider trajectories of matter points. They at least constitute efficient models for the behaviour of so many objects around us. Furthermore, physicists, when they are students, have been trained, when learning classical mechanics, to the study of the mechanical behaviour of matter points, before entering the realm of the mechanics of extended solids. There is also this famous theorem telling us *that the force between two uniform spheres of masses M_1, M_2 is equal to the force between two point masses M_1, M_2 at their respective centers* [69].

We therefore feel very comfortable with the concept of trajectories in the Newton's formulation of classical mechanics and, as a consequence, we may feel very uncomfortable with the Hamilton-Jacobi's formulation in which trajectories are dressed by a field. What could physically be this field S, which extends in the space around the object under motion, that no one has ever seen and which is effectively unobservable? At the best, it is a convenient intermediary for computations and, at the worst, it is some kind of metaphysical artificial excrescence. At least, this is what we could instinctively believe.

But, actually, although this way of thinking seems very reasonable, it may be dramatically misleading. Let us effectively consider the special non-singularity principle and its application to classical scattering. It leads us to the rejection of classical mechanics, as being inadmissible, on the basis that, in some cases, a classical problem may produce singularities, in particular the rainbow singularity when the differential cross-section becomes infinite. If we hold fast on to this conclusion and use it as an ampliative argument, we reach an immediate consequence: Newtonian trajectories of matter points do not exist (and cannot exist).

This statement is actually in agreement with sense data. Effectively, no one ever observed the trajectory of a matter point. Simply try to imagine, at the most fundamental level of understanding, what could be the trajectory of a matter point, that is to say the trajectory of an object of mass m, confined to a vanishingly small volume. Such an object would have an infinite density, and Newtonian mechanics would have to be merged with the most extraordinary predictions of general relativity. There would not be any domain of validity for classical mechanics.

A special kind of trajectory is the motionless trajectory of a particle at rest. Then, we have the particle (the matter point) standing still (in some frame of reference). From this, we can see that the concept of a matter point, independently of the concept of trajectory, cannot be accepted. Beside what we have already said, including the idea of a collapse of a mass m to a geometrical point to generate a matter point, just think, for instance, of the *unbelievable* behaviour of potentials of point charges varying as $1/r$, and therefore diverging when the point charge is indefinitely approached (producing an actual infinity). Obviously, from a pure logical point of view, we might as well start to refute first the existence of matter points invoking the special non-singularity principle, and thereafter the existence of trajectories of matter points, since something which does not exist cannot have any trajectory.

Nevertheless, what about trajectories of cars, balls ... and planets? The answer is that there should not be any deep and definitive conflict between the nonexistence of trajectories of matter points, and the observability of trajectories of extended solids. Cars, balls, ... planets are not matter points that we cannot observe, but macroscopic objects that we do observe. Their existence, their properties, the fact that we can observe them, and by which processes we observe them, must emerge as the consequences of a quantum theory, whatever its ultimate formulation will be. In other words, classical Newtonian trajectories of extended objects have to be accepted, but such extended objects, rather than being considered as a collection of matter points, in the Newton's style, must emerge from a more fundamental description of nature, e.g. from the underlying quantum mechanical level (e.g. decoherence theory). Hence, the rejection of matter points, and of their trajectories, does not offend our everyday intuition.

3.4.6 Deciding Between Undecidables

The fact that Newtonian trajectories of matter points do not exist is the ultimate ampliative argument to discriminate between Newton's and Hamilton-Jacobi's formulation of classical mechanics. In contrast with our naive expectation in which the field S was viewed as a simple intermediary tool for computations, without any physical significance, it is the Newtonian formulation of classical mechanics which is to be rejected, and the Hamilton-Jacobi's formulation which is to be given a due privilege, let us say which is "closer to truth". We are then left with a dressing field S without any trajectory to be dressed. To understand such a weird situation, it is sufficient to remark that, not only do Newtonian trajectories of matter points not exist but also, it is classical mechanics as a whole which collapsed. Hence, Hamilton-Jacobi's formulation of classical mechanics is also to be rejected. Nevertheless, the field S still remains physically meaningful. It actually appears to be related to the phase of Ψ in quantum mechanics. Indeed [29], we may write Ψ under the form $\Psi = \exp(iS/\hbar)$ in which, in the classical limit, the quantum field S of the previous expression identifies with the classical field S of Hamilton-Jacobi's formulation. The physical meaning of the classical S of the inadmissible classical mechanics is therefore that it constitutes a formal anticipation of the Ψ of quantum mechanics, more precisely of its phase. The constant $\hbar$, which has the dimension of an action, has the virtue of changing S, which has the dimension of an action too, to a dimensionless phase $iS/\hbar$.

Concerning the second case study (pilot wave versus usual quantum mechanics), the relevance of ontological underdetermination for a discussion of Bohm's pilot wave has been noticed by Cushing [46, 70]. His point of view is that [70] "*one formalism, with two different interpretations, counts as two different theories*" and that [46] "*the physical interpretation refers to what the theory tells us about the underlying structure of... phenomena, i.e. the corresponding story about the furniture of the world.*" This furniture of the world is that he called an ontology (the point of view also adopted in this paper). Therefore, quantum mechanics and Bohm's pilot wave, although experimentally equivalent (implying that both of them *should* be accepted by positivists), provided two different ontologies. Bohm himself was aware of the issue at least implicitly [39].

However, we may ampliatively discriminate between the pilot wave and the usual quantum mechanics. For this, we just need to remark that the same ampliative arguments (non-singularity principle, inexistence of Newtonian trajectories of matter points) which have been used to discriminate between Newton's and Hamilton-Jacobi's formulations of classical mechanics may be used to discriminate between Bohm's pilot wave and quantum mechanics. If objective, deterministic, Newtonian trajectories of matter points (and matter points themselves) do not exist, it is nonsensial to introduce them in quantum mechanics as done by Bohm. The attempt to propulse the classical concepts of matter points and matter point trajectories in the quantum domain leads definitively to a failure. Trajectories of matter points do not exist in classical mechanics; hence, they should not be reintroduced in quantum mechanics, even if they are hidden. This statement applies to all causal theories (pilot wave of Louis de Broglie and Bohm, and double solution of Louis de Broglie).

The rejection of inadmissible theories is not contradictory with the fact that they can make very decent models, much useful in practice. And, after all, our best theories are very far from being perfect and completed: from this point of view, they are all models. In particular, the fact that singularities in the behaviour of Newtonian trajectories are rare, occurring only occasionally, demonstrate that the concept of trajectory of matter points still remains useful for many practical purposes, in the same way that optical ray computing and tracing will forever remain invaluable tools. But, if you need to dig deep into the mysteries of the world, it is definitely of good advice, and even compulsory, to abandon inadmissible theories, when possible. The idea, expressed by Bohm in its auto appraisals, and by other authors, that the pilot wave could provide complementary insights in to a better, more thorough, and deeper, understanding of quantum mechanics, is then erroneous and even dangerous, for there could not be complementary insights in erroneous ideas, only an opportunity to spoil the clarity of the mind. Any theoretician in quantum field theory, for whom the description of reality in terms of particles being permanent entities with fixed numbers is more than naive, would agree with this statement.

The two case studies discussed above are strongly correlated. One pertains to classical mechanics, the other to quantum mechanics, but both of them were anchored on the existence of the Hamilton-Jacobi's formulation of classical mechanics, and undecidables were made decidable by relying on similar ampliative arguments. We are now going to briefly discuss a third case study, related to another physical framework, as discussed in the introduction, and solve a question that I asked myself about 30 years ago, a long time before I heard from Quine's underdetermination and from ampliative arguments, when I started to work on light scattering theories with the building of a generalised Lorenz–Mie theory (GLMT) in mind [71].

For this third case study, we may again rely on an ampliative argument to discriminate between Lorenz and Mie. This argument simply invokes the well-founded rejection of ether by Einsteinian relativity. Hence, Mie was "closer to truth" than Lorenz. Considering the effort accomplished by Lorenz, he certainly deserved to have his name associated to what I always preferred to call Lorenz–Mie theory. Nevertheless, the two theories, although empirically equivalent, are ontologically different (conflicting), and better have to be viewed as two genuine different theories.

3.5 Conclusion

Quine's underdetermination thesis, loosely speaking, states that theories are underdetermined by experiments. I also introduced an ontological version in which we may have several empirically equivalent theories, with however different ontologies. We have examined three exemplifying case studies. Two of them (Newton's formulation versus Hamilton-Jacobi's formulation of classical mechanics, causal theories versus the orthodox interpretation in quantum mechanics) are closely related to the debates on the foundations of quantum mechanics. The third case study (Mie's theory versus Lorenz' theory) is borrowed from electromagnetic theory, more specificaly from light scattering theory. Each case study exhibits a couple of theories which lead to identical experimental predictions but are contradictory insofar as they do provide conflicting visions of the world. The existence of conflicting empirically equivalent theories implies, in principle, strong limitations to any realistic interpretation of science. However, in each case study discussed in this chapter, we have been able to invoke ampliative arguments allowing one to decide between undecidables.

References

1. G. Mie, Beiträge zur Optik trüber Medien speziell kolloidaler Metalösungen. Annalen der Physik **25**, 377–452 (1908)
2. H. Horvath (ed.), Light scattering, Mie and more. J. Quant. Spectrosc. Radiat. Transf.(special issue) 2009
3. J.H. Hough (ed.), 11th conference on electromagnetic and light scattering by nonspherical particles. J. Quant. Spectrosc. Radiat. Transf.(special issue) 2009
4. M.I. Mischenko, L.D. Travis, Gustav Mie and the evolving discipline of electromagnetic scattering by particles. Bull. Am. Meteorol. Soc. **89**(12), 1853–1861 (2008)
5. L. Lorenz, Lysbevaegelsen i og uden for en af plane lysbolger belyst kulge. Vidensk. Selk. Skr. **6**, 1–62 (1890)
6. L. Lorenz, *Sur la lumière réfléchie et réfractée par une sphère transparente*. Lib. Lehmann et Stage, Oeuvres scientifiques de L.Lorenz, revues et annotées par H.Valentiner, 1898
7. N.A. Logan. Survey of some early studies of the scattering of plane waves by a sphere, reprinted in selected papers on light scattering by M. Kerker. SPIE, p. 951, 1965
8. N.A. Logan. Survey of some early studies of the scattering of plane waves by a sphere, ed. by D. Hirleman. in Proceedings of the Second International Symposium on Optical Particle Sizing, (Arizona State University, Arizona, 1990), pp. 7–15
9. H. Kragh. Ludwig Lorenz: his contributions to optical theory and light scattering by spheres, ed. by D. Hirleman, in Proceedings of the Second International Symposium on Optical Particle Sizing, (Arizona State University, Arizona, 1990), pp.1–6
10. H. Kragh, Ludwig Lorenz and the nineteenth century optical theory. the work of a great Danish scientist. Appl. Opt. **30**(33), 4688–4695 (1991)
11. W.V. Quine, On empirically equivalent systems of the world. Erkenntnis **9**, 313–328 (1975)
12. R. Harré, *Varieties of realism, a rationale for the natural sciences* (Basil Blackwell, Oxford, 1986)
13. A. Koestler, *Les somnambules, French translation of "The sleepwalkers"* (Calmann-Lévy, Paris, 1960)
14. O. Gingerich, *Le livre que nul n'avait lu, à la poursuite du "De Revolutionibus" de Copernic. French translation of: "The book nobody read-chasing the revolution of Nicolaus Copernicus"* (Dunod, Paris, 2008)
15. Galileo Galilei, *Dialogue sur les deux grands systemes du monde, French translation of: Dialogo sopra i due massimi sistemi del mondo tolemaico e copernicano (Editions du Seuil, Paris, 1992)*
16. P. Duhem, *La théorie physique, son objet, sa structure. English version: the aim and structure of physical theory* (Librairie philosophique J. Vrin, Paris, 1997)
17. B. Van Fraassen, *Quantum mechanics: an empiricist view* (Clarendon Press, Oxford, 2000)
18. R. Descartes, *Les principes de la philosophie, troisième partie: pirouettes et tourbillons des cieux* (Paleo, France, 2000)

19. R. Descartes, *Les principes de la philosophie, quatrième partie: la Terre et son histoire* (Paleo, France, 1999)
20. B. Pascal, *Oeuvres complètes, tome 1*, chapter: Lettre de Blaise Pascal au très Révérend Père Noël (Gallimard, Paris, 1998), pp. 377–386
21. W.V. Quine, *Le mot et la chose, French translation of "Word and Object"* (MIT Press/Flammarion, Cambridge/Paris, 1960/1977)
22. W.V. Quine, *La poursuite de la vérité, French Translation of "Pursuit of Truth"* (Harvard University Press/Editions du Seuil, Cambridge/Paris, 1990/1993)
23. Sous la direction de Pierre Wagner, editor. *Les philosophes et la science* (Gallimard, Paris, 2002)
24. S. Laugier-Rabaté, *L'anthropologie logique de Quine, l'apprentissage de l'obvie* (Librairie philosophique J Vrin, Paris, 1992)
25. B. Van Fraassen, *Lois et symétries, French translation of "Laws and symmetry"* (Oxford University Press/Librairie philosophique J Vrin, Oxford/Paris, 1989/1994)
26. J.T. Cushing, *Wave-particle duality*, ed. by F. Selleri, Causal quantum theory, why a nonstarter?, (Plenum Press, New York, 1992), pp. 37–63
27. E. Squires, *The mystery of the quantum world* (Adam Hilger, Bristol, 1986)
28. Louis de Broglie, *Introduction à l'étude de la mécanique ondulatoire* (Hermann et Cie, Paris, 1930)
29. D.I. Blokhintsev, *Mécanique quantique* (Masson et Cie, Paris, 1967)
30. L.D. Landau, E.V. Lifchitz, *Mécanique, English translation: Mechanics* (Pergamon Press/Editions Mir, Oxford,/Moscow, 1969/1966)
31. P.R. Holland, *The quantum theory of motion, an account of the de Broglie-Bohm causal interpretation of quantum mechanics* (Cambridge University Press, Cambridge, 1993)
32. Louis de Broglie, Sur la possibilité d'une interprétation causale et objective de la mécanique ondulatoire. Comptes-Rendus de l'Académie des Sciences **234**, 265–268 (1952)
33. Louis de Broglie, *Louis de Broglie, physicien et penseur* Vue d'ensemble sur mes travaux scientifiques, Collection dirigée par André George (Editions Albin Michel, Paris, 1953) pp. 457–486
34. Louis de Broglie, *La physique quantique restera-t-elle indéterministe?* (Gauthier-Villars, Paris, 1953)
35. Louis de Broglie, *Une tentative d'interprétation causale et non linéaire de la mécanique ondulatoire (la théorie de la double solution). English translation: Nonlinear Wave Mechanics* (Elsevier/Gauthier-Villars, Amsterdam/Paris, 1960/1956)
36. Louis de Broglie, *La théorie de la mesure en mécanique ondulatoire (interprétation usuelle et interprétation causale)* (Gauthier-Villars, Paris, 1957)
37. Louis de Broglie, *Etude critique des bases de l'interprétation actuelle de la mécanique ondulatoire* (Gauthier-Villars, Paris, 1963)
38. D. Bohm, A suggested interpretation of the quantum theory in terms of "hidden" variables. part 1. Phys. Rev. **85**, 166–179 (1952)
39. D. Bohm, *Causality and chance in modern physics* (Routledge and Paul Kegan, London, 1957)
40. D. Bohm and B.J. Hiley, *The undivided universe, an ontological interpretation of quantum theory* (Routledge and Paul Kegan, London, 1993)
41. J.P. Vigier, *Structure des micro-objets dans l'interprétation causale de la théorie des quanta* (Gauthier-Villars, Paris, 1956)
42. M. Jammer, *The philosophy of quantum mechanics. The interpretations of quantum mechanics in historical perspective* (Wiley, New York, 1974)
43. M. Jammer, *The conceptual development of quantum mechanics. The History of Modern Physics, 1800–1950*, vol 12 (Tomash Publishers, Los Angeles, 1989)
44. T.J. Pinch, in *What does a proof do if it does not prove, a study of the social conditions and metaphysical divisions leading to David Bohm and John von Neumann failing to communicate in quantum physics*, ed. by E. Mendelson, P. Weingart and R. Whitley. Social production of scientific knowledge, sociology of the sciences, Vol 1 (D. Reidel, Dordrecht, 1977), pp. 171–215

45. J.A. Wheeler, W.H. Zurek (eds.), Quantum theory and measurement. Princeton ser. phys. (Princeton University Press, Princeton, 1983)
46. J.T. Cushing, Quantum mechanics, historical contingency and the Copenhagen hegemony (The University of Chicago Press, Hyde Park, 1994)
47. H. Freistadt, The causal formulation of the quantum mechanics of particles (the theory of de Broglie, Bohm and Takabayasi). Supplemento al Nuovo Cimento, Ser. **10**(5), 1–70 (1957)
48. F.J. Belinfante, *A survey of hidden-variable theories* (Pergamon, Oxford, 1973)
49. F.M. Pipkin, in *Atomic physics tests of the basic concepts in quantum mechanics*, ed. by D.R. Bates and B. Bederson. *Advances in Atomic and Molecular Physics* (Academic Press, New York, 1978) pp. 281–340
50. B. d'Espagnat, *Traité de physique et philosophie* (Fayard, Paris, 2002)
51. *Electrons et photons, Rapports et discussions du cinquième congrès de physique tenu à Bruxelles du 24 au 29 octobre 1927* (Gauthier-Villars, Paris, 1928)
52. P. Marage and G. Wallenborn, *Les conseils Solvay et les débuts de la physique moderne* (Université libre de Bruxelles, Brussels, 1995)
53. Louis de Broglie, *Electrons et photons, Rapports et Discussions du cinquième conseil de physique tenu à Bruxelles du 24 au 29 octobre 1927*, chapter: Nouvelle dynamique des quanta, (Gauthier-Villars, Paris, 1928) pp. 105–141
54. N. Bohr *Electrons et Photons, Rapports et Discussions du cinquième conseil de physique tenu à Bruxelles du 24 au 29 octobre 1927*, chapter: Le postulat des quanta et le nouveau dèveloppement de l'atomistique, (Gauthier-Villars, Paris, 1928), pp. 215–287
55. E. Schrödinger, *Electrons et Photons, Rapports et Discussions du cinquième conseil de physique tenu à Bruxelles du 24 au 29 ocotbre 1927*, chapter: La mécanique des ondes, (Gauthier-Villars, Paris, 1928), pp. 185–213
56. W. Heisenberg, *La partie et le tout, le monde de la physique atomique, French translation of: Der Teil und das Ganze, Gespräche im Umkreis der Atomphysik* (Editions Albin Michel, Paris, 1972)
57. W. Heisenberg, *La nature dans la physique contemporaine, translated from German*, collection Idées (Gallimard, Paris, 1962)
58. M. Born, *Louis de Broglie, physicien et penseur, collection dirigée par André George*, chapter: La grande synthèse, (Editions Albin Michel, Paris, 1953) pp. 165–170
59. J. von Neumann, *Les fondements de la mécanique quantique, French translation of "Mathematische Grundlagen der Quanten-mechanik", 1932. English version: Mathematical foundations of quantum mechanics* (Princeton University Press/Librairie Félix Alcan, Princeton/Paris, 1955/1946)
60. D. Bohm, A suggested interpretation of the quantum theory in terms of "hidden" variables, part 2. Phys. Rev. **85**, 180–193 (1952)
61. W. Pauli, *Louis de Broglie, physicien et penseur, collection dirigée par André George*, chapter: Remarques sur le problème des paramètres cachés dans la mécanique quantique et sur la théorie de l'onde pilote (Editions Albin Michel, Paris, 1953) pp. 33–42
62. B. d'Espagnat, *Le réel voilé, analyse des concepts quantiques* (Fayard, Paris, 1994)
63. M. Born, *The Born-Einstein letters, correspondence between Albert Einstein and Max and Hedwig Born, from 1916 to 1955 with commentaries by Max Born*, ed. by M.Born (MacMillan, London, 1971)
64. R.G. Newton, *Scattering theory of waves and particles* (Dover Publications, New York, 2002)
65. H.M. Nussenzveig, *Diffraction effects in semiclassical scattering* (Cambridge University Press, Cambridge, 1992)
66. W.V. Quine, *Philosophie de la logique, présentation par Denis Bonnay et Sandra Laugier* (Aubier-Montaigne, Paris, 1975)
67. W.V. Quine, *Relativit de l'ontologie et autres essais, présentation par Sandra Laugier* (Aubier-Montaigne, Paris, 1977)
68. W.V. Quine, *Du point de vue logique, neuf essais logico-philosophiques, French translation of: from a logical point of view*, bibliothèque des textes philosophiques, (Editions Vrin, Paris, 2003)

69. C. Kittel, W.D. Knight, and M.A. Ruderman, *Mechanics* (McGraw-Hill Book Company, New York, 1962)
70. J.T. Cushing, Bohm's theory: common sense dismissed. Stud. Hist. Philos. Sci. **24**(5), 815–842 (1993)
71. G. Gouesbet, Generalized Lorenz-Mie theories, the third decade: a perspective, invited review paper. J. Quant. Spectrosc. Radiat. Transf. **110**, 1223–1238 (2009)

Chapter 4
Predicting the Appearance of Materials Using Lorenz–Mie Theory

Jeppe Revall Frisvad , Niels Jørgen Christensen and Henrik Wann Jensen

Computer graphics systems today are able to produce highly realistic images. The realism has reached a level where an observer has difficulties telling whether an image is real or synthetic. The exception is when we try to compute a picture of a scene that really exists and compare the result to a photograph of the real scene. In this direct comparison, an observer quickly identifies the synthetic image. One of the problems is to model all the small geometrical details correctly. This is a problem that we will not consider. But even if we pick a simple experimental set up, where the objects in the scene have few geometrical details, a graphics system will still have a hard time predicting the result of taking a picture with a digital camera. The problem here is to model the optical properties of the materials correctly. In this chapter, we show how Lorenz–Mie theory enables us to compute the optical properties of turbid materials such that we can predict their appearance. To describe the entire process of predicting the appearance of a material, we include a description of the mathematical models used in realistic image synthesis.

4.1 Introduction

The appearance of an object is a result of light reaching the eye after interacting with matter. One way to capture the appearance of an object is to take a picture of it. This means that a digital camera is not only an inexpensive consumer product, it is also a fairly advanced device for measuring light. As we know from spectroscopy, light can tell us a great deal about the material that it is illuminating. The question is then: what can we learn about a material by taking a picture of it using a digital camera? The

J. R. Frisvad (✉) · N. J. Christensen
Technical University of Denmark, Kongens Lyngby, Denmark
e-mail: jrf@imm.dtu.dk

H. W. Jensen
University of California, San Diego, California

W. Hergert and T. Wriedt (eds.), *The Mie Theory*, Springer Series in Optical Sciences 169, DOI: 10.1007/978-3-642-28738-1_4,

immediate response is: not a lot. Unless we have a lot of pictures of similar materials in the same lighting environment which we can use to make a statistical model (this is image analysis). Or unless we have a physical model of the material and lighting environment that we can use to predict the picture (this is computer graphics). The purpose of this chapter is to show how to construct a model which predicts the result of taking a picture of a material in a controlled lighting environment. In short, we describe how to predict the appearance of materials. Once we have an appearance model that predicts the result of taking a picture of a material, we can take a picture of a similar material and fit the parameters in the model such that the picture of this similar material is obtained. In this way, we can use a picture of a material to estimate the properties of the material (analysis by synthesis).

The advantage of using a camera instead of a spectrometer is that it is inexpensive and that it does not require small samples to be prepared for analysis inside a larger device. Taking a picture using a digital camera is a simple, non-invasive process. This means that it is particularly useful for food quality control and medical diagnosis.

To construct an appearance model that predicts the result of taking a picture of a material, we need (a) a model of the camera, (b) a geometrical model of the material sample, (c) a model of the light sources, (d) a model for light propagation and (e) a model for light scattering. The problematic model is the one for light scattering because the material properties are unknown. The macroscopic optical properties of materials are so diverse that it is impossible to measure and tabulate the properties of all materials directly. This is where Lorenz–Mie theory comes in handy. Lorenz–Mie theory provides the link between the particle composition of a material and its macroscopic optical properties. The particle composition of a material is a very flexible starting point. There are many different ways to combine small particles into one macroscopic bulk material. To limit the potentially large number of input parameters, we use empirical models to describe the size distribution and optical properties of the particles. When available, these empirical models enable us to construct material appearance models that depend on only few essential inputs. A list of ingredients, for example, where the relative contents are specified in weight percent. The example used in this chapter is cow's milk with fat and protein contents of the milk (in weight percent) as input parameters.

4.2 Realistic Image Synthesis

The development of models for computing realistic images is a branch of computer graphics known as *realistic image synthesis* [1, 2]. This field of research provides the link between physical models of light and materials and the process of computing an image as if it were taken with a real digital camera or as if it were seen with the eyes of a human observer. Light transport and scattering is modelled in realistic image synthesis using the mathematical model developed for radiative transfer [3, 4]. The process of generating an image using a computer is referred to as *rendering*. Thus realistic image synthesis is also known as realistic rendering.

The basics of realistic image synthesis have been described in many textbooks. A useful reference which provides both detailed descriptions and source code is the book by Pharr and Humphreys [5]. The following sections provide brief descriptions of the five models that we need.

4.2.1 Camera

Think about taking a picture. The first thing you do is to position the camera and orient it towards the objects that you wish to photograph. To model this set up, we need a camera position $\boldsymbol{p}$, a viewing direction $\mathbf{v}$, and an up-vector $\mathbf{u}$. A digital camera is built around a photoactive charge-coupled device chip (CCD chip) which has a limited number of bins across its area. This means that output images from a digital camera have a specific *resolution* (image width W and height H) measured in numbers of pixels. Each *pixel* corresponds to a small area of the CCD chip and has an associated vector which describes the amount of light (of different colours) that reached this area during exposure. To *capture* an image is to expose the chip for a short amount of time and record the vectors for all the pixels. We model the light sensitive area of the CCD chip by a rectangle in the *image plane*, and we will call it the *film*. The viewing direction $\mathbf{v}$ is normal to the image plane and the up direction of the image plane is given by the up-vector $\mathbf{u}$. To position the film in space, we centre it around the camera position $\boldsymbol{p}$ and move it a distance d in the viewing direction. The distance d depends on the lens system of the camera and is called the *camera constant*. The size of the film (width by height, $w \times h$) is usually specified by an angle in the vertical direction called the *field of view*, fov and an aspect ratio a such that

$$
\begin{aligned}
w &= ah \\
h &= 2d \tan(\mathrm{fov}/2).
\end{aligned}
$$

Using the camera resolution ($W \times H$), we can divide the film into pixels. To compute an image, we trace a number of rays from $\boldsymbol{p}$ through each pixel of the film.

4.2.2 Geometry of Objects

Representation of arbitrarily shaped objects is a difficult problem with many solutions. The computer graphics community is with little doubt the most extensive user of surface models. The mainstream approach is to use a large number of small triangles. If we let some triangles have common edges with others, we obtain a *triangle mesh*. If there are no holes in the mesh, it represents the surface of a volume which could be an object in our scene.

The triangle mesh poses some problems. Unless we want to make objects with sharp edges, we need an awfully large number of triangles to make an object look smooth. Even more triangles are needed if we want smooth reflections and refractions as well. This problem is usually solved by computing a normal for each vertex in the triangle mesh as originally suggested by Phong [6]. Such a *vertex normal* is an average of the normals associated with the neighbouring triangle faces. When we want to find the *surface normal* where a ray intersects a triangle, we interpolate the vertex normals across the triangle using trilinear interpolation. In this way, we have a number of points (the vertices in the triangle mesh) which make out a discrete representation of a smooth surface.

4.2.3 Light Sources

A light source is typically modelled by letting an object in the scene emit light equally in all directions from every surface point. Another option is to have a spherical map around the scene which acts as a background source or environment lighting. This is called an *environment map* and it is modelled as if it were infinitely far away. It is particularly useful for outdoor sceneries, where one can use a model of the atmosphere to compute a realistic environment map [7–9].

Another important type of light source is laser. As has been discovered in biomedical optics [10], shining a laser into a turbid material (skin, for example) tells us surprisingly much about the material. To model a laser source, we use a circular disk which emits light weighted by a Gaussian function diminishing with the distance to the centre of the disk. Laser is also nearly collimated. This is modelled by letting the source emit light only in the direction normal to the surface of the disk.

4.2.4 Light Propagation

Pictures most often capture light in the visible part of the electromagnetic spectrum. Visible wavelengths are very short, ranging from 380 to 780 nm, and it was shown by Sommerfeld and Runge [11] that plane electromagnetic waves propagating in a non-absorbing, homogeneous dielectric are equivalent to *rays of light* for $\lambda \to 0$, where λ denotes the wavelength in vacuum. This means that rays of light are often a good model of visible light. According to Fermat's principle, rays of light follow the path of least time [12, pp. 457–463]. We can use this principle to trace rays of light through a digitally modelled scene. The main limitation is that wave phenomena such as interference and diffraction will not be captured. The visual effects from such phenomena must then be incorporated into the scattering properties of the materials in the scene. Diffraction by particles is incorporated into the macroscopic scattering properties of a material using Lorenz–Mie theory. This is covered in Sect. 4.2.5. Interference and diffraction effects caused by microfacets in the surface of an object are incorporated into surface reflection models. We will not discuss microfacetted surface reflection models.

Rays of light are traced in straight lines as long as they travel in a homogeneous medium, since, in this case, the shortest path is also the path of least time. Having settled on a ray theory of light, we trace light backwards from the camera position $\boldsymbol{p}$ through a pixel into our scene to see if it intersects one of the triangles that the objects are composed of. The number of triangles is large, so some kind of spatial data structure is needed to find the triangles that are most likely to be intersected [13–16]. An efficient ray-triangle intersection algorithm has been developed by Möller and Trumbore [17]. Only the closest intersected triangle is considered, other intersected triangles are hidden behind it. Once a point of intersection is identified and a surface normal has been computed, the next step is to estimate the light that reaches this point. Assuming a smooth surface, light reflects off the surface and refracts into the material as described by Fresnel's formulae for reflectance [18]. New rays are traced from the point of intersection in the directions of reflection and refraction (as dictated by the laws of reflection and refraction). Light reflects and refracts recursively until it reaches a light source. This is called *ray tracing*, and this recursive version, where new rays are traced from the point of intersection, was first described for graphics by Whitted [19]. If light refracts into a turbid material, it scatters as described in Sect. 4.2.5.

What we trace along a ray is electromagnetic energy. The rays follow the direction of the time-averaged Poynting vector as closely as possible.

4.2.5 Light Scattering

In the previous section, we found a way to trace energy through matter. This is all we need to render materials with a continuous interior. However, if a material is composed of millions of microscopic particles, it is infeasible to model the surface of every particle. Therefore, we introduce a model for macroscopic scattering. The model used to describe light scattering is the radiative transfer equation [3]:

$$(\boldsymbol{\omega} \cdot \nabla) L(\boldsymbol{x}, \boldsymbol{\omega}) = -\sigma_t(\boldsymbol{x}) L(\boldsymbol{x}, \boldsymbol{\omega}) + \sigma_s(\boldsymbol{x}) \int_{4\pi} p(\boldsymbol{x}, \boldsymbol{\omega}', \boldsymbol{\omega}) L(\boldsymbol{x}, \boldsymbol{\omega}') \, \mathrm{d}\omega' + L_e(\boldsymbol{x}, \boldsymbol{\omega}), \tag{4.1}$$

where $L(\boldsymbol{x}, \boldsymbol{\omega})$ is the *radiance* at $\boldsymbol{x}$ in the direction $\boldsymbol{\omega}$, the subscript e denotes emission, and σ_s, σ_a, and $\sigma_t = \sigma_s + \sigma_a$ are the scattering, absorption and extinction coefficients, respectively. The phase function p specifies the normalised distribution of the scattered light. Radiance is a radiometric quantity measured in energy flux per solid angle per projected area. The equation splits the directional derivative (left-hand side), that is, the change in radiance along a ray, into three terms (right-hand side): The first term denotes the exponential attenuation, the second denotes the in-scattering from all directions and the third is an emission term.

The radiometric quantity *radiance* is fundamental in radiative transfer. It is what the radiative transfer equation (Eq. 4.1) is concerned with and is defined (radiometrically) by

$$L = \frac{\mathrm{d}^3 Q}{\mathrm{d}t\, \mathrm{d}\omega\, \mathrm{d}A_\perp} = \frac{\mathrm{d}^2 \Phi}{\mathrm{d}\omega\, \mathrm{d}A \cos\theta},$$

where Q is energy, t is time, $\Phi = \mathrm{d}Q/\mathrm{d}t$ is energy flux, ω is solid angle, $A_\perp$ is projected area, A is area and θ is the angle between the normal to the area $\mathrm{d}A$ and the considered direction. Radiance describes the flow of energy through a differential area $\mathrm{d}A$. The energy flows in a directional differential volume $\mathrm{d}\omega$ which is not necessarily normal to the area. The purpose of radiance is thus to describe the energy flow in a ray of light incident on a surface.

Since radiance is a quantity based on energy, we assume that it follows the flow of energy through a medium. Thus we trace rays using Fermat's principle. However, we have to be careful and take into account that radiance denotes flux per solid angle per projected area. When we follow a ray of light through a material, we have to take into account that the radiance may change along the ray. Let us see how radiance changes upon refraction.

Consider a ray of light in a medium with refractive index $n_1 = n_1' + i\, n_1''$ incident on a surface patch $\mathrm{d}A$ of a medium with refractive index $n_2 = n_2' + i\, n_2''$. Due to energy conservation at the boundary, the following condition must hold:

$$L_i \cos\theta_i\, \mathrm{d}A\, \mathrm{d}\omega_i = L_r \cos\theta_r\, \mathrm{d}A\, \mathrm{d}\omega_r + L_t \cos\theta_t\, \mathrm{d}A\, \mathrm{d}\omega_t,$$

where L_i, L_r and L_t are the incident reflected and transmitted radiances and likewise θ_i, θ_r, and θ_t are the angles of incidence, reflection, and refraction. In spherical coordinates, the solid angles are defined by

$$\begin{aligned}
\mathrm{d}\omega_i &= \sin\theta_i\, \mathrm{d}\theta_i\, \mathrm{d}\phi_i \\
\mathrm{d}\omega_r &= \sin\theta_r\, \mathrm{d}\theta_r\, \mathrm{d}\phi_r \\
\mathrm{d}\omega_t &= \sin\theta_t\, \mathrm{d}\theta_t\, \mathrm{d}\phi_t\,.
\end{aligned}$$

Using the law of reflection $\theta_i = \theta_r$ and the fact that both the reflected and transmitted rays lie in the plane of incidence $\phi_i = \phi_r = \phi_t$, the boundary condition becomes

$$L_i \cos\theta_i \sin\theta_i\, \mathrm{d}\theta_i = L_r \cos\theta_i \sin\theta_i\, \mathrm{d}\theta_i + L_t \cos\theta_t \sin\theta_t\, \mathrm{d}\theta_t\,.$$

To find the transmitted angle, we have to consider the direction of the transmitted ray. This is given by the law of refraction. We find

$$\begin{aligned}
\sin\theta_t &= \frac{n_1'}{n_2'} \sin\theta_i \\
\cos\theta_t\, \mathrm{d}\theta_t &= \frac{n_1'}{n_2'} \cos\theta_i\, \mathrm{d}\theta_i\,.
\end{aligned}$$

With this result, the boundary condition is $L_i = L_r + \left(\frac{n'_1}{n'_2}\right)^2 L_t$. Or, in terms of outgoing radiance,

$$L_0 = RL_i + (1 - R)\left(\frac{n'_1}{n'_2}\right)^2 L_i, \tag{4.2}$$

where R is the Fresnel reflectance.

These equations show that radiance is not constant along a ray of light. If we move along a ray through a heterogeneous medium, the radiance should be modified. Generalising the result above to the interior of a medium, we approximately have [20, 21]

$$\frac{L_1}{n_1'^2}\mathrm{d}A_1 = \frac{L_2}{n_2'^2}\mathrm{d}A_2,$$

where the subscripts 1 and 2 denote two locations along a ray of light. This means that the quantity $L_1/n_1'^2$ is (approximately) constant along the ray. If we store $L_1/n_1'^2$ before tracing a ray from one point in a medium, then the radiance at the destination point is $L_2 = n_2'^2 L_1/n_1'^2$. Now that we know how radiance behaves as we move along a ray, we are ready to evaluate the radiative transfer Eq. (4.1) using ray tracing.

Rendering realistic images using the radiative transfer equation, was first proposed by Kajiya and Von Herzen [22]. Most rendering algorithms use approximate evaluation schemes to gain speed. Here, we will only describe the general way of evaluating the radiative transfer equation. The remainder of this section is a short account of *Monte Carlo path tracing*, which is a sampling-based rendering algorithm that works in general. More information is available in the book by Pharr and Humphreys [5]. Unfortunately, Pharr and Humphreys stop their treatment at single scattering. Evaluation of the general case has been described by Pattanaik and Mudur [23].

The emission term in the radiative transfer equation is just an added constant. It is not difficult to include it, but it makes the equations rather long. Therefore, we leave out the emission term in the following. The general approach is as follows. We first parameterise the radiative transfer equation using the distance s' that light has travelled along a path into a medium. We have (for a non-emitter):

$$\frac{\mathrm{d}L(s')}{\mathrm{d}s'} + \sigma_t(s')L(s') = \sigma_s(s') \int_{4\pi} p(s', \boldsymbol{\omega}', \boldsymbol{\omega})L(s', \boldsymbol{\omega}')\,\mathrm{d}\omega', \tag{4.3}$$

where $\boldsymbol{\omega}$ denotes the tangential direction of the path at the distance s' along the path.

The parameterised Eq. (4.3) is a linear, first-order, ordinary differential equation (where $\sigma_t(s')$ is a variable coefficient). One way to solve such an equation is by means of an integration factor:

$$T_r(s', s) = \exp\left(-\int_{s'}^{s} \sigma_t(t)\,\mathrm{d}t\right). \tag{4.4}$$

With this factor, Eq. (4.3) becomes

$$\frac{\mathrm{d}}{\mathrm{d}s'}\left(T_r(s', s)L(s')\right) = T_r(s', s)\sigma_s(s')\int_{4\pi} p(s', \boldsymbol{\omega}', \boldsymbol{\omega})L(s', \boldsymbol{\omega}')\,\mathrm{d}\omega'. \tag{4.5}$$

Note that $T_r(s, s) = 1$. Then by integration along the ray from the surface $s' = 0$ to the considered location in the medium $s' = s$, Eq. (4.5) attains the form:

$$L(s) = T_r(0, s)L(0) + \int_0^s T_r(s', s)\sigma_s(s')\int_{4\pi} p(s', \boldsymbol{\omega}', \boldsymbol{\omega})L(s', \boldsymbol{\omega}')\,\mathrm{d}\omega'\,\mathrm{d}s'. \tag{4.6}$$

For convenience, some of the different mathematical quantities encountered in this derivation have been given appropriate names in radiative transfer theory [3]. The distance s (or s') travelled along the ray inside the medium, is referred to as the *depth*. The integral in Eq. (4.4) is called the *optical thickness* and is denoted by the symbol τ, that is,

$$\tau(s', s) = \int_{s'}^{s} \sigma_t(t)\,\mathrm{d}t.$$

The integration factor itself $T_r(s', s) = e^{-\tau(s', s)}$ is sometimes referred to as the *beam* (or path) *transmittance*. Finally, the first term on the right-hand side of the formal solution (4.6) for the radiative transfer equation (Eq. 4.1) is referred to as the *direct transmission term*, whereas the second term is called the *diffusion term*. Realistic rendering is all about evaluating these terms in various kinds of ways.

To evaluate Eq. (4.6) using Monte Carlo path tracing, we first take a look at the direct transmission term $T_r(0, s)L(0)$. For this term we need to estimate the value of the optical thickness $\tau(0, s)$. If the medium is homogeneous, this optical thickness is simply given by

$$\tau(0, s) = \sigma_t s. \tag{4.7}$$

For heterogeneous media, we use sampling. The optical thickness is found quite efficiently by the estimator:

$$\sum_{i=0}^{N-1} \sigma_t(t_i)\Delta t, \tag{4.8}$$

where Δt is the step size (given as user input) and t_i are locations along the ray found using a single uniform random variable $\xi \in [0, 1]$:

$$t_i = \frac{\xi + i}{N} s. \tag{4.9}$$

The number N of locations along the ray is found using the depth s which is the distance to the next surface, that is, $N = \lfloor s/\Delta t \rfloor$. Having estimated the optical thickness $\tau(0, s)$, the beam transmittance $T_r(0, s)$ is easily found and multiplied by the amount of radiance $L(0)$ to reveal the direct transmission term. The radiance $L(0)$ is the radiance which contributes to the ray at the surface of the medium.

Evaluating the diffusion term is more involved. What we need is an estimator of the form

$$\frac{1}{N} \sum_{j=0}^{N-1} \frac{T_r(s'_j, s)\sigma_s(s'_j) J(s'_j)}{\text{pdf}(s'_j)}, \tag{4.10}$$

where the probability distribution function (pdf) preferably cancels out the transmittance. The source function

$$J(s') = \int_{4\pi} p(s', \omega', \omega) L(s', \omega') \, \mathrm{d}\omega' \tag{4.11}$$

is evaluated using a distribution of samples over the entire unit sphere.

The following pdf is a good choice for the diffusion term estimator (4.10):

$$\text{pdf}(s'_j) = \sigma_t(s'_j) T_r(s'_j, s),$$

since it has the simple complementary cumulative distribution function

$$\text{ccdf}(s'_j) = \int_{s'_j}^{s} \sigma_t(t) T_r(t, s) \, \mathrm{d}t = T_r(s, s) - T_r(s'_j, s) = 1 - T_r(s'_j, s).$$

Using the inverse transformation method [5, e.g.], we get the following equation for sampling an interaction along the ray according to this pdf:

$$\xi_j = 1 - T_r(s'_j, s) = 1 - e^{-\tau(s'_j, s)} \quad \text{or} \quad \ln(\xi_j) + \tau(s'_j, s) = 0,$$

where $\xi_j \in [0, 1]$ is a uniform random variable for sample j. An interaction is either scattering according to the source function (4.11) or absorption. The depth of the sample is easily found for homogeneous media, where we have $\tau(s'_j, s) = (s - s'_j)\sigma_t$, which gives

$$s'_j = s + \frac{\ln(\xi_j)}{\sigma_t}. \tag{4.12}$$

If $s'_j < 0$, there is no interaction for sample j. For heterogeneous media we have to step along the ray to find out where the optical thickness matches the event. Starting at

t_1 in $t_i = s - i\Delta t$, we step along the ray by incrementing i, and if $\ln(\xi_j) + \tau(t_i, s) > 0$, we stop and compute the location of the interaction as follows [23]:

$$s'_j = s - (i-1)\Delta t + \frac{\tau(t_{i-1}, s) + \ln(\xi_j)}{\sigma_t(t_i)}. \tag{4.13}$$

Here, it is assumed that $\sigma_t(t_i)$ is approximately constant in steps of size Δt along the ray, such that $\sigma_t(s'_j) \approx \sigma_t(t_i)$. If we end up with $t_i \le 0$ before the other criteria is fulfilled, there is no interaction for sample j. The optical thicknesses needed in order to find s'_j are evaluated in the same way as when we evaluated $\tau(0, s)$ only with $s - t_i$ in Eq. (4.9) instead of s.

Finally, the scattering coefficient $\sigma_s(s'_j)$ in the estimator (4.10) and the extinction coefficient $\sigma_t(s'_j)$ in the probability distribution function (pdf) are cancelled out by means of a Russian roulette. At every interaction a Russian roulette is carried out using the *scattering albedo*, which is defined by

$$\alpha(s'_j) = \sigma_s(s'_j)/\sigma_t(s'_j),$$

as the probability of a scattering event.

Using the sampling scheme described above, the estimator (4.10) becomes:

$$\frac{1}{N}\sum_{j=0}^{N-1} J(s'_j) = \frac{1}{NM}\sum_{j=0}^{N-1}\sum_{k=0}^{M-1} \frac{p(s'_j, \omega'_k, \omega)L(s'_j, \omega'_k)}{\text{pdf}(\omega'_k)} \tag{4.14}$$

for $\xi < \alpha(s'_j)$ and 0 otherwise. For an isotropic phase function, sampling a uniform distribution over the unit sphere leaves only $L(s'_j, \omega'_k)$ in the sum.

To summarise the algorithm, a ray is traced from an observer through a scene, when it refracts into a turbid material, we do the following:

1. Trace a refracted ray. The radiance carried along the refracted ray is corrected according to the boundary condition (Eq. 4.2).
2. The tracing of the refracted ray gives the depth s to the next surface.
3. The radiance $L(0)$ which contributes to the ray at the surface is found by tracing new rays in the directions of reflection and refraction. If $L(0)$ is not too small, we evaluate the direct transmission term:

   a. The optical thickness $\tau(0, s)$ is estimated using Eqs. 4.8 and 4.9 (or Eq. 4.7 for homogeneous media).
   b. The direct transmission term $T_r(0, s)L(0)$ is found using the optical thickness $\tau(0, s)$ (see Eq. 4.4) and the radiance which contributes to the ray at the surface $L(0)$.

4. For every diffusion term sample $j = 1, \ldots, N$, a sample depth s'_j is found using Eq. 4.13 (or Eq. 4.12 for homogeneous media).

5. For the samples $s'_j > 0$, a Russian roulette is done using the scattering albedo $\alpha(s'_j)$. For $\xi < \alpha(s'_j)$, where $\xi \in [0, 1]$ is a random variable, there is a scattering event.
6. For every scattering event, the phase function $p(s'_j, \boldsymbol{\omega}'_k, \boldsymbol{\omega})$ is evaluated in M sampled directions $\boldsymbol{\omega}'_k$ with $k = 1, \ldots, M$. Likewise, M new rays are traced at the position s'_j in the directions $\boldsymbol{\omega}'_k$ to obtain the radiances $L(s'_j, \boldsymbol{\omega}_\mathrm{k})$. Using Eq. 4.14 this gives an estimate of the diffusion term.
7. Finally, the direct transmission term and the diffusion term are added to get the radiance emergent at the surface $L(s)$.

The number of samples chosen are often $N = 1$ and $M = 1$. Then, we get a very noisy sample image rather quickly. Another sample is then rendered and this is averaged with the previous one. The next sample is weighted by one-third and added to the other two samples which are weighted by two-thirds and so on. In this way the image will improve itself over time. Even so, the Monte Carlo path tracing procedure spawns a formidable number of rays. It is very slow, but it converges to the intended result in an unbiased manner. This is nice in a predictive rendering context, where we want to predict the appearance of real-world materials.

4.3 Predicting Appearance

In order to predict the result of taking a picture with a digital camera, we need to digitally model the 3D scenery that is going to be in the picture as explained in Sect. 4.2.2. For example, in order to compute a realistic image of a glass of milk, we need a geometric model of the glass and the milk inside it. The most practical way of modelling such geometry is to use a computer aided design (CAD) system (e.g. AutoCAD® or Pro/ENGINEER®) or a 3D modelling system (e.g. 3ds Max®, Softimage® or Blender™). A scene modelled using one of these systems can be exported to a text file and imported into a rendering system which implements a camera model (see Sect. 4.2.1) and the models for light propagation and scattering described in Sects. 4.2.4 and 4.2.5. This could be our own rendering system or another physically based rendering system such as Maxwell Render™ or Indigo Renderer. Once the geometry of the scene is in place, we need information about the materials and the light sources in the scene. More specifically, we need the optical properties of the materials and the emission profiles of the light sources.

To acquire the optical properties of a material, one option is to measure them. Another option is to compute them from the particle contents of the material. The second option is made possible by the Lorenz–Mie theory, and it provides us with a very flexible way of modelling how the appearance of a material changes when we change its contents. Nevertheless, the Lorenz–Mie theory has only reluctantly been adopted in graphics. In the two papers [24, 25] where the theory was first considered for graphics applications, it was found to be either too complicated [24] or too restricted [25] to be useful. The problematic restrictions are that the material should consist of nearly spherical particles embedded in a non-absorbing host medium. This significantly limits the number of materials that can be modelled by the original version of the theory. Even so, the original theory has proven useful for modelling the appearance of some special materials: Callet [26] used the theory to model pigmented materials (such as paints, plastics, inks and cosmetics which consist of pigmented particles in a transparent solvent). Jackèl and Walter [8] used it to model the atmosphere and rainbows. Nishita and Dobashi [27] used the theory to model various other materials consisting of particles in air (clouds, smoke/gas, fog/haze, snow and sand). Most recently, we have seen a development towards generalisation of the Lorenz–Mie theory such that it becomes useful for a wider range of materials. Such new developments were adapted for graphics by Frisvad et al. [28]. In the following section, we describe the link between material contents and macroscopic optical properties of the material as provided by the Lorenz–Mie theory.

4.3.1 Computing Optical Properties

The input parameters for the radiative transfer equation (Eq. 4.1) are the phase function $p(\boldsymbol{x}, \boldsymbol{\omega}', \boldsymbol{\omega})$, the scattering coefficient $\sigma_s(\boldsymbol{x})$ and the extinction coefficient $\sigma_t(\boldsymbol{x})$ (or the absorption coefficient since $\sigma_t = \sigma_a + \sigma_s$). Together with the index of refraction $n_{\text{bulk}}(\boldsymbol{x})$, these parameters constitute the optical properties of a material. In the following, we will omit the dependency on the position $\boldsymbol{x}$ in the material. Then we just have to remember that this dependency should be inserted if the material is heterogeneous.

The phase function and the scattering coefficient are collectively referred to as the scattering properties of the material. The direction of the incoming light $\boldsymbol{\omega}'$ is called the *forward direction* while $\boldsymbol{\omega}$ is the direction of the scattered light. The plane spanned by these two vectors is called the *scattering plane* and the angle between them θ is called the *scattering angle*.

Scattering under the surface of a material is typically caused by particles. For simplicity, let us assume that the particles are small, randomly distributed throughout the material, not too densely packed, and approximately spherical. A particle is considered small when it is not directly visible to the human eye from the distance that it is observed. If every particle were visible, we would have to model the surface of each individual particle. If the particles were not randomly distributed, we would again need to make a very precise model taking into account the ordered placement of the particles in the material. When we assume that the particles are not too densely packed, we assume that the distance between them is considerably larger than the wavelength of the light. Under this assumption, we say that the particles scatter light independently of each other. The assumption about approximately spherical particles ensures that scattering is symmetric around the forward direction and that the two polarisation components do not affect each other. In other words, we only need two scattering components to describe the scattering of a spherical particle: $S_1(\theta)$ for the polarisation component with the electric vector perpendicular to the scattering plane and $S_2(\theta)$ for the polarisation component with the electric vector parallel to the scattering plane. For unpolarised light, these two scattering components define the phase function of a single particle by

$$p_r(\theta) = \frac{|S_1(\theta)|^2 + |S_2(\theta)|^2}{2|k|^2 C_s}, \tag{4.15}$$

where C_s is the scattering cross section of the particle, $k = 2\pi n_{\text{med}}/\lambda$ is the wave number and n_{med} is the refractive index of the host medium. If the host medium is absorbing, n_{med} is a complex number. The scattering components (S_1 and S_2) and the scattering cross section of a particle change depending on the radius r of the particle. To indicate this, we let p_r denote the phase function of a single particle of radius r.

The scattering cross section C_s of a particle is the area that would receive the same amount of energy as the particle scatters if we subtend it normal to the incident light. If we multiply the scattering cross section of a particle with the number density of this type of particle in the material, we get the amount of light that will be scattered away from a ray of light per unit distance as it propagates through the material, that is, we get the scattering coefficient σ_s. Since the particles may have many sizes with different cross sections and scattering components, the scattering coefficient is an integral over particle radii r:

$$\sigma_s = \int_{r_{\text{min}}}^{r_{\text{max}}} C_s(r) N(r)\, \mathrm{d}r. \tag{4.16}$$

The number density in this expression is $N(r)\,\mathrm{d}r$ while $N(r)$ itself is a number density distribution. It is a distribution because it denotes the number of particles of radius r in the range of particle sizes $\mathrm{d}r$. The justification for this simple combination of scattering cross sections into the macroscopic scattering coefficient is the assumption about independent scattering of the particles.

To keep the phase function normalised, the expression for the macroscopic phase function, often called the *ensemble phase function*, is:

$$p(\theta) = \frac{\int_{r_{\text{min}}}^{r_{\text{max}}} \left(|S_1(\theta)|^2 + |S_2(\theta)|^2\right) \mathrm{d}r}{2|k|^2 \sigma_s} = \frac{1}{\sigma_s} \int_{r_{\text{min}}}^{r_{\text{max}}} C_s(r) p_r(\theta)\, \mathrm{d}r.$$

As the equation shows, another way to find the ensemble phase function is as a scattering cross section weighted average of the single particle phase functions p_r (defined by Eq. 4.15).

Now we have an idea about how to find the scattering due to one type of particle in a medium. To deal with several different types of particles, we let A denote the set of homogeneous substances appearing as particles in the host medium. Then $p_i, \sigma_{s,i}$ and

$\sigma_{t,i}$ denote the phase function and the scattering and extinction coefficients for every individual particle inclusion $i \in A$. Once the phase function has been determined for each individual particle inclusion, the ensemble phase function is computed using a scattering coefficient weighted average:

$$p(\theta) = \frac{1}{\sigma_s} \sum_{i \in A} \sigma_{s,i} \, p_i(\theta). \tag{4.17}$$

Considering the number of single particle phase functions required to approximate the ensemble phase function $p(\theta)$, it is only practical to either tabulate the phase function or to use the ensemble asymmetry parameter g (see below) with one of the standard phase functions, e.g. the Henyey-Greenstein phase function which is defined by [29]

$$p(\theta) = \frac{1}{4\pi} \frac{1 - g^2}{(1 + g^2 - 2g \cos\theta)^{3/2}}.$$

A more exact option is to use a multi lobed phase function where the Henyey-Greenstein function replaces $p_i(\theta)$ in Eq. 4.17. The *asymmetry parameter* is defined by the integral over all solid angles of the cosine weighted phase function:

$$g = \int_{4\pi} p(\theta) \cos\theta \, \mathrm{d}\omega.$$

If the asymmetry parameter is computed for single particles and individual particle inclusions, we can compute the *ensemble asymmetry parameter* as a weighted average using the same weights as for the ensemble phase function.

Because we assume that particles scatter light independently, not only scattering cross sections are additive (see Eq. 4.16), but also scattering coefficients (and extinction coefficients) are additive. Finding the bulk scattering coefficient is straightforward:

$$\sigma_s = \sum_{i \in A} \sigma_{s,i}.$$

Note that volume fractions are *not* included in this formula, because they are a part of the number density distributions.

In a transparent medium, the extinction coefficient is defined by an equivalent sum, but in an absorbing medium an important correction must be made. Since the host medium is a part of the extinction process, a non-absorbing particle will reduce the extinction of the bulk medium. This means that the extinction cross sections can be negative [30]. The extinction cross section resulting from the Lorenz–Mie theory is, in other words, relative to the absorption of the host medium and the necessary correction is to include the host medium absorption $\sigma_{a,\mathrm{med}}$ in the sum. For this purpose, we compute the bulk extinction coefficient for particles in an absorbing medium by

$$\sigma_t = \sigma_{a,\mathrm{med}} + \sum_{i \in A} \sigma_{t,i},$$

and the bulk absorption coefficient is given by the simple relation $\sigma_a = \sigma_t - \sigma_s$. These bulk coefficients are never negative.

To compute the refractive index of the bulk medium n_{bulk}, we follow van de Hulst's [31] derivation of a formula for the effective index of refraction, but we remove the assumptions of non-absorbing media and particles of only one radius. This gives the following approximate relation for the real part of the bulk refractive index:

$$\mathrm{Re}(n_{\mathrm{bulk}}(\lambda)) = \mathrm{Re}(n_{\mathrm{med}}(\lambda)) + \lambda \sum_{i \in A} \int_{r_{\min}}^{r_{\max}} \mathrm{Im}\left(\frac{S_{i,r,\lambda}(0)}{k^2}\right) N_i(r)\,\mathrm{d}r,$$

where $S_{i,r,\lambda}(0) = S_1(0) = S_2(0)$ is the amplitude in the forward direction of the wave of wavelength λ scattered by a type i particle of radius r, $N_i(r)$ is the number density distribution and k is the wave number. The imaginary part is found by its relation to the bulk absorption coefficient:

$$\mathrm{Im}(n_{\mathrm{bulk}}(\lambda)) = \sigma_a(\lambda)\frac{\lambda}{4\pi}. \tag{4.18}$$

4.3.2 Lorenz–Mie Theory

Lorenz [32] and Mie [33] showed that when the scattering components are expanded using spherical functions they are defined, for a homogeneous plane wave, by [31–34]

$$S_1(\theta) = \sum_{n=1}^{\infty} \frac{2n+1}{n(n+1)} \left(a_n \pi_n(\cos\theta) + b_n \tau_n(\cos\theta)\right) \tag{4.19}$$

$$S_2(\theta) = \sum_{n=1}^{\infty} \frac{2n+1}{n(n+1)} \left(a_n \tau_n(\cos\theta) + b_n \pi_n(\cos\theta)\right), \tag{4.20}$$

where the functions π_n and τ_n are related to the Legendre polynomials P_n as follows:

$$\pi_n(\cos\theta) = \frac{P_n^1(\cos\theta)}{\sin\theta} = \frac{\mathrm{d}P_n(\cos\theta)}{\mathrm{d}(\cos\theta)}$$

$$\tau_n(\cos\theta) = \frac{\mathrm{d}P_n^1(\cos\theta)}{\mathrm{d}\theta} = \cos\theta\,\pi_n(\cos\theta) - \sin^2\theta\,\frac{\mathrm{d}\pi_n(\cos\theta)}{\mathrm{d}(\cos\theta)}.$$

Their numeric evaluation can be found in standard references on Lorenz–Mie theory [35, 36].

It remains to give expressions for the Lorenz–Mie coefficients a_n and b_n. Suppose the refractive index of the particle is n_p. Then the coefficients are given, in the far field, by [31–34]

$$a_n = \frac{n_{\text{med}}\psi_n'(y)\psi_n(x) - n_p\psi_n(y)\psi_n'(x)}{n_{\text{med}}\psi_n'(y)\zeta_n(x) - n_p\psi_n(y)\zeta_n'(x)} \tag{4.21}$$

$$b_n = \frac{n_p\psi_n'(y)\psi_n(x) - n_{\text{med}}\psi_n(y)\psi_n'(x)}{n_p\psi_n'(y)\zeta_n(x) - n_{\text{med}}\psi_n(y)\zeta_n'(x)}, \tag{4.22}$$

where the primes $'$ denote derivative. The spherical functions $\psi_n(z)$ and $\zeta_n(z)$ are known as Riccati-Bessel functions. They are related to the spherical Bessel functions $j_n(z)$ and $y_n(z)$ as follows:

$$\psi_n(z) = z j_n(z)$$
$$\zeta_n(z) = z(j_n(z) - i y_n(z)).$$

The argument z is an arbitrary complex number, the arguments x and y used for the Lorenz–Mie coefficients are related to particle and host media as follows:

$$x = \frac{2\pi r n_{\text{med}}}{\lambda} \quad \text{and} \quad y = \frac{2\pi r n_p}{\lambda},$$

where λ is the wavelength in vacuum and r is the radius of the spherical particle.

When computers came around, it turned out to be quite difficult to find a numerically stable way of evaluating the spherical functions ψ_n and ζ_n for complex arguments. Eventually, the numerical difficulties were solved for complex y [35, 37, 38]. This is sufficient for the traditional Lorenz–Mie theory with a non-absorbing host medium. When people started considering spheres in an absorbing host, starting with Mundy et al. [39], it became necessary to find a robust way of evaluating the Lorenz–Mie coefficients for complex x as well. This is considerably more difficult. The following describes a robust evaluation scheme proposed by Frisvad et al. [28].

In the case of an absorbing host medium, n_{med} has an imaginary part and then the parameter x is complex. The consequence is that most numerical evaluation schemes become unstable because the Riccati-Bessel functions enter the exponential domain and run out of bounds.

To avoid the ill-conditioning of the Riccati-Bessel functions ψ_n and ζ_n, the Lorenz–Mie coefficients (4.21–4.22) are rewritten in a form involving only ratios between them [37]

$$a_n = \frac{\psi_n(x)}{\zeta_n(x)} \frac{n_{\text{med}}A_n(y) - n_p A_n(x)}{n_{\text{med}}A_n(y) - n_p B_n(x)} \tag{4.23}$$

$$b_n = \frac{\psi_n(x)}{\zeta_n(x)} \frac{n_p A_n(y) - n_{\text{med}}A_n(x)}{n_p A_n(y) - n_{\text{med}}B_n(x)}. \tag{4.24}$$

Here $A_n(z)$ and $B_n(z)$ denote the logarithmic derivatives of $\psi_n(z)$ and $\zeta_n(z)$, respectively:

$$A_n(z) = \frac{\psi_n'(z)}{\psi_n(z)} \quad \text{and} \quad B_n(z) = \frac{\zeta_n'(z)}{\zeta_n(z)}.$$

The ratio A_n is only numerically stable with downward recurrence. Therefore, the following formula is employed for its evaluation [37]

$$A_n(z) = \frac{n+1}{z} - \left(\frac{n+1}{z} + A_{n+1}(z)\right)^{-1}. \tag{4.25}$$

This formula is also valid for the ratio B_n, but then it is unfortunately unstable for both upward and downward recurrences [40]. Instead, we use a different formula for B_n which has been developed by Mackowski et al. [41] in the field of multilayered particles embedded in a non-absorbing medium. It is numerically stable with upward recurrence for any complex argument [41]:

$$B_n(z) = A_n(z) + \frac{i}{\psi_n(z)\zeta_n(z)} \tag{4.26}$$

$$\psi_n(z)\zeta_n(z) = \psi_{n-1}(z)\zeta_{n-1}(z)\left(\frac{n}{z} - A_{n-1}(z)\right)\left(\frac{n}{z} - B_{n-1}(z)\right). \tag{4.27}$$

It remains to give a recurrence relation for the ratio $\psi_n(z)/\zeta_n(z)$ in Eqs. 4.23 and 4.24. Recent developments in the context of multilayered particles, provide a recurrence relation that works well for small $\mathrm{Im}(z)$ [42, 43]:

$$\frac{\psi_n(z)}{\zeta_n(z)} = \frac{\psi_{n-1}(z)}{\zeta_{n-1}(z)} \frac{B_n(z) + n/z}{A_n(z) + n/z}. \tag{4.28}$$

The restriction to small $\mathrm{Im}(z)$ is not a problem in graphics applications, as a larger $\mathrm{Im}(z)$ means that the host medium is highly absorbing, and then we would not be able to see the effect of particle scattering anyway.

The amplitude functions (4.19– 4.20) are defined by an infinite sum, and in order to get a decent approximation, we must find an appropriate number of terms M to sum. This is also necessary for initialisation of the downward recurrence (4.25) which computes $A_n(x)$ and $A_n(y)$. A formula determining M, which has both an empirical [38, 41] and a theoretical [40] justification, is

$$M = \left\lceil |x| + p|x|^{1/3} + 1 \right\rceil,$$

where $p = 4.3$ gives a maximum error of 10^{-8}. It is possible to calculate an approximate initial value for the downward recurrence (4.25), but, as explained by Dave [35], the recurrence is not sensitive to the initial value, and therefore we can arbitrarily choose $A_M(z) = 0$.

Once $A_0(z), \ldots, A_M(z)$ have been computed for both $z = x$ and $z = y$, we are able to find the ratios $B_n(x)$ and $\psi_n(x)/\zeta_n(x)$ as well as the Lorenz–Mie coefficients, a_n and b_n, step by step. Note that there is no need to store $B_n(x)$ and $\psi_n(x)/\zeta_n(x)$ since they are computed using upward recurrences (4.26– 4.28). These recurrences should be initialised by

$$\begin{aligned} B_0(z) &= i \\ \psi_0(z)\zeta_0(z) &= \tfrac{1}{2}\left(1 - e^{i2z}\right) \\ \psi_0(z)/\zeta_0(z) &= \tfrac{1}{2}\left(1 - e^{-i2z}\right). \end{aligned}$$

Recall that there is a direct relationship between wavelength λ and the size parameters x and y. This tells us that the Lorenz–Mie coefficients are spectrally dependent and should preferably be sampled at different wavelengths. They also depend on the particle radius r and are valid for spherical particles of arbitrary size as long as they do not exhibit diffuse reflection (which is only possible if the particle size greatly exceeds the wavelength and, even so, the surface of the particle might still be smooth) [31]. Furthermore, the equations provided in this section reveal that the complex refractive index of each particle inclusion, as well as that of the host medium, are needed as input parameters for computing the optical properties of a scattering material.

This robust way of computing the Lorenz–Mie coefficients enables us to compute the scattering amplitudes S_1 and S_2 (using Eqs. 4.19 and 4.20). With these, we are able to find the extinction and scattering cross sections as well as the phase function of the particle. These are all well-defined quantities for particles in a non-absorbing medium. For a particle in an absorbing medium, the scattering cross section is a problematic quantity because the resulting formula depends on the distance to the observer.

When particles are embedded in an absorbing host, the extinction cross section C_t is the only well-defined observable quantity [30]. It is computed using an optical theorem first presented by van de Hulst [31, 44]. The original theorem by van de Hulst is valid for particles of arbitrary shape and size, but it only applies to a non-absorbing host medium. To account for an absorbing host, we use a slightly modified equation presented by Bohren and Gilra [30]:

$$C_t = 4\pi \,\mathrm{Re}\left(\frac{S(0)}{k^2}\right), \tag{4.29}$$

where $S(0) = S_1(0) = S_2(0)$ is the amplitude in the forward direction of the scattered wave and $k = 2\pi n_{\text{med}}/\lambda$ is the wave number. Since the host medium was assumed by van de Hulst to be non-absorbing, n_{med} and therefore also k were assumed real and moved outside the Re operator (which takes the real part of a complex number). This is not allowed if the host medium is absorbing as the result would be a meaningless complex extinction coefficient. Correction by discarding the imaginary part of the result would not be a good approximation (except when particle absorption is considerably stronger than that of the host medium [30]). Inserting the expression for $S(0)$ in this optical theorem (4.29), we get

$$C_t = \frac{\lambda^2}{2\pi} \sum_{n=1}^{\infty} (2n+1) \mathrm{Re}\left(\frac{a_n + b_n}{n_{\mathrm{med}}^2} \right).$$

A form has not been found for the scattering cross section C_s which is independent of the distance to the observer, but we still have to approximate C_s to evaluate the radiative transfer Eq. (4.1). We use a far-field approximation which has been reported to be consistent with measured data [45, 46]. The chosen formula is identical to the scattering cross section for transparent media except for two correction terms: an exponential term and a geometrical term γ. The formula is

$$C_s = \frac{\lambda^2 e^{-4\pi r \,\mathrm{Im}(n_{\mathrm{med}})/\lambda}}{2\pi \gamma |n_{\mathrm{med}}|^2} \sum_{n=1}^{\infty} (2n+1) \left(|a_n|^2 + |b_n|^2 \right), \tag{4.30}$$

where r in the exponential term is the uncertain part of the equation because it ought to be the distance to where the scattered wave is observed. This distance is unknown, and consequently it has been projected to the particle surface, such that r denotes the particle radius.

The geometrical term γ accounts for the fact that the incident wave changes over the surface of the particle as a consequence of the absorbing host medium. It is defined by [39]

$$\gamma = \frac{2(1 + (\alpha - 1)e^{\alpha})}{\alpha^2}, \tag{4.31}$$

where $\alpha = 4\pi r \,\mathrm{Im}(n_{\mathrm{med}})/\lambda$ and $\gamma \to 1$ for $\alpha \to 0$. Note that α is 0 when the medium is transparent and close to 0 for small particles in a weakly absorbing medium. To avoid numerical errors, one should use $\gamma = 1$ for $\alpha < 10^{-6}$.

The precision of the far-field approximation (4.30, 4.31) has recently been reviewed [47] and compared to experimental data [45, 46]. The conclusion is that it (as expected) does not give entirely accurate results, but it does give physically plausible results. It is also concluded that significant errors can result if the absorption of the host medium is ignored (this is especially true when the size parameter x is large).

In the same way that it is possible to formulate expressions for the scattering and extinction cross sections of a spherical particle using Lorenz–Mie theory, it is also possible to derive the following formula for the asymmetry parameter of a single spherical particle [31]:

$$g = \frac{\sum_{n=1}^{\infty} \left\{ \frac{n(n+2)}{n+1} \mathrm{Re}(a_n a_{n+1}^* + b_n b_{n+1}^*) + \frac{2n+1}{n(n+1)} \mathrm{Re}(a_n b_n^*) \right\}}{\frac{1}{2} \sum_{n=1}^{\infty} (2n+1) \left(|a_n|^2 + |b_n|^2 \right)},$$

where the asterisks * denotes the complex conjugate.

This concludes the robust scheme for computing the scattering properties of a sphere in a host medium. When using the Lorenz–Mie theory to compute macroscopic scattering properties of a medium (as described in Sect. 4.3.1), some information about the particle shapes and sizes is needed. We will look at this in Sects. 4.3.3 and 4.3.4.

4.3.3 Number Density Distributions

Particle *size distribution* is the common term for distributions that we can use to find the number densities of particles of different sizes. One type of size distribution, which is often encountered in the literature, is the *volume frequency distribution* $r^3 N(r)$. Such distributions typically follow a log-normal distribution. Log-normal distributions are often described by a mean particle size μ and a coefficient of variation $c_v = \sigma/\mu$, where σ is the standard deviation.

If we find that the volume frequency of some type of particle in a medium follows the log-normal distribution with mean value μ and standard deviation σ, the volume frequency distribution is given by

$$r^3 N(r) = \frac{1}{r\beta\sqrt{2\pi}} e^{-\frac{1}{2}\left(\frac{\ln r - \alpha}{\beta}\right)^2}, \tag{4.32}$$

where r is the particle radius and

$$\alpha = \ln \mu - \frac{1}{2} \ln\left(\frac{\sigma^2}{\mu^2} + 1\right) \quad \text{and} \quad \beta = \sqrt{\ln\left(\frac{\sigma^2}{\mu^2} + 1\right)}.$$

This type of distribution is commonly observed for small particles. Distributions of larger particles tend to follow a *power law*. If the particle number density of some type of particle in a medium follows a power law, the distribution is given by

$$N(r) = N_* r^{-\alpha},$$

where α is usually determined empirically and N_* is a constant which is determined by the relationship between number density and the volume fraction of the medium occupied by the considered type of particle (Eq. 4.33).

Measured data are sometimes available which specify the volume fraction v_i, $i \in A$, of each particle inclusion that is present in the bulk medium. In any case, volume fractions are a reasonable choice of input parameters. The number density distribution $N(r)$ specifies the number of particles per unit volume with radii in the interval $[r, r + dr]$. This means that the volume fraction occupied by a particle inclusion, which consists of spherical particles, is

$$v = \frac{4\pi}{3} \int_{r_{\min}}^{r_{\max}} r^3 N(r)\,\mathrm{d}r. \tag{4.33}$$

Suppose we measure the particle size distributions for some sample of material. Then we would have empirical functions or tabulated data that fit the volume fractions of the particles in the original sample. Most probably the original volume fractions are not the volume fractions we desire in our medium. Equation (4.33) is important because it explains how we find the original volume fraction $v_{\text{original},i}$ of particle type i. If the volume fraction v_i is desired rather than $v_{\text{original},i}$, the measured number density distribution should be scaled by $v_i/v_{\text{original},i}$.

4.3.4 Non-Spherical Particles

Particles are not always spherical. There exist a number of theories for the scattering of other perfect mathematical shapes like cylinders, hexagonal columns and plates, etc. They are useful because some materials actually do consist of particles approximately of these shapes. Halos are, for example, the result of scattering by hexagonal ice crystals in the atmosphere. Instead of the mathematical approach, let us use a more practical approach for non-spherical particles.

With the theory we have already developed, we are able to approximate non-spherical particles by an appropriate collection of spherical particles. It is not obvious what set of spheres we should choose to model a non-spherical particle in the best way. Many different concepts have been tried: equal-volume spheres, equal-area spheres, etc. The best approach we are aware of is that of Grenfell and Warren [48]. They use volume-to-area equivalent spheres. As opposed to equal-volume and equal-area spheres, the volume-to-area equivalent spheres have proven to be quite exact. They have been tested for cylinders [48], hexagonal columns and plates [49], and hollow columns and plates [50]. In most cases the error is less than 5 %. At least this is true for scattering and extinction coefficients. The approximation is, as could be expected, less accurate with respect to the phase function.

To represent a particle of volume V and surface area A by a collection of spheres, the radius of the equivalent spheres is found simply using the volume to surface area ratio of a sphere [48]:

$$r_{\text{eq}} = 3\frac{V}{A}.$$

Since the number of equivalent spheres is not equal to the number of non-spherical particles, the number density must be adjusted accordingly [48]:

$$\frac{N_{\text{eq}}}{N} = \frac{3V}{4\pi r_{\text{eq}}^3}.$$

The equivalent radius r_{eq} and the equivalent number density N_{eq} are then used for computing the optical properties of the material with Lorenz–Mie theory for computing the cross sections of the equivalent spheres. This is a simple and practical approach which gives rather good results. It has been used with the theory presented here to develop an appearance model for natural ice. which contains both cylindrically and spheroidally shaped particles [28, 51]. In the milk case study (Sect. 4.4), the volume-to-area equivalent size distribution is reported directly in the literature, so the radius and number density adjustments are not necessary.

4.3.5 Scattering of a Gaussian Beam

In wave optics, a generalised version of the Lorenz–Mie theory [52, 53] would be needed to model the scattering by a particle of a shaped beam such as laser. However, we are using the scattering by particles in a geometrical optics context. The scattering of a ray, which traces an infinitely thin part of the wavefront, is adequately modelled by the scattering of a plane wave. Thus we do not employ the generalised Lorenz–Mie theory, but model a Gaussian beam (a laser source) as described in Sect. 4.2.3.

When we render an image using the algorithm described in Sect. 4.2.5, the laser source poses a problem. Rays are traced from the observer and new directions are sampled at every scattering event. Since the laser source is small and collimated, the chance of a ray hitting the laser source from the right direction is almost non-existing. To solve this problem, we use bidirectional path tracing [54]. As described in Sect. 4.2, rays are traced from the observer through each pixel in the image into the scene. To account for a laser source, we also sample a position on the laser source and trace a ray from this position to the first scattering event in a medium. For every ray from the observer that reaches this medium, we compute the contribution from the scattered laser light. Every time all pixels have been sampled once by rays from the observer, a new sample ray is traced from the laser source. This is how we accommodate a laser source in the path tracing algorithm. In the following, we will use the theory to predict the appearance of milk.

4.4 Milk as a Case Study

Milk consists roughly of an emulsion of milkfat globules; a colloidal suspension of protein particles; and lactose, soluble proteins, minerals, vitamins, acids, enzymes and other components dissolved in water [55]. About 80 % of the protein in milk is casein protein. Most of this casein, about 95 % [56], exists in colloidal particles known as casein micelles. From an optical point of view, milk can then be treated as two different types of nearly spherical particles, namely fat globules and casein micelles, suspended in a host medium with almost the same optical properties as pure water. The absorption spectrum of the host medium needs to be adjusted because

Table 4.1 The coefficients for the empirical formula (4.34) by Quan and Fry [57]

$n_1 = 1.779 \cdot 10^{-4}$	$n_4 = -2.02 \cdot 10^{-6}$	$n_7 = -0.00423$
$n_2 = -1.05 \cdot 10^{-6}$	$n_5 = 15.868$	$n_8 = -4382$
$n_3 = 1.6 \cdot 10^{-8}$	$n_6 = 0.01155$	$n_9 = 1.1455 \cdot 10^6$

of dissolved vitamin B2 (riboflavin) which exhibits absorption in the visible range of the spectrum. In fact riboflavin is also fluorescent, but we will not take that into account. What we use is, in other words, a simplified model of milk, but it should be sufficient for considering the appearance of milk.

4.4.1 Particle Composition

The host medium is water in which many different components are dissolved. For the real part of the refractive index of the milk host, we use the refractive index for pure fresh water ($S = 0‰$ in Eq. 4.34).

Quan and Fry [57] have developed an empirical formula for computing the real part of the refractive index of pure water or brine as a function of salinity S, temperature T, and wavelength λ. It is as follows [57]:

$$n'_{\text{water}}(\lambda, T, S) = 1.31405 + (n_1 + n_2 T + n_3 T^2)S + n_4 T^2 + \frac{n_5 + n_6 S + n_7 T}{\lambda} + \frac{n_8}{\lambda^2} + \frac{n_9}{\lambda^3}. \tag{4.34}$$

The coefficients are listed in Table 4.1. This formula describes the dependency of n'_{water} on salinity in the range $0‰ < S < 35‰$, temperature in the range $0\,°\text{C} < T < 30\,°\text{C}$, and wavelength in the range $400\,\text{nm} < \lambda < 700\,\text{nm}$. Moreover, Huibers [58] has reported that the same formula is valid over a broader spectrum of wavelengths ($200\,\text{nm} < \lambda < 1100\,\text{nm}$) than originally assumed.

To find the imaginary part of the refractive index of the milk host, we make a correction for the imaginary part of the refractive index of pure water. The dissolved component exhibiting the most significant absorption in the visible range is vitamin B2 (riboflavin). Spectral data for the absorption of vitamin B2 are available in the PhotochemCAD application[1] [59]. The absorption coefficient is not measured directly, instead the absorbance D is measured. The absorption coefficient is calculated from the absorbance using a molar absorption coefficient ε for some wavelength. A molar absorption coefficient of riboflavin is $\varepsilon(266.5\,\text{nm}) = 3.3 \cdot 10^6\,\text{M}^{-1}\text{m}^{-1}$ [60]. The following formula finds the remaining molar absorption coefficients for riboflavin using the absorbance data:

[1] http://omlc.ogi.edu/spectra/PhotochemCAD/abs_html/riboflavin.html.

Table 4.2 Imaginary part of the refractive index for pure water [61], riboflavin [59] (0.17 mg pr. 100 g solution), milk host n''_{milk}, and milk fat n''_{fat}

λ(nm)	n''_{water}	$n''_{\text{riboflavin}}$	n''_{milk}	n''_{fat}
375	$3.393 \cdot 10^{-10}$	$2.927 \cdot 10^{-7}$	$2.93 \cdot 10^{-7}$	$4.0 \cdot 10^{-6}$
400	$2.110 \cdot 10^{-10}$	$2.603 \cdot 10^{-7}$	$2.60 \cdot 10^{-7}$	$6.4 \cdot 10^{-6}$
25	1.617	3.363	3.36	8.6
50	3.302	4.096	4.10	$1.1 \cdot 10^{-5}$
75	4.309	3.323	3.33	1.1
500	$8.117 \cdot 10^{-10}$	$1.076 \cdot 10^{-7}$	$1.08 \cdot 10^{-7}$	$1.0 \cdot 10^{-5}$
25	$1.742 \cdot 10^{-9}$	$5.470 \cdot 10^{-8}$	$5.64 \cdot 10^{-8}$	$4.7 \cdot 10^{-6}$
50	2.473	5.772	6.02	4.6
75	3.532	7.554	7.91	4.7
600	$1.062 \cdot 10^{-8}$	$6.889 \cdot 10^{-8}$	$7.95 \cdot 10^{-8}$	$4.9 \cdot 10^{-6}$
25	1.410	7.169	8.58	5.0
50	1.759	7.563	9.32	5.0
75	2.406	4.967	7.37	5.1
700	$3.476 \cdot 10^{-8}$	$7.937 \cdot 10^{-8}$	$1.14 \cdot 10^{-7}$	$5.2 \cdot 10^{-6}$
25	8.591	4.683	1.33	5.2
50	$1.474 \cdot 10^{-7}$	7.287	2.20	5.2
775	$1.486 \cdot 10^{-7}$	$8.626 \cdot 10^{-8}$	$2.35 \cdot 10^{-7}$	$5.2 \cdot 10^{-6}$

The milk host spectrum is $n''_{\text{water}} + n''_{\text{riboflavin}}$. The milk fat spectrum is from Michalski et al. [62]

$$\varepsilon(\lambda) = \frac{\varepsilon(266.5\,\text{nm})}{D(266.5\,\text{nm})} \ln(10) D(\lambda).$$

The natural content in milk of riboflavin is 0.17 mg per 100 g milk [56]. Using the molar mass of riboflavin which is 376.3682 g/mol, we find that the natural concentration of riboflavin in milk is

$$c = \frac{0.17\,\text{mg}}{376.3682\,\text{g/mol}} \bigg/ \frac{100\,\text{g}}{1.03\,\text{g/mL}} = 4.65 \cdot 10^{-6}\,\text{mol/L}.$$

By multiplication of the molar absorption coefficient with this concentration, we obtain the absorption coefficient of riboflavin which is converted into the imaginary part of a refractive index (Eq. 4.18) and added to the imaginary part of the refractive index for pure water to obtain the imaginary part for the milk host, n''_{milk}. See Table 4.2.

The Fat Inclusion

Walstra and Jenness [63] have found experimentally that the real part of the refractive index of milk fat approximately follows the function

$$n'_{\text{fat}}(\lambda, T) = \sqrt{\frac{(b(T) + 2)\lambda^2 - 0.03}{(b(T) - 1)\lambda^2 - 0.03}}, \tag{4.35}$$

where wavelength is measured in μm and b is found using a measurement of the refractive index at the temperature T. We use measurements by Michalski et al. [62] since they also present the wavelength dependent imaginary part of the refractive index. They find $n'_{\text{fat}}(0.589\,\mu\text{m}, 20\,^\circ\text{C}) = 1.461$ which gives $b(20\,^\circ\text{C}) = 3.73$. This corresponds well to the $b(40\,^\circ\text{C}) = 3.77$ reported by Walstra and Jenness [63]. Table 4.2 includes the imaginary part of the refractive index for milk fat n''_{fat} as we read it from the curve reported by Michalski et al. [62].

The volume frequency of the fat globules follows a log-normal distribution [64] (cf. Eq. 4.32). The mean of the volume-to-area equivalent sphere radii $r_{\text{eq,fat}}$ of the fat globules change depending on the volume fraction of the globules in the milk. By a least-squares, two-piece fit to measured data reported by Olson et al. [65], we have found a functional expression describing this relationship:

$$r_{43,\text{fat}} = \begin{cases} -0.2528\, w_f^2 + 1.419\, w_f & \text{for } w_f < 2.0 \\ 1.456\, w_f^{0.36} & \text{otherwise} \end{cases},$$

where $r_{43,\text{fat}}$ is measured in micrometres. The relationship between $r_{43,\text{fat}}$ and $r_{\text{eq,fat}}$ is [64]

$$r_{\text{eq,fat}} = r_{43,\text{fat}}/(c_{v,\text{fat}}^2 + 1).$$

The radius $r_{43,\text{fat}}$ is used since it can be estimated empirically with good accuracy [64]. The coefficient of variation $c_{v,\text{fat}}$ is usually between 0.4 and 1.2 in normal milk. Reasonable limits for the range of fat globule radii are $r_{\text{min,fat}} = 0.005\,\mu\text{m}$ and $r_{\text{max,fat}} = 10\,\mu\text{m}$.

The Protein Inclusion

The refractive index of casein micelles is not readily available in the literature. For comparison to goat's milk it has been determined to be the following for cow's milk [66]:

$$n_{\text{casein}} = 1.503.$$

This value is assumed to be constant in the visible range and absorption of the casein micelles is neglected.

Structure and size distribution of casein micelles is still being disputed in the literature. Recent research on the matter is discussed by Gebhardt et al. [67]. Most investigations are based on either light scattering or electron microscopy. Light scattering approaches find micelles of large average size while electron microscopy report a large number of very small casein particles in addition to the larger micelles. Sometimes, these very small particles are excluded from the reported size distribution since they are regarded to represent non-micellar casein or single submicelles. No matter what we call these very small particles, they scatter light as do the larger

Table 4.3 A summary of the microscopic properties of cow's milk

Property	Milk host	Fat globules	Casein micelles
n'	Eq. 4.34	Eq. 4.35	1.503
n''	Table 4.2	Table 4.2	0.00
r		[0.005 μm, 10 μm]	[0 nm, 150 nm]
$N(r)$		Eq. 4.32	Eq. 4.32

Table 4.4 Densities for computing of milk fat and casein volume fractions using weight percents

ρ_{fat}	ρ_{protein}	ρ_{milk}
1.11 g/mL	0.915 g/mL	1.03 g/mL

Measured by Walstra and Jenness [63] at 20 °C

aggregates and therefore should be included in the size distribution employed for the Lorenz–Mie calculations.

A size distribution based on electron microscopy, which includes the single submicelles in the distribution, was reported by Schmidt et al. [68]. They found the mean $r_{\text{eq,casein}} = 43$ nm and showed that a log-normal distribution (4.32) of $r/(r_{\text{max,casein}} - r)$ is a good fit of the measured volume frequency distribution. The limits for the casein micelle radii are $r_{\text{min,casein}} = 0$ nm and $r_{\text{max,casein}} = 150$ nm.

The microscopic properties of milk are summarised in Table 4.3.

4.4.2 Appearance Model

To model the concentration of fat and protein we use wt% (g per 100 g milk), since this value is used on content declarations on the side of milk cartons. In the remainder of this chapter we let w_f and w_p denote the wt% of fat and protein respectively. To translate wt% into volume fractions, we use the densities given by Walstra and Jenness [63]. They are summarised in Table 4.4. Casein micelles make up about $80\% \cdot 95\% = 76\%$ of the protein volume fraction (see above). This means that

$$v_{\text{fat}} = \frac{w_f/\rho_{\text{fat}}}{100\,\text{g}/\rho_{\text{milk}}} \quad \text{and} \quad v_{\text{casein}} = 0.76\frac{w_p/\rho_{\text{protein}}}{100\,\text{g}/\rho_{\text{milk}}}.$$

This simple translation from fat and protein contents to volume fractions of the particle inclusions in the milk means that we have an appearance model with the following parameters:

- Fat content w_f
- Protein content w_p.

These two parameters are all we need to model most types of milk.

Fig. 4.1 Rendered images of the components in milk (*top row*) as well as mixed concentrations (*bottom row*). From *top left* to *bottom right* the glasses contain: pure water, water and vitamin B2, water and protein, water and fat, skimmed milk, regular milk, and whole milk

4.4.3 Results

The protein content of the milk we buy in a grocery store is usually around $w_p = 3.4$ wt.% while the content of fat is what we use to distinguish between different milk products: skimmed milk, $w_f = 0.1$ wt.%; low fat milk, $w_f = 1.5$ wt.%; whole milk $w_f = 3.5$ wt.%. This information, and the appearance model described above, enables us to visualise different types of milk. We are also able to show the visual significance of each component in the milk. This is a particular strength of the approach that we take. Knowing the visual significance of the different ingredients in a material is important if we would like to design the appearance of the material or if we would like to interpret the appearance of the material. Figure 4.1 shows the visual significance of different components in milk as well as the predicted appearance of different milk products. Starting leftmost in the top row, the first glass

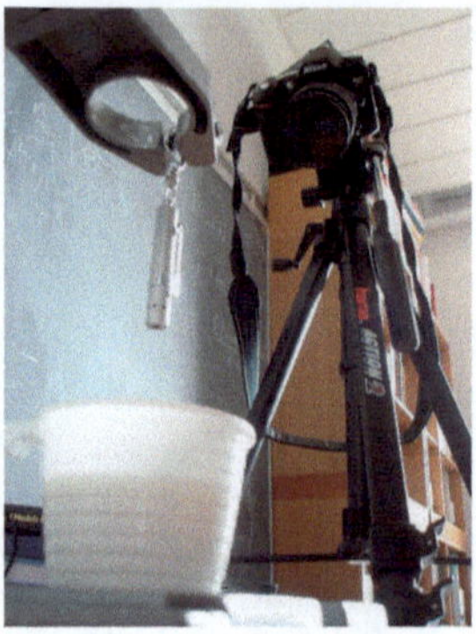

Fig. 4.2 A simple experimental setup where we can take a picture of a laser pointer shining light into a cup of milk

contains pure water, the second includes the absorption of vitamin B2 and is the host medium of the particles in the milk, the third glass contains casein micelles in water, the fourth contains fat globules in water, and the three glasses in the bottom row contain skimmed milk, low fat milk and whole milk, respectively. The contents of these glasses have all been rendered using the appearance model described in this section. Spectral optical properties (sampled at every 25 nm) were computed as described in Sect. 4.3 for all the milk materials. For rendering, we computed RGB representations of the optical properties by weighted averages using the RGB colour matching functions of a standard human observer [69]. We used homogeneous scattering properties for the milk and the approximative Henyey-Greenstein phase function with an ensemble asymmetry parameter.

The results provided in Fig. 4.1 (bottom row) compare only qualitatively to pictures of real milk. Our eyes are the instruments in such a comparison. To find out if our model correctly predicts the appearance of real milk, some sort of quantitative comparison is necessary. We can do a quantitative comparison by constructing a simple experimental setup which we can easily model and render. The set-up we choose is photographed in Fig. 4.2. It consists of a digital camera on a tripod pointing at the surface of milk in a cup, and some arbitrary device for hanging up a laser pointer over the cup. The experiment is to take a picture of the milk while the laser pointer shines light into it directly from above. The room should be darkened as much as possible, but it is usually not possible nor practical to black it out completely. For this reason, a picture was first taken with the laser pointer turned *off* and one was then taken with the laser pointer turned *on*. Subtracting the first picture from the second removes the background illumination to reveal the scattering of the laser in the milk alone. To avoid scattering of light in the cup itself, the white cup in Fig. 4.2 was replaced by a black, opaque cup when the pictures were taken. In addition, the digital camera was and should be configured to process the picture as little as possible.

To model the experimental set-up digitally, all we need is a correctly sized cylinder of milk, a laser source placed directly above it, and a camera placed at the right distance from the milk surface. The laser pointer we used had a diameter of 3 mm, power of $\Phi = 1$ mW, and wavelength of $\lambda = 650$ nm. This means that all the laser

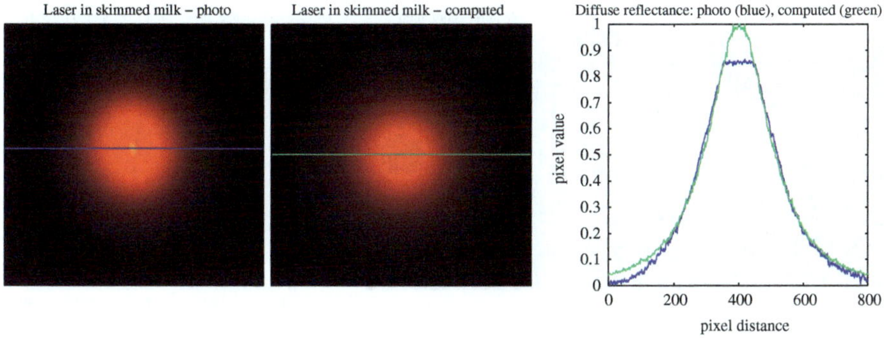

Fig. 4.3 Photographed laser in skimmed milk ($w_f = 0.1$ wt.%, $w_p = 3.4$ wt.%), simulated laser in skimmed milk, and comparison of the diffusion curves, that is, the middle lines through the images

light should go into the red colour band. The horizontal line in the image through the laser spot centre is referred to as the *diffuse reflectance curve*. We only use the diffuse reflectance curve of the red colour band in our quantitative comparison. This line is perpendicular to the direction towards the camera, so only the distance to the camera is important not the angle. For this reason, we place both camera and laser pointer directly above the milk in our digital scene. To save computations, we only render pixels from the laser spot centre to the right border of the film (half the diffuse reflectance curve), and then we revolve this result around the centre to get a full synthesised image of the laser spot.

The photographed milk, the rendered milk, and the quantitative comparison are provided in Fig. 4.3. There are a few differences between theory and experiment that we should discuss. The real camera is overexposed by the reflected laser light while the synthetic camera is not (the top of the blue curve has been cut off). The reason is the background illumination which was subtracted. In this experiment, the background illumination was just enough to overexpose the laser spot in the photograph. Another difference is that the photographed laser spot is not symmetric. It is more elliptically shaped. The reason for this deviation in the rendered image is that the laser source was modelled as a circular disc (see Sect. 4.2.3), and this is not the shape of the light emitted by a standard laser pointer. Finally, the central part of the photographed laser spot is not only red. The most likely explanation is that light penetrates the colour filters in the CCD chip when it is overexposed. This means that some of the red light spills into the other colour bands in the overexposed area of the image. Apart from these explainable differences, there is surprisingly good quantitative agreement between theory and experiment (the diffuse reflectance curves are close to each other).

From a qualitative perspective, the rotational symmetry of the simulated image is a giveaway which makes it easy for the human eye to spot the synthetic image in comparison to the real one. This might not be as easy if we had not revolved a single line in the image around the centre to save computations. Although it is not physically

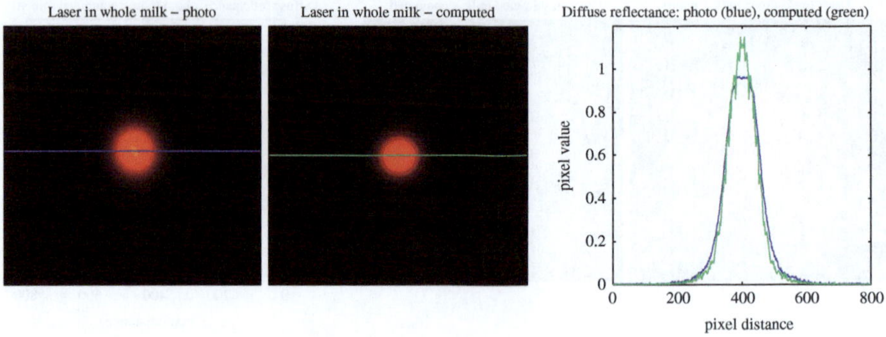

Fig. 4.4 Photographed laser in whole milk ($w_f = 3.5$ wt.%, $w_p = 3.4$ wt.%), simulated laser in whole milk, and comparison of the diffusion curves, that is, the middle lines through the images

accurate, the red light spilling into the other colour bands where the light is intense in the photograph convinces the eye that the photograph is more real. This is because the human eye is able to perceive a more intense red colour than what the camera can capture (and what a standard computer screen can display). The added intensity from the other colour bands therefore seems more realistic than a flat red colour. As a final example, to illustrate that the appearance model is reliable for different types of milk, Fig. 4.4 compares photographed and simulated laser in whole milk.

4.5 Conclusion

The milk example demonstrates that we are able to show the visual significance of the different components in a material by computing synthetic images. It also demonstrates that we are able to *predict* the appearance for various ratios between the contents. This makes a number of interesting applications possible:

- If you want to *design* materials with a specific appearance, the appearance model can help you choose the right components to obtain the desired appearance.
- If you want to *detect* whether a component is present in a material or not, the appearance model can help you visualise the material as it would look with and without the component.

In other words, synthesised images with a connection to the contents and the physical conditions of a material enable us to learn a lot about the reasons for the appearance of materials.

Considering the input parameters that brought us to the simulated diffuse reflectance curves in Figs. 4.3 and 4.4, the presented method is definitely useful for analysis by synthesis. If we simulate a large number of diffuse reflectance curves for milks with different particle inclusions, we can construct numerical methods for analysing the properties of these inclusions from photographed diffuse reflectance curves. The slope of the diffuse reflectance curve, for example, reveals a lot about the fat content of the milk (as is obvious from the figures). Further investigation can lead to more precise models for determining the fat content of milk from a digital photograph.

The Lorenz–Mie theory provides the link from the particle composition of a material to the macroscopic scattering properties that we can use with radiative transfer theory to compute a realistic image. In this way, a century after its formulation, the Lorenz–Mie theory is the key component when we want to predict the appearance of materials using a physical model.

References

1. R.A. Hall, D.P. Greenberg, IEEE Comput. Graph. Appl. **3**(8), 10 (1983)
2. A.S. Glassner, *Principles of Digital Image Synthesis* (Morgan Kaufmann Publishers, Inc., San Francisco, California, 1995). Two-volume set
3. S. Chandrasekhar, *Radiative Transfer* (Oxford, Clarendon Press, 1950). Unabridged and slightly revised version published by Dover Publications, Inc., in 1960
4. R. Siegel, J.R. Howell, *Thermal Radiation Heat Transfer*, 4th edn. (Taylor & Francis, New York, 2002)
5. M. Pharr, G. Humphreys, *Physically Based Rendering: From Theory to Implementation* (Morgan Kaufmann Publishers, an imprint of Elsevier Inc., 2004)
6. B.T. Phong, Commun. ACM **18**(6), 311 (1975)
7. R.V. Klassen, ACM Trans. Graph. **6**(3), 215 (1987)
8. D. Jackèl, B. Walter, Comput. Graph. Forum **16**(4), 201 (1997)
9. A.J. Preetham, P. Shirley, B. Smits, in *Proceedings of ACM SIGGRAPH 1999* (ACM Press, 1999), pp. 91–100
10. L.V. Wang, H.I. Wu, *Biomedical Optics: Principles and Imaging* (John Wiley & Sons, Inc., Hoboken, New Jersey, 2007)
11. A. Sommerfeld, J. Runge, Annalen der Physik **340**, 277 (1911)
12. P.d. Fermat, OEuvres de Fermat: Correspondance, vol. 2 (Gauthier-Villars et fils, 1894)
13. A. Glassner, IEEE Comput. Graph. Appl. **4**(10), 15 (1984)
14. F.W. Jansen, in *Data Structures for Raster Graphics: Proceedings of a Workshop Held at Steensel*, the Netherlands, from 24–28 June 1985. Eurographics seminars, (Springer, 1986), pp. 57–73
15. A. Fujimoto, T. Tanaka, K. Iwata, IEEE Comput. Graph. Appl. **6**(4), 16 (1986)
16. K. Sung, P. Shirley, in *Graphics Gems III*, ed. by D. Kirk (Academic Press, 1992), pp. 271–274
17. T. Möller, B. Trumbore, J. Graph. Tools **2**(1), 21 (1997)
18. A. Fresnel, Mémoires de l'Académie des sciences de l'Institut de France **11**, 393 (1832). Presented 7 January 1823
19. T. Whitted, Commun. ACM **23**(6), 343 (1980). Presented at SIGGRAPH 79
20. R.W. Preisendorfer, *Radiative Transfer on Discrete Spaces* (Pergamon Press, 1965)
21. A. Ishimaru, *Wave Propagation and Scattering in Random Media* (Academic Press, New York, 1978). Reissued by IEEE Press and Oxford University Press 1997
22. J.T. Kajiya, B.P. Von Herzen, Computer graphics (Proc. ACM SIGGRAPH 84) **18**(3), 165 (1984)
23. S.N. Pattanaik, S.P. Mudur, J. Vis. Comput. Animat. **4**(3), 133 (1993)
24. T. Nishita, Y. Miyawaki, E. Nakamae, Computer graphics (Proc. ACM SIGGRAPH 87) **21**(4), 303 (1987)

25. H. Rushmeier, in *Realistic Input for Realistic Images*. ACM Press (ACM SIGGRAPH 95 Course Notes, 1995). Also appeared in the ACM SIGGRAPH 98 Course Notes—A Basic Guide to Global Illumination
26. P. Callet, Comput. Graph. Forum **15**(2), 119 (1996)
27. T. Nishita, Y. Dobashi, in *Proceedings of Computer Graphics International 2001* (IEEE Computer Society, 2001), pp. 149–156
28. J.R. Frisvad, N.J. Christensen, H.W. Jensen, ACM Trans. Graph. **26**(3) (2007). Article 60
29. L.G. Henyey, J.L. Greenstein, Annales d'Astrophysique, **3**, 117, (1940) Also in. Astrophys. J. 93, (1941)
30. C.F. Bohren, D.P. Gilra, J. Colloid Interface Sci. **72**(2), 215 (1979)
31. H.C. van de Hulst, *Light Scattering by Small Particles* (John Wiley & Sons, Inc., New York, 1957), Unabridged and corrected version of the work published by Dover Publications, Inc., in 1981
32. L. Lorenz, Det kongelig danske Videnskabernes Selskabs Skrifter **6**(1) (6. Række, naturvidenskabeligog mathematisk Afdeling, 1890), pp. 2–62
33. G. Mie, Annalen der Physik **25**(3), 377 (1908). IV. Folge
34. M. Kerker, *The Scattering of Light and Other Electromagnetic Radiation* (Academic Press, New York, 1969)
35. J.V. Dave, IBM J. Res. Dev. **13**(3), 302 (1969)
36. C.F. Bohren, D.R. Huffman, *Absorption and Scattering of Light by Small Particles* (John Wiley & Sons, Inc., 1983)
37. G.W. Kattawar, G.N. Plass, Appl. Opt. **6**(8), 1377 (1967)
38. W.J. Wiscombe, Appl. Opt. **19**(9), 1505 (1980)
39. W.C. Mundy, J.A. Roux, A.M. Smith, J. Opt. Soc. Am. **64**(12), 1593 (1974)
40. V.E. Cachorro, L.L. Salcedo, J. Electromagn. Waves Appl. **5**(9), 913 (1991)
41. D.W. Mackowski, R.A. Altenkirch, M.P. Menguc, Appl. Opt. **29**(10), 1551 (1990)
42. Z.S. Wu, Y.P. Wang, Radio Sci. **26**(6), 1393 (1991)
43. W. Yang, Appl. Opt. **42**(9), 1710 (2003)
44. H.C. van de Hulst, Physica **15**(8–9), 740 (1949)
45. J. Randrianalisoa, D. Baillis, L. Pilon, J. Opt. Soc. Am. A **23**(7), 1645 (2006)
46. J. Yin, L. Pilon, J. Opt. Soc. Am. A **23**(11), 2784 (2006)
47. Q. Fu, W. Sun, J. Quant. Spectr. Radiat. Transf. **100**(1–3), 137 (2006)
48. T.C. Grenfell, S.G. Warren, J. Geophys. Res. **104**(D24), 31697 (1999)
49. S.P. Neshyba, T.C. Grenfell, S.G. Warren, J. Geophys. Res. **108**(D15, 4448), 6 (2003)
50. T.C. Grenfell, S.P. Neshyba, S.G. Warren, J. Geophys. Res. 110(D17203), 1 (2005)
51. J.R. Frisvad, *Light, Matter, and Geometry: The Cornerstones of Appearance Modelling (VDM* (Verlag Dr. Müller, 2008)
52. J.A. Lock, G. Gouesbet, J. Quant. Spectr. Radiat. Transf. **110**(11), 800 (2009). Review
53. G. Gouesbet, J. Quant. Spectr. Radiat. Transf. **110**(14–16), 1223 (2009). Review
54. E.P. Lafortune, Y.D. Willems, in *Proceedings of the 7th Eurographics Workshop on Rendering (1996)*, pp. 91–100
55. H.D. Goff, A.R. Hill, in *Dairy Science and Technology Handbook: Principles and Properties*, vol. 1, ed. by Y.H. Hui (VCH Publishers, Inc., New York, 1993), Chap. 1, pp. 1–81
56. P.F. Fox, P.L.H. McSweeney, *Dairy Chemistry and Biochemistry* (Blackie Academic & Professional, London, 1998)
57. X. Quan, E.S. Fry, Appl. Opt. **34**(18), 3477 (1995)
58. P.D.T. Huibers, Appl. Opt. **36**(16), 3785 (1997)
59. H. Du, R.C.A. Fuh, J. Li, L.A. Corkan, J.S. Lindsey, Photochem. Photobiol. **68**(2), 141 (1998)
60. J. Koziol, Photochem. Photobiol. **5**, 41 (1966)
61. G.M. Hale, M.R. Querry, Appl. Opt. **12**(3), 555 (1973)
62. M.C. Michalski, V. Briard, F. Michel, Lait **81**, 787 (2001)
63. P. Walstra, R. Jenness, *Dairy Chemistry and Physics* (John Wiley & Sons, New York, 1984)
64. P. Walstra, Neth. Milk Dairy J. **29**, 279 (1975)
65. D.W. Olson, C.H. White, R.L. Richter, J. Dairy Sci. **87**(10), 3217 (2004)

66. R. Attaie, R.L. Richtert, J. Dairy Sci. **83**, 940 (2000)
67. R. Gebhardt, W. Doster, J. Friedrich, U. Kulozik, Eur. Biophys. J. **35**, 503 (2006)
68. D.G. Schmidt, P. Walstra, W. Buchheim, Neth. Milk Dairy J. **27**, 128 (1973)
69. A. Stockman, L.T. Sharpe, Vis. Res. **40**(13), 1711 (2000)

Chapter 5
Dipole Re-Radiation Effects in Surface Enhanced Raman Scattering

Logan K. Ausman and George C. Schatz

Abstract In this chapter we use extensions of Mie theory to study electromagnetic enhancement factors associated with surface enhanced Raman scattering (SERS) from molecules adsorbed onto metal sphere array structures, comparing results from the more rigorous dipole re-radiation (DR) expression for Raman enhancement with the commonly used plane-wave (PW) enhancement formula. The DR and PW calculations are based on the T-matrix method for determining optical scattering from multiple spheres. In the PW expression, the enhancement is considered to be equal to the product of the squares of the local electric fields, $|\mathbf{E}_{\mathrm{loc}}(\omega)|^2|\mathbf{E}_{\mathrm{loc}}(\omega')|^2$ or $|\mathbf{E}_{\mathrm{loc}}(\omega)|^4$ for zero Stokes shift, obtained from plane wave Mie scattering. In the DR calculation, the induced dipole in a molecule that is located at the surface of one of the particles serves as a dipole source at the Stokes-shifted frequency that scatters from the particles to define an overall enhancement factor. The SERS enhancement factors are determined for chains of 100 nm diameter Ag spheres and for chains of 100 nm Ag sphere dimers for various sphere and dimer separations and for various chain lengths, with the dimer gap fixed at 6.25 nm. We compare the PW and DR results for two different detector locations, a backscattering configuration normal to the axis of the chain, and a 135° scattering direction that includes the plane of the chain axis, in order to highlight far-field phase interference effects that are incorporated in the DR result but not PW. We find that the DR and PW results have negligible differences for the backscattering geometry, but far-field effects play a significant role in the overall enhancement factor for the non-backscattered location. This demonstrates the importance of including DR effects in the interpretation of SERS experiments.

L. K. Ausman · G. C. Schatz (✉)
Department of Chemistry Northwestern University, 2145 Sheridan Road,
Evanston, IL, 60208-311 USA
e-mail: schatz@chem.northwestern.edu

W. Hergert and T. Wriedt (eds.), *The Mie Theory*, Springer Series in Optical Sciences 169, DOI: 10.1007/978-3-642-28738-1_5,

5.1 Introduction

Surface enhanced Raman scattering (SERS) was discovered in the mid-1970s when it was realised that Raman scattering from molecules like pyridine adsorbed in monolayer concentrations on anodised (roughened) silver electrode surfaces was enhanced by a factor of 10^6 compared to the same scattering process for molecules in solution [1–3]. Shortly after this discovery, it was postulated that this enhancement could be understood based on the enhancement in local fields on the surfaces of the nanoparticles being irradiated as a result of plasmon excitation in these particles [4]. Approximate theories that include plasmon excitation were then formulated, as reviewed in many places in the mid-1980s [5–10], in which the electromagnetic (EM) contribution to the enhancement factor is given by $|\mathbf{E}_{\text{loc}}(\omega)|^2|\mathbf{E}_{\text{loc}}(\omega')|^2$, where $\mathbf{E}_{\text{loc}}$ is the electric field at the position of the molecule and ω and ω' are the incident and Stokes shifted frequencies, respectively. This so-called plane wave (PW) approximation assumes that the incident and scattered field enhancements can both be derived from the scattering of plane waves from the metal nanostructure; however, it was noted in 1980 by Kerker [11] that this is only an approximation to the rigorous EM enhancement factor. Kerker used concepts derived from Mie theory [12] which he had extended in the mid-1970s to describe fluorescence from molecules inside polymer nanoparticles [13–15] to show that although the incident photon enhancement factor is given by $|\mathbf{E}_{\text{loc}}(\omega)|^2$, the emitted photon enhancement factor should be derived from emission by the dipole induced in the adsorbed molecule at the Stokes frequency. The emitted radiation interacts with the nanostructure, leading to plasmon enhancement in the scattered field. Kerker demonstrated that this dipole re-radiation (DR) enhancement factor matches the plane wave (PW) enhancement factor for spherical particles in the quasistatic (small particle limit) approximation for molecules in special locations. However, the more general evaluation of DR effects has received scant attention over the years, with the few exceptions being work done by Kerker and others in the mid-1980s [10, 16–18], and a recent study by Ausman and Schatz [19]. Instead the vast majority of the SERS models are based on the PW approximation [20–29]. A further restriction is often made where the difference between ω and ω' is neglected, leading to the zero Stokes shift enhancement factor $|\mathbf{E}_{\text{loc}}(\omega)|^4$.

The lack of interest in using the dipole re-radiation expression for the SERS enhancement factor has arisen in part because of the technical difficulty associated with evaluation of the enhancement factor (few codes provide this capability and the computational effort is higher), and in part because Kerker's early estimates of DR effects [18] suggested that the PW approximation was generally adequate. However, in the recent work by Ausman and Schatz [19], it has been found that the DR/PW comparison can show a factor of 2–3 differences for large enough particles or for dimers of particles when averaged over detector locations. In addition, there can be much larger differences for specific molecule/detector locations. These differences are typically due to interference effects that are important in the DR enhancements but which are completely neglected by PW. These interference effects typically become important when multipole resonances are important, or when the molecule is located in a position that does not lead to large SERS enhancement based on PW excitation. Such effects can be significant for many kinds of SERS measurements, particularly single molecule SERS measurements [27, 29–34] and for other measurements that are designed to yield the highest possible SERS enhancements.

Many SERS substrates are currently being produced with the goal of being highly sensitive and at the same time highly reproducible in applications to biomolecule [29, 35–38] and chemical sensing [39–41]. Array structures are one substrate of interest due to fabrication capabilities that are routinely available, along with the added benefit that plasmon resonances in each particle can couple to give photonic enhancement at wavelengths that can be specified by the array structure [20–25, 42–44]. Experiments on nanoparticle aggregates and array structures have generated interest in the incorporation of dipole re-radiation into SERS models [19, 45–49]; however, the comparison between DR and PW results for these structures has not been assessed. Therefore, knowledge concerning the limitations of the PW approximation for these structures will aid in its proper employment and also hopefully spur new advances in numerical methods (i.e. DDA, FDTD, etc) for describing SERS.

This chapter will highlight the incorporation of dipole re-radiation in the modeling of SERS due to silver particles and silver particle array structures where all particles are modeled as spheres so that Mie theory (with extensions) can be used. We will only briefly consider applications to an isolated sphere and to a dimer of spheres as this has been addressed in our previous work. Instead, most of our development will refer to larger array structures, with the aim of determining the validity of the PW approximation and to assess the conditions under which it breaks down. We do this by considering SERS enhancements for several molecule locations and two different detector locations. These applications will show that far-field phase interference effects can result in SERS enhancement factors that differ significantly from the PW approximation.

5.2 Methods

5.2.1 Theory

There is a vast literature that presents Mie theory and its extensions to coupled spheres; however, our focus will be on the inclusion of dipole re-radiation. For thorough treatments of single sphere plane wave scattering see Bohren and Huffman [50]. The extension of this to multiple spheres using the T-matrix method can be found in papers by Mackowski and Mishchenko [51–53], along with the accompanying references.

The starting point for all of these theories is the single sphere scattering problem. This is treated through an expansion of incident and scattered electric fields in a basis of vector spherical harmonics. A general incident electric field has the form

$$\mathbf{E}_i(\mathbf{r},\omega)=\sum_{l=0}^{\infty}\sum_{m=-l}^{l}\left[p_{lm}\mathbf{N}_{lm}^{(1)}(k\mathbf{r})+q_{lm}\mathbf{M}_{lm}^{(1)}(k\mathbf{r})\right], \tag{5.1}$$

and the resulting scattered electric field is

$$\mathbf{E}_{sc}(\mathbf{r},\omega)=\sum_{l=0}^{\infty}\sum_{m=-l}^{l}\left[a_{lm}\mathbf{N}_{lm}^{(3)}(k\mathbf{r})+b_{lm}\mathbf{M}_{lm}^{(3)}(k\mathbf{r})\right], \tag{5.2}$$

where $\mathbf{N}_{lm}$ and $\mathbf{M}_{lm}$ are the vector spherical harmonic basis functions that can be found in [11, 50, 51, 54]. The coefficients a_{lm} and b_{lm} are related to the general incident electromagnetic field coefficients p_{lm} and q_{lm} through the Lorenz–Mie single sphere coefficients α_l and β_l by the following:

$$a_{lm}=\alpha_l p_{lm}, \tag{5.3}$$

$$b_{lm}=\beta_l q_{lm}. \tag{5.4}$$

Explicit expressions for the Lorenz–Mie single sphere coefficients α_l and β_l for a sphere of radius a and permittivity ε_1 in a medium with ε_0 are

$$\alpha_l=-\frac{\varepsilon_1 j_l(k_1a)\left[k_0aj_l(k_0a)\right]'-\varepsilon_0 j_l(k_0a)\left[k_1aj_l(k_1a)\right]'}{\varepsilon_1 j_l(k_1a)\left[k_0ah_l^{(1)}(k_0a)\right]'-\varepsilon_0 h_l^{(1)}(k_0a)\left[k_1aj_l(k_1a)\right]'}, \tag{5.5}$$

$$\beta_l=-\frac{j_l(k_1a)\left[k_0aj_l(k_0a)\right]'-j_l(k_0a)\left[k_1aj_l(k_1a)\right]'}{j_l(k_1a)\left[k_0ah_l^{(1)}(k_0a)\right]'-h_l^{(1)}(k_0a)\left[k_1aj_l(k_1a)\right]'}. \tag{5.6}$$

In (5.5) and (5.6) $j_l(x)$ is the spherical Bessel function and $h_l^{(1)}(x)$ is the spherical Hankel function of the first kind. The coefficients p_{lm} and q_{lm} when the incident field is a plane wave (the traditional Lorenz–Mie theory), and the resulting scattered field coefficients a_{lm} and b_{lm} can be found in [50], in [51] and in Ausman and Schatz [54].

In order to describe dipole re-radiation, the coefficients p_{lm} and q_{lm} in (5.1) are modified from their usual plane wave expressions by making use of the fact that the electric dipole field can be expressed as the product of the free-space tensor Green's function, G_s, with the dipole moment induced by plane wave scattering, $\mathbf{p}=\alpha_R\cdot\left[\mathbf{E}_{\mathrm{pw}}(\mathbf{r}_0)+\mathbf{E}_{\mathrm{sc,pw}}(\mathbf{r}_0)\right]$, through the relation

$$\mathbf{E}_{\mathrm{dp}}(\mathbf{r};\omega_s)=\frac{4\pi k_s^2}{\varepsilon_s}\mathsf{G}_s(\mathbf{r},\mathbf{r}_0;\omega_s)\cdot\mathbf{p}. \tag{5.7}$$

The terms in the induced dipole moment are the Raman polarisability tensor, α_R, and the local field at the molecule location, $\mathbf{r}_0$, due to plane wave scattering. The field $\mathbf{E}_{\text{dp}}$ can be expressed in a vector spherical harmonic basis by noting that the tensor Green's function can be expanded in a basis of vector spherical harmonics by way of the Ohm-Rayleigh method as described by Tai [55]. With the modified p_{lm} and q_{lm} for a molecule located at $\mathbf{r}_0$, the resulting scattered field coefficients a_{lm} and b_{lm} have the form [11, 16, 19, 54]

$$a_{lm} = (-1)^m \alpha_l \frac{ik_s^3}{\varepsilon_s} \frac{2l+1}{l(l+1)} \mathbf{N}^{(3)}_{l(-m)}(k_s \mathbf{r}_0) \cdot \mathbf{p}, \tag{5.8}$$

$$b_{lm} = (-1)^m \beta_l \frac{ik_s^3}{\varepsilon_s} \frac{2l+1}{l(l+1)} \mathbf{M}^{(3)}_{l(-m)}(k_s \mathbf{r}_0) \cdot \mathbf{p}. \tag{5.9}$$

In the above equations ε_s is the dielectric function of the medium at the frequency ω_s, $k_s = \omega_s \sqrt{\varepsilon_s \mu_s}/c$ and $\sqrt{\varepsilon_s \mu_s}$ is the complex refractive index of the medium. Note that the expressions in (5.8) and (5.9) will generally be nonzero for all values of m between $-l$ and $+l$. In the case of plane wave scattering m will only take on the values of ± 1 which means that only specific symmetry components of any multipole l can be induced in a sphere. This is no longer the case for the scattering of an electric dipole field. As the l value increases the difference between accessible modes for a plane wave compared to those of a dipole field increases.

Now we extend this theory to dipole re-radiation that arises from emission in the presence of many interacting spheres, following the notation used in [51]. By making use of (5.9) and (5.10), sphere i is then coupled to sphere j by use of the addition theorem for vector spherical harmonics in the same way as plane wave scattering is treated. These coefficients are [19, 51–53]

$$a^{\mathrm{i}}_{lm} = \alpha^{\mathrm{i}}_l \left(\frac{ik_s^3}{\varepsilon_s} \mathbf{N}^{(3)}_{l(-m)} \left(k_s \mathbf{r}^{\mathrm{i}}_0\right) \cdot \mathbf{p} + \sum_{\substack{\mathrm{j}=1 \\ \mathrm{j}\neq \mathrm{i}}}^{N_s} \sum_{l'=1}^{\infty} \sum_{m'=-l'}^{l'} \left[a^{\mathrm{j}}_{l'm'} A^{(3)}_{lml'm'} \left(k_s \mathbf{R}^{\mathrm{ji}}\right) + b^{\mathrm{j}}_{l'm'} B^{(3)}_{lml'm'} \left(k_s \mathbf{R}^{\mathrm{ji}}\right) \right] \right), \tag{5.10}$$

$$b^{\mathrm{i}}_{lm} = \beta^{\mathrm{i}}_l \left(\frac{ik_s^3}{\varepsilon_s} \mathbf{M}^{(3)}_{l(-m)} \left(k_s \mathbf{r}^{\mathrm{i}}_0\right) \cdot \mathbf{p} + \sum_{\substack{\mathrm{j}=1 \\ \mathrm{j}\neq \mathrm{i}}}^{N_s} \sum_{l'=1}^{\infty} \sum_{m'=-l'}^{l'} \left[b^{\mathrm{j}}_{l'm'} A^{(3)}_{lml'm'} \left(k_s \mathbf{R}^{\mathrm{ji}}\right) + a^{\mathrm{j}}_{l'm'} B^{(3)}_{lml'm'} \left(k_s \mathbf{R}^{\mathrm{ji}}\right) \right] \right). \tag{5.11}$$

$A^{(3)}_{lml'm'}$ and $B^{(3)}_{lml'm'}$ are vector harmonic addition coefficients, expressions for which can be found in [51–53]. The superscript (3) signifies that the coefficients are generated using spherical Hankel functions and $\mathbf{R}^{\mathrm{ji}} = \mathbf{r} - \mathbf{r}_0$.

When the individual sphere coefficients are determined, the total field, $\mathbf{E}^{\mathrm{tot}}_{\mathrm{sc}}$, at a point $\mathbf{r}$ is generated by taking the sum of the individual scattered fields from each sphere in Cartesian space

$$\mathbf{E}^{\mathrm{tot}}_{\mathrm{sc}} = \sum_{\mathrm{i}=1}^{N_s} \mathbf{E}^{\mathrm{i}}_{\mathrm{sc}}. \tag{5.12}$$

This field is then added to the field $\mathbf{E}_{\mathrm{dp}}$ produced by the radiating dipole, and for a sufficiently large r_s is multiplied by $r_s/\exp(ikr_s)$ to obtain the far-field amplitude $\mathbf{F}_R(\theta_s, \phi_s) = r_s(\mathbf{E}_{\mathrm{dp}} + \mathbf{E}^{\mathrm{tot}}_{\mathrm{sc}})/\exp(ikr_s)$. With this it is possible to obtain the dipole re-radiation Raman enhancement factor, G, using

$$G(\theta_s, \phi_s) = \frac{|\mathbf{F}_R(\theta_s, \phi_s)|^2}{\left|\mathbf{F}_{R,0}(\theta_s, \phi_s)\right|^2}, \tag{5.13}$$

with $\mathbf{F}_R(\theta_s, \phi_s)$ being the far-field amplitude of the molecule/particle system and $\mathbf{F}_{R,0}(\theta_s, \phi_s)$ being the corresponding far-field amplitude of the scattered radiation from an isolated molecule. We note that $\mathbf{F}$ implicitly depends on r_s but we simply assume that this distance is large with respect to the wavelength of light and fixed.

5.2.2 Computational Details

As is mentioned elsewhere, determining the coefficients a^{i}_{lm} and b^{i}_{lm} would formally be done by fully inverting a $(2N_{\mathrm{o,i}}(N_{\mathrm{o,i}}+2))^2$ matrix where the l value of each sphere is truncated at $N_{\mathrm{o,i}}$; however, in practice these are solved iteratively by employing a biconjugate gradient method. The computational cost is further reduced by taking advantage of various rotational and translational transformations and symmetry properties of the $A^{(3)}_{lml'm'}$ and $B^{(3)}_{lml'm'}$ with the details of these available in the previously mentioned references.

In this chapter we use the dielectric data for Ag as reported by Johnson and Christy [56]. We also mention that in all calculations performed in this work the incident electric field is assumed to be plane polarised in the $\hat{\mathbf{x}}$ direction and propagating in the $\hat{\mathbf{z}}$ direction. Additionally, the external medium is taken to be a vacuum, and the reported enhancement factors are evaluated for zero Stokes shift (so that the PW enhancement factor is $|\mathbf{E}_{\mathrm{loc}}|^4$). The molecule location is denoted by the coordinates r_0, θ_0 and ϕ_0 while the detector location is denoted by the coordinates r_s, θ_s and ϕ_s; depictions of these are shown in Fig. 5.1. Also in Fig. 5.1 we show the arrangement of the sphere chains and dimer chains with respect to the incident electric field. In the single sphere chains the spheres are aligned along the axis that coincides with

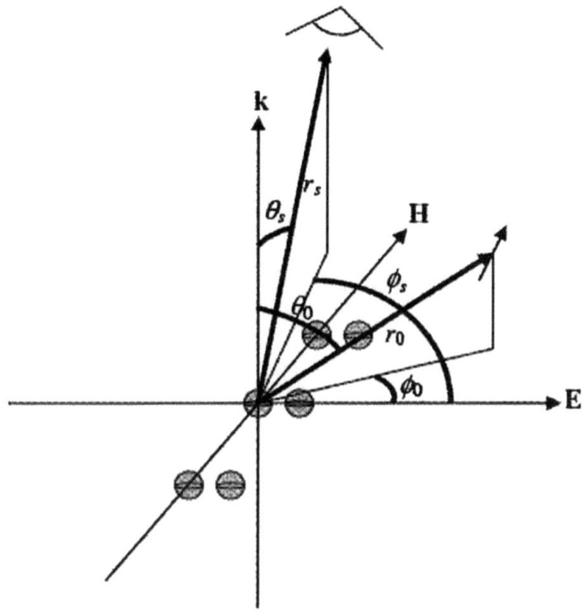

Fig. 5.1 Scheme of the relevant molecule and detector coordinates and the arrangement of the spheres

H (the y-axis). The dimer chains are similarly arranged, with the second sphere in the dimer being offset along the axis that coincides with **E**. The molecule is located along the **E**-axis (i.e., x-axis) a distance 0.5 nm from the surface of the sphere that is at the origin. This is the central sphere in the chain,and the central dimer in the chain of dimers. Note that the classical point dipole is assumed to have an isotropic response. While there are likely to be many interesting effects due to anisotropic responses, the choice of an isotropic α_R is done for simplicity and we note that other responses could be considered. Indeed, a quantum mechanical molecular response can be included (still keeping the particle as a classical scatterer) by employing the many-body theory of SERS developed by Masiello and Schatz [57].

5.3 Dipole Re-Radiation Effects in SERS from Isolated Particles and Dimers

The most extensive studies of dipole re-radiation have been done for isolated spheres. In 1980, Kerker and coworkers examined SERS in the presence of Ag spheres and dielectric spheres [11], considering distance dependence, frequency dependence and sphere size dependence of the SERS enhancement factor. Later, Kerker studied sensitivity of the estimated enhancement factors to the dielectric data used [17] and also briefly compared the average enhancement over the surface of a sphere as calculated by dipole re-radiation with the average local electric field enhancement [18].

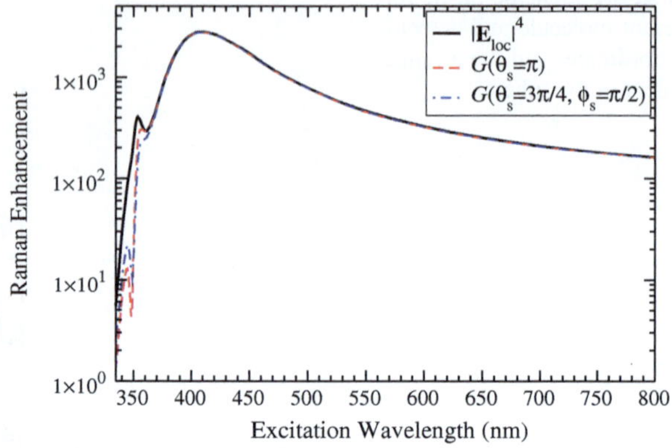

Fig. 5.2 SERS excitation profile for a single 100 nm diameter Ag sphere with a molecule located 0.5 nm from the sphere surface along the axis parallel with the incident E-field polarisation. The *solid black line* is the PW approximation result, the *dashed red line* is the backscattering DR result and the *blue dash–dotted line* is the DR result for a detector at $\theta_s = 3\pi/4$ and $\phi_s = \pi/2$

However, Kerker fixed the detector location in these studies at a position that is orthogonal to the **kE**-plane. Recently, we compared the dipole re-radiation G value and the PW approximation for other detector locations [19]. In that work we found that the agreement between the PW approximation and G is dependent upon both molecule and detector location. Here, we only consider the influence of the detector location.

In Fig. 5.2 we present an isolated sphere result so that a comparison with arrays made up of a chain of spheres can be made later. The sphere under consideration is a 100 nm diameter Ag sphere in vacuum. We choose two different detector locations: (1) a backscattering configuration at $\theta_s = \pi$ and (2) an off axis direction $\theta_s = 3\pi/4$ and $\phi_s = \pi/2$. Neither of these locations was considered in our earlier study. The first location corresponds to a common detector direction while the second is somewhat arbitrarily chosen but is typical of directions used for SERS measurements. As was observed in [19] for other detector locations, we see that the PW approximation is accurate for the dipole plasmon part of the spectrum but there are significant differences for higher order multipoles. For the detector positions considered here there is no difference between the G values except in the blue region of the spectrum. This behaviour is indicative of far-field interference effects due to the angular structure of higher order multipoles. These effects have been explained previously [19], and we will return to this in more detail later.

As can be seen in Fig. 5.2 the peak SERS enhancement factor from an isolated sphere is rather small (approximately 3×10^3). However, the junction of a sphere dimer can produce large local electric field enhancements that lead to considerably larger SERS enhancement factors. Therefore, we also will briefly examine SERS enhancements due to sphere dimers in order to make later comparisons with chains of dimers analogous to the sphere chains. In Fig. 5.3 we present the PW enhancement factor along with the dipole re-radiation enhancement factor at the same two detector locations as used in the isolated sphere case for two 100 nm diameter Ag spheres separated by 6.25 nm. It has been shown previously that the absolute maximum enhancement factors occur when the interparticle axis of the dimer system lies parallel

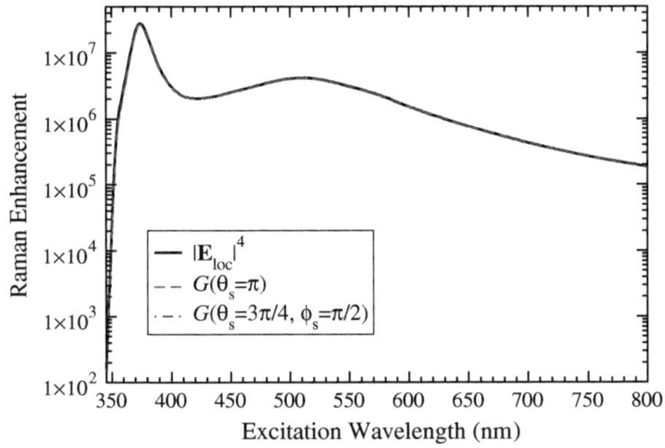

Fig. 5.3 SERS excitation profile for a dimer of 100 nm diameter Ag spheres with a gap distance of 6.25 nm. The molecule is located in the junction between the two spheres at a distance of 0.5 nm from the surface of the sphere that is at the coordinate origin and the inter-sphere axis lies along the polarisation direction of the incident E-field. The *solid black line* is the PW approximation result, the *dashed red line* is the backscattering DR result, and the *blue dash-dotted line* is the DR result for a detector at $\theta_s = 3\pi/4$ and $\phi_s = \pi/2$

to the incident electric field polarisation. In this case we see in Fig. 5.3 that the peak enhancement factor for the sphere dimer is around 3×10^7 which is approximately 4 orders of magnitude larger than the isolated sphere case.

For the dimer system phase interference effects in the far-field are small at the two detector locations chosen in this work and therefore the SERS excitation profiles are qualitatively similar to the PW enhancements. We note that this is not necessarily true for other detector locations as has been described previously in [19]. What is interesting about the dimer excitation profile is that the most intense peak occurs for the quadrupole plasmon rather than the dipole plasmon, whereas the opposite is the case for smaller spheres. This is likely due to the fact that the near-zone quadrupole field varies as $1/r^4$ which means that at close enough distances it can play a more significant role than the electric dipole field.

5.4 Dipole Re-Radiation Effects in SERS from Single Sphere Chains

Having introduced some ideas concerning the differences between dipole re-radiation and the locally enhanced field approximation for an isolated sphere and a sphere dimer we now turn our attention to systems that are more complex. Earlier work has examined the role of linear arrays of spheres separated by various distances. It was shown by Zou and Schatz [20–24] that 1D arrays with sufficiently many spheres, specific sphere separations and sphecific choices of polarization and wave-

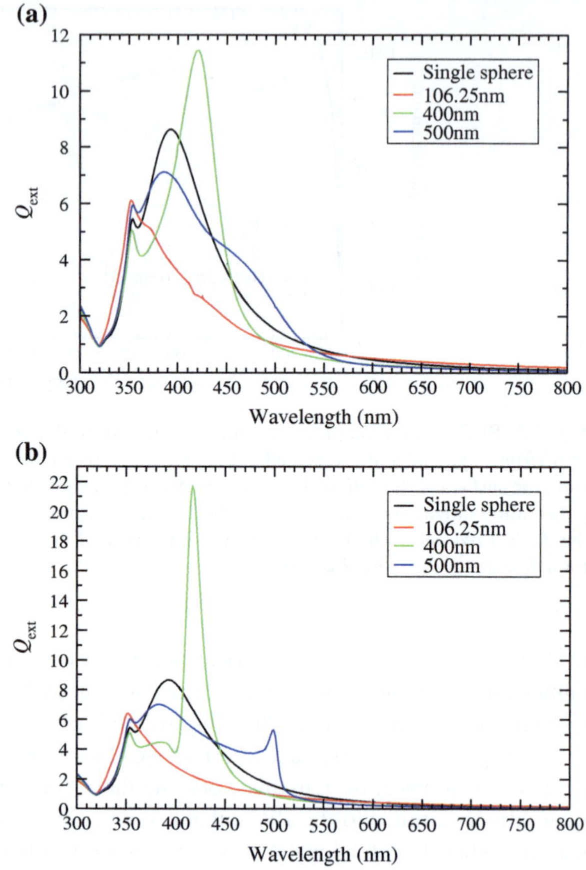

Fig. 5.4 Extinction spectra for sphere chains made up of 5 spheres (**a**) and 45 spheres (**b**) for different sphere separations. The separations given in the legend are the center-to-center distances

vector could exhibit long range interactions that can increase the isolated sphere enhancement factor by at least one order of magnitude at specific wavelengths. These photonic effects are characterised by a sharp peak in the extinction spectrum that occurs at a wavelength that is roughly equal to the center-to-center particle spacing (and higher order modes also exist). In this section, we examine the role of DR effects for these structures.

Figure 5.4 presents the extinction spectra of 1D arrays of either 5 or 45 diameter spheres measuring 100 nm, where the array axis is taken to be perpendicular to both the **k** and **E** directions (and therefore is along the y-direction). The figure shows that for short separation distances the dipole plasmon in the isolated sphere spectrum near 390 nm is suppressed (by destructive interference of interacting dipole near-fields) while it becomes enhanced at longer separations (due to the different dipolar interactions for far-field interaction compared to near-field interaction). As has been described [20–24], we see that longer separations with a larger number of spheres can result in sharp peaks in the extinction spectrum at wavelengths that roughly

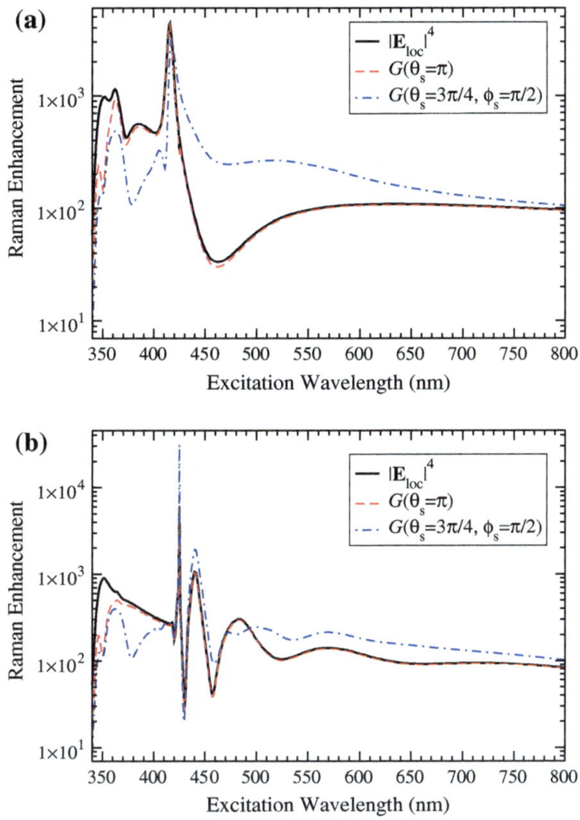

Fig. 5.5 SERS excitation profile for a linear chain made of 100 nm diameter Ag spheres with a center-to-center separation of 106.25 nm. The molecule is located 0.5 nm from the surface of the sphere at the coordinate origin along the axis parallel with the incident E-field polarisation. The case of 5 spheres is presented in **a** and the case of 45 spheres is presented in **b**. In both **a** and **b** the *solid black line* is the PW approximation result, the *dashed red line* is the backscattering DR result, and the *blue dash–dotted line* is the DR result for a detector at $\theta_s = 3\pi/4$ and $\phi_s = \pi/2$

match the sphere separation. These are photonic resonance peaks, and we note that the peaks here are not as sharp as were observed previously due to the fact that our chains are truncated at 45 spheres (because of computational limitations), instead of hundreds.

The trends observed in the extinction spectra are also seen in the frequency dependence of the PW enhancement factor for a molecule that is located along the x-axis. Figure 5.5 shows sharp peaks near 425 nm that do not correlate with features in the extinction spectrum, reflecting the contributions of localised photonic enhancement at the position of the molecule that are obscured by the more global nature of extinction. Figure 5.6 also shows a peak near 415 nm, but in this case the resonance peak reflects a photonic resonance at that wavelength. Figure 5.7 shows similar photonic resonance behaviour near 500 nm. The largest enhancement factors occur for sphere to sphere separations around 400 nm. For backscattered polarisation, there is generally good agreement between DR and PW results except at very short wavelengths where higher multipoles are expected to be important. This good agreement indicates that for a detector that is located in a nearly symmetric position with respect to the chain, mismatches between the phases of the scattered fields in the far-field, an effect

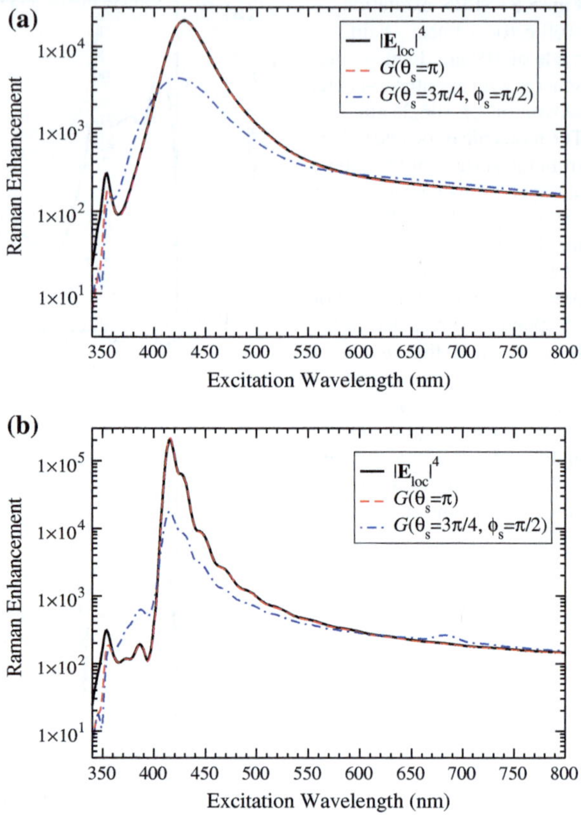

Fig. 5.6 SERS excitation profile for a linear chain made of 100 nm diameter Ag spheres with a center-to-center separation of 400 nm. The molecule is located 0.5 nm from the surface of the sphere at the coordinate origin along the axis parallel with the incident E-field polarisation. The case of 5 spheres is presented in **a** and the case of 45 spheres is presented in **b**. In both **a** and **b** the solid black line is the PW approximation result, the dashed red line is the backscattering DR result, and the *blue dash–dotted line* is the DR result for a detector at $\theta_s = 3\pi/4$ and $\phi_s = \pi/2$

that we will explain later, are largely balanced and cancel out. This leaves the major contribution to the enhancement factor being from the local environment near the molecule.

The blue dash–dot lines in Figs. 5.5–5.7 are for the case when the detector is at $\theta_s = 3\pi/4$ and $\phi_s = \pi/2$. For this detector location there are significant differences between the PW and DR results, demonstrating far-field phase interference effects that are significant for an array structure that is not symmetrically positioned with respect to the detector. The enhancement factors for this detector location are generally lower than for the backscattering arrangement in the blue region of the spectrum where the maximum enhancement occurs, and are larger in the red region of the spectrum if the minimum separation between the spheres in the chain is below 300 nm. The change for longer separation distances arises because cavity-like photonic resonance effects begin to occur in the visible region of the spectrum for separation distances larger than 400 nm. One thing to note is that for the largest separations the local field from each sphere closely approximates the radiation profile of a point dipole when evaluated at the position of other spheres. In other words, the point-dipole approximation works well since the spheres are far-enough apart and the wavelengths long

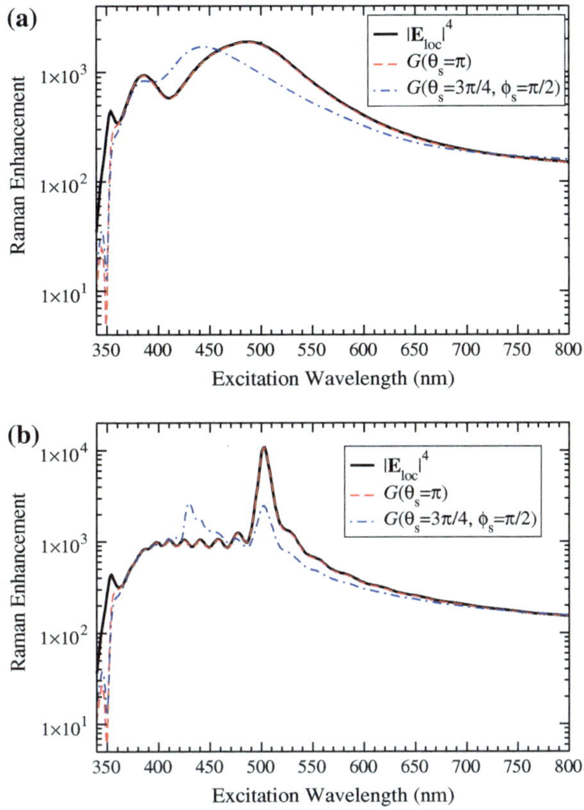

Fig. 5.7 SERS excitation profile for a linear chain made of 100 nm diameter Ag spheres with a center-to-center separation of 500 nm. The molecule is located 0.5 nm from the surface of the sphere at the coordinate origin along the axis parallel with the incident E-field polarisation. The case of 5 spheres is presented in **a** and the case of 45 spheres is presented in **b**. In both **a** and **b** the *solid black line* is the PW approximation result, the *dashed red line* is the backscattering DR result and the *blue dash–dotted line* is the DR result for a detector at $\theta_s = 3\pi/4$ and $\phi_s = \pi/2$

enough that higher multipolar contributions are unimportant. We will expand on this point later when examining chains of sphere dimers.

Since the most interesting features of the excitation profiles are the maximum enhancement factors, we now note the absolute errors in the PW approximation for the two detector positions. For the backscattering position the errors in the 5–sphere chains are 1.74, 0.31 and 0.88% at separations of 106.25, 400 and 500 nm in Figs. 5.5a, 5.6a, and 5.7a. The error is 0.52, 2.37 and 0.13% for these same separations in Figs. 5.5b, 5.6b and 5.7b when the number of spheres is increased to 45. Of course all these errors would be negligible in practical applications, which means that the PW approximation can be used with confidence. When the detector is moved to the second position at $\theta_s = 3\pi/4$ and $\phi_s = \pi/2$ the errors become considerably larger. With the same separations as before the 5–sphere PW errors are 63.5, 416 and 64.5% while the 45–sphere PW errors are 80.6, 16.2 and 334%. Looking at Figs. 5.5–5.7 we note that the wavelength at which the maximum enhancement occurs can change depending on the detector location. In these results we have taken the maximum to be the location at which the PW is largest.

The reason that we see significant differences between the PW approximation and dipole re-radiation results can be explained by phase interference effects as follows. We recall from Sect. 5.2.1 that the electric field in the far-field has the form

$$\mathbf{E} = \frac{\mathbf{F}\exp(ikr)}{r}$$

and therefore each field $\mathbf{E}_\mathrm{i}$ for sphere i has an oscillatory term that behaves as $\exp(ikr_s^i)/r_s^i$, where r_s^i is the distance from the center of sphere i to the detector location. Therefore, the overall far-field amplitude behaves as

$$\mathbf{F}_\mathrm{tot} = \sum_\mathrm{i} \frac{r_s}{r_s^\mathrm{i}} \mathbf{F}_\mathrm{i} \exp\left(ik\left[r_s^\mathrm{i} - r_s\right]\right),$$

where r_s is defined in Fig. 5.1. The result of this is that for array structures where spheres have larger separations there can be interference patterns in the far-field between amplitudes of the form

$$Re\left\{\mathbf{F}_\mathrm{i}^* \cdot \mathbf{F}_\mathrm{j}\right\} \cos\left(k\left[r_s^\mathrm{j} - r_s^\mathrm{i}\right]\right) - Im\left\{\mathbf{F}_\mathrm{i}^* \cdot \mathbf{F}_\mathrm{j}\right\} \sin\left(k\left[r_s^\mathrm{j} - r_s^\mathrm{i}\right]\right),$$

where i$\neq$j. These phase interference effects contribute to differences between the PW and dipole re-radiation enhancement factors. In our earlier study, it was demonstrated that phase interferences are smallest when the nanospheres are arranged symmetrically with respect to the detector location such that emission from the different particles is all in-phase. This also applies in the present application to the backscattering results, which explains the small errors we find for that geometry. In the nonsymmetric geometry that we have studied (detector along the y-axis), the scattering from different particles is no longer in-phase, so much larger errors are found.

5.5 Dipole Re-Radiation Effects in SERS from Dimer Chains

As was mentioned earlier, the largest observed enhancement factors in SERS have been found to occur at the junction between two particles. In an analogous fashion to the construction of arrays made up of single sphere chains, arrays made of two chains next to each other, in effect a chain of dimers, can be constructed. This structure was also examined by Zou and Schatz [22] and it was found that long range photonic effects similar to those observed with single sphere chains can be combined with plasmonic hotspots between the particles to produce even higher enhancement factors. The main difference is that the location of the most intense photonic resonances are now red shifted from the isolated sphere case due to the red shift in the dipole plasmon mode of a sphere dimer system. In this work we study models similar to Zou and Schatz, but now including for DR effects in the enhancement.

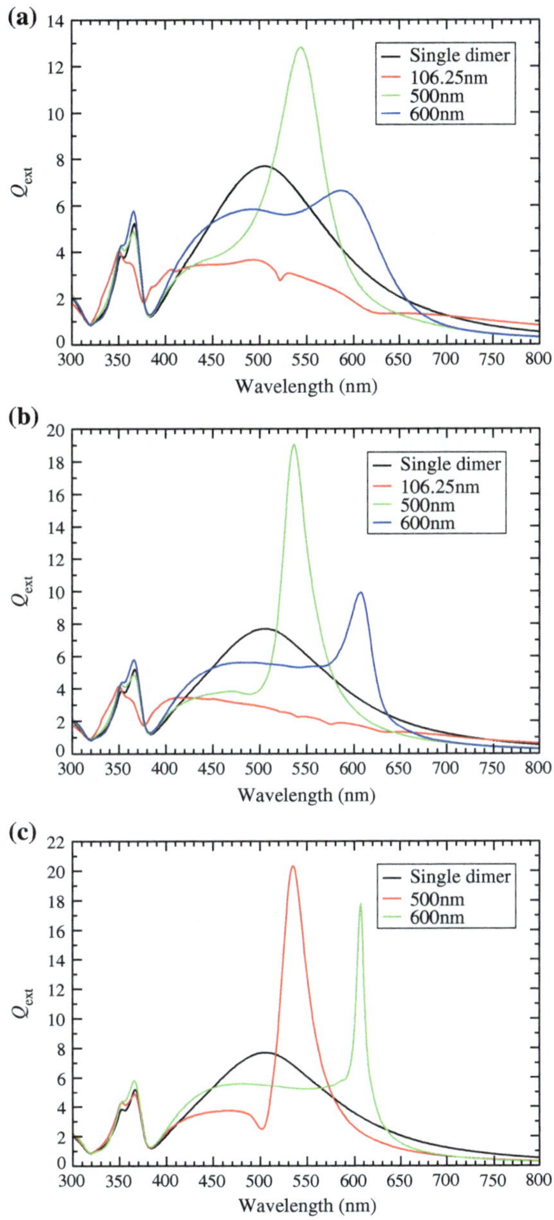

Fig. 5.8 Extinction spectra for dimer chains made up of 5 dimers (**a**), 15 dimers (**b**), and 45 dimers (**c**) for different dimer separations. The dimer-to-dimer separations given in the legend are the center-to-center distances. The gap between the two spheres in the dimers is 6.25 nm

In Fig. 5.8 we present the extinction spectra of arrays of sphere dimer chains with a dimer gap of 6.25 nm for various chain lengths and dimer-to-dimer separation distances. Note that the dimer chain is taken to be along the y-axis, but with each dimer oriented along the x-axis. Also, the spheres have a diameter of 100 nm, and the mole-

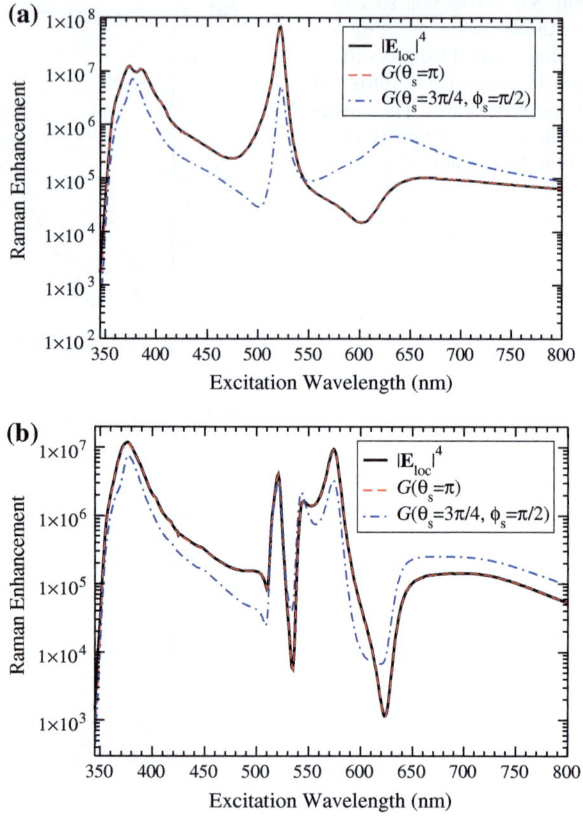

Fig. 5.9 SERS excitation profile for a linear chain made of a dimer of 100 nm diameter Ag spheres with a dimer center-to-center separation of 106.25 nm. The gap between the two chains is 6.25 nm. The molecule is located 0.5 nm from the surface of the sphere at the coordinate origin along the axis parallel with the incident E-field polarisation. The case of 5 dimers is presented in **a** and the case of 15 dimers is presented in **b**. In both **a** and **b** the *solid black line* is the PW approximation result, the *dashed red line* is the backscattering DR result and the *blue dash-dotted line* is the DR result for a detector at $\theta_s = 3\pi/4$ and $\phi_s = \pi/2$

cule is located 0.5 nm from the surface of one sphere in the gap of the central dimer. An examination of the extinction spectra of the arrays of dimers shows trends similar to the single sphere chains in Fig. 5.4. For small dimer-to-dimer separations the dipole peak is suppressed and then becomes enhanced as the separation between the dimers becomes larger. Increasing the number of components in the chains for both single spheres and dimers increases the extinction, and an examination of Figs. 5.5–5.7 and Figs. 5.9–5.11 indicates that this also leads to an increase in the PW enhancement factor. At separations greater than 400 nm we see intense enhancements that are due to long range photonic resonances. These become more prominent as the number of dimers is increased, just as is observed for single sphere chains. There is little change in the maximum observed enhancement factor when the number of dimers is increased from 5 to 15, but the larger number of dimers results in a maximum in the dipole plasmon region that is not as prominent in the 5–dimer case.

In Figs. 5.9a, 5.10a and 5.11a the SERS excitation profiles for chains made up of 5 dimers are shown. We see that chains made up of 5 dimers where the detector is located in a backscattering configuration have SERS enhancement factors that are accurately approximated by the PW enhancement factor. In fact comparisons with

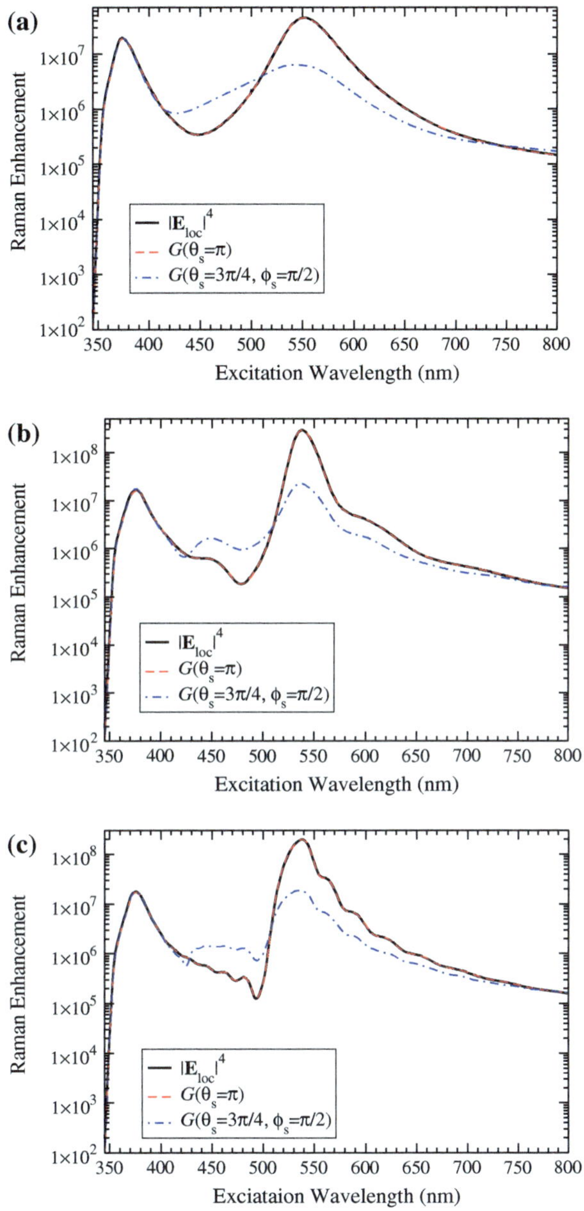

Fig. 5.10 SERS excitation profile for a linear chain made of a dimer of 100 nm diameter Ag spheres with a dimer center-to-center separation of 500 nm. The gap between the two chains is 6.25 nm. The molecule is located 0.5 nm from the surface of the sphere at the coordinate origin along the axis parallel with the incident E-field polarisation. The case of 5 dimers is presented in **a**, the case of 15 dimers is presented in **b**, and the case of 45 dimers is presented in **c**. In **a**, **b**, and **c** the *solid black line* is the PW approximation result, the *dashed red line* is the backscattering DR result and *the blue dash-dotted line* is the DR result for a detector at $\theta_s = 3\pi/4$ and $\phi_s = \pi/2$

Figs. 5.5a, 5.6a and 5.7a show that the PW approximation is better in the dimer case than is observed in the case of single sphere chains. This is because for single sphere chains, higher order multipoles are important at certain frequencies (towards the blue)

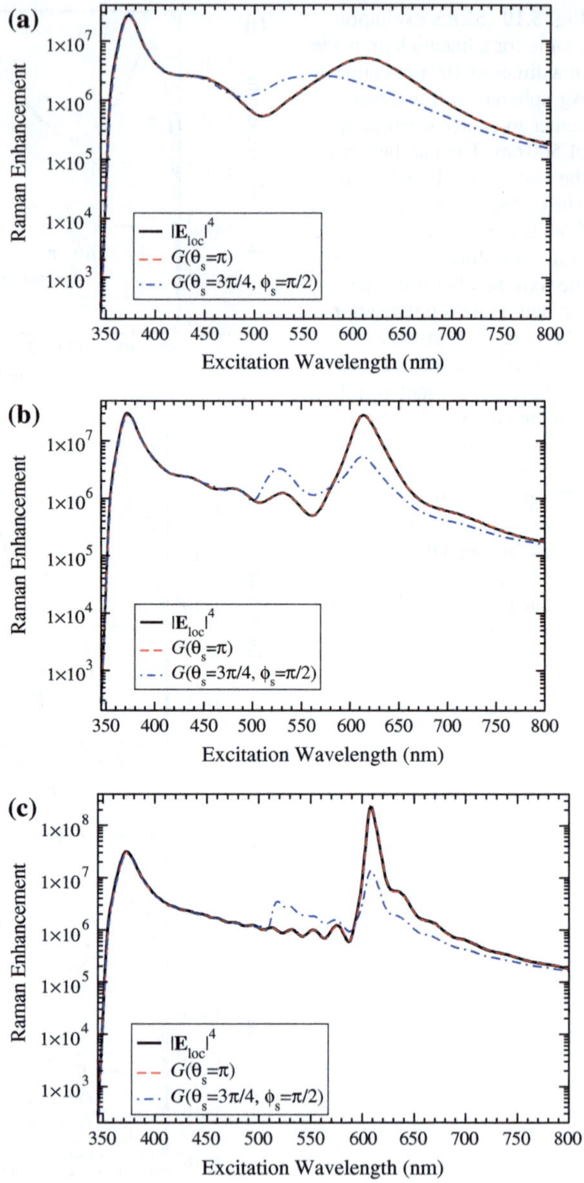

Fig. 5.11 SERS excitation profile for a linear chain made of a dimer of 100 nm diameter Ag spheres with a dimer center-to-center separation of 600 nm. The gap between the two chains is 6.25 nm. The molecule is located 0.5 nm from the surface of the sphere at the coordinate origin along the axis parallel with the incident E-field polarisation. The case of 5 dimers is presented in **a**, the case of 15 dimers is presented in **b**, and the case of 45 dimers is presented in **c**. In **a**, **b**, and **c** the *solid black line* is the PW approximation result, the *dashed red line* is the backscattering DR result and the *blue dash-dotted line* is the DR result for a detector at $\theta_s = 3\pi/4$ and $\phi_s = \pi/2$

which are not accessed by plane wave scattering. However the dimer resonances are dipolar in character so the influence of higher multipoles is less important.

Figures 5.9b, 5.10b and 5.11b show SERS excitation profiles for 15–dimer chains at dimer-to-dimer separations of 106.25, 500 and 600 nm while Figs. 5.10c and 5.11c show excitation profiles of 45–dimer chains with dimer-to-dimer separations of 500 and 600 nm. Again there is good agreement of PW and DR results for the backscattering arrangement. In Figs. 5.10b and 5.11b we see the influence of long range photonic resonances which become more prominent in Figs. 5.10c and 5.11c. These photonic resonances combine with the large field enhancement associated with the dimer to give peak enhancement factors above 10^8, thus confirming the previous conclusions of Zou and Schatz concerning the effect of combining plasmonic and photonic resonance contributions to the electromagnetic enhancement.

As seen in Sect. 5.4 with single sphere chains, Figs. 5.9b, 5.10b and 5.11b show that changing the detector position so that it is no longer symmetric with respect to the dimer chain results in significant differences between the PW approximation and dipole re-radiation. These differences are due to phase interference effects as discussed earlier. There are certain frequencies where the PW and DR results are in good agreement for this detector location, especially for larger dimer separations and for longer dimer chains. This is especially noticeable for shorter wavelengths, where there are no important photonic resonances, so only localised plasmonic excitatiton is important.

We now examine the absolute errors associated with the PW approximation at the maximum enhancement for the dimer chains. The reported errors are for separations of 106.25, 500, and 600 nm for the 5–dimer chains in Figs. 5.9a, 5.10a, and 5.11a, the 15–dimer chains in Figs. 5.9b, 5.10b, and 5.11b and separations of 500 and 600 nm for the 45–dimer chains in Figs. 5.10c and 5.11c. In a similar fashion to the single sphere chains, the backscattering detector location results in smaller errors which for 5–dimer chains are 0.077, 0.22 and 3.7 %. For 15–dimer chains these errors become 0.59, 0.23 and 3.84% and for 45 dimers the errors are 0.89 and 0.56%. Larger errors are again seen when the detector location is at $\theta_s = 3\pi/4$ and $\phi_s = \pi/2$. The 5–dimer chain has errors of 1205, 636 and 5.78% for this detector location while the 15–dimer chain has errors of 56.2, 1192 and 7.72%. The errors associated with this detector location in the 45–dimer chain are 986 and 1555%.

5.6 Conclusions

Dipole re-radiation effects can play a significant role in the modeling of SERS enhancement factors. In single particle chains and in dimer chains we observe that the main differences between dipole re-radiation calculations and the PW approximation are the result of phase interference effects that result from scattering of the emitted dipole wave from multiple spheres. Such interferences are relatively unimportant for cases where the detector location is in a location that is symmetric with respect to the array structure; however, they can produce major differences in peak SERS enhancement factors for other detector geometries. Differences between PW and DR also occur in certain cases for isolated spheres due to multipole excitation effects, as multipole resonances can be more easily excited by a dipole field than by plane waves.

The maximum enhancement factors for single spheres are around 10^3, and for sphere dimers around 10^7. Photonic resonances in sphere chains and dimer chains with the right orientation can increase these enhancement factors by 1–2 orders of magnitude when the center-to-center separation distance is close to the wavelength of light. We note, however, that these effects will be smaller when nonzero Stokes shifts are included in the Raman intensity evaluation.

Acknowledgments This research was supported by grant DE-SC0004752 funded by the US Department of Energy, Office of Science and Office of Basic Energy Sciences.

References

1. M.G. Albrecht, J.A. Creighton, J. Am. Chem. Soc. **99**, 5215 (1977)
2. D.L. Jeanmaire, R.P. van Duyne, Electroanal. Chem. **84**, 1 (1977)
3. M. Fleischmann, P.J. Hendra, A.J. McQuillian, Chem. Phys. Lett. **26**, 163 (1974)
4. M. Moskovits, J. Chem. Phys. **69**, 4159 (1978)
5. M.R. Philpott, J. Phys. Colloques **44**, 295 (1983)
6. H. Metiu, P. Das, Ann. Rev. Phys. Chem. **35**, 507 (1984)
7. G.C. Schatz, Acc. Chem. Res. **17**, 370 (1984)
8. M. Moskovits, Rev. Mod. Phys. **57**, 783 (1985)
9. A. Wokaun, Mol. Phys. **56**, 1 (1985)
10. M. Kerker, *Enhanced Raman scattering in colloidal systems*. No. 45 in Studies in Physical and Theoretical Chemistry (Elsevier, Amsterdam, 1987)
11. M. Kerker, D.S. Wang, H. Chew, App. Opt. **19**, 4159 (1980)
12. G. Mie, Ann. Phys. **25**, 377 (1908)
13. H. Chew, P.J. McNulty, M. Kerker, Phys. Rev. A **13**, 396 (1976)
14. H. Chew, M. Kerker, P.J. McNulty, J. Opt. Soc. Am. **66**, 440 (1976)
15. H. Chew, M. Kerker, D.D. Cooke, Phys. Rev. A **16**, 320 (1977)
16. R. Ruppin, J. Chem. Phys. **76**, 1681 (1982)
17. M. Kerker, J. Opt. Soc. Am. B **2**, 1327 (1985)
18. M. Kerker, J. Colloid Interface Sci. **118**, 417 (1987)
19. L.K. Ausman, G.C. Schatz, J. Chem. Phys. **131**, 084708 (2009)
20. S. Zou, N. Janel, G.C. Schatz, J. Chem. Phys. **120**, 10871 (2004)
21. S. Zou, G.C. Schatz, J. Chem. Phys. **121**, 12606 (2004)
22. S. Zou, G.C. Schatz, Chem. Phys. Lett. **403**, 62 (2005)
23. S. Zou, G.C. Schatz,*Coupled Plasmonic Plasmon/Photonic Resonance Effects in SERS*. No. 103 in Topics in Applied Physics (Springer, Berlin, 2006)
24. S. Zou, G.C. Schatz, Nanotechnology **17**, 2813 (2006)
25. S. Zou, G.C. Schatz, Isreal J. Chem. **46**, 293 (2006)
26. L. Quin, S. Zou, C. Xue, A. Atkinson, G.C. Schatz, C.A. Mirkin, Proc. Natl. Acad. Sci. USA **103**, 13300 (2006)
27. J.P. Camden, J.A. Dieringer, Y. Wang, D.J. Masiello, L.D. Marks, G.C. Schatz, R.P. van Duyne, J. Am. Chem. Soc. **130**, 12616 (2008)
28. W. Wei, S. Li, J.E. Millstone, M.J. Banholzer, X. Chen, X. Xu, G.C. Schatz, C.A. Mirkin, Angew. Chem. Int. Edit. **48**, 4210 (2009)
29. J.A. Dieringer, K.L. Wustholz, D.J. Masiello, J.P. Camden, S.L. Kleinman, G.C. Schatz, R.P. van Duyne, J. Am. Chem. Soc. **131**, 849 (2009)
30. K. Kneipp, W. Wang, H. Kneipp, L.T. Perelman, I. Itzkan, R.R. Dasari, M.S. Feld, Phys. Rev. Lett. **78**, 1667 (1997)
31. S. Nie, S. Emory, Science **275**, 1102 (1997)

32. M. Moskovits, L.L. Tay, J. Yang, T. Haslett, *SERS and the single molecule*. No. 82 in Topics in Applied Physics (Springer, Berlin, 2002)
33. B. Vlčková, M. Moskovits, I. Pavel, K. Šišková, M. Sládková, M. Šlouf, Chem. Phys. Lett. **455**, 131 (2008)
34. Y. Fang, N.H. Seong, D.D. Dlott, Science **321**, 388 (2008)
35. K.E. Shafer-Peltier, C.L. Haynes, M.R. Glucksberg, R.P. van Duyne, J. Am. Chem. Soc. **125**, 588 (2003)
36. X. Zhang, M.A. Young, O. Lyandres, R.P. van Duyne, J. Am. Chem. Soc. **127**, 4484 (2005)
37. N.C. Shah, O. Lyandres, C.R. Yonzon, X. Zhang, R.P. van Duyne, ACS Symposium Series **963**, 107 (2007)
38. S.D. Hudson, G. Chumanov, Anal. Bioanal. Chem. **394**, 679 (2009)
39. D.A. Stuart, K.B. Biggs, R.P. van Duyne, Analyst **131**, 568 (2006)
40. A.V. Whitney, F. Casadio, R.P. van Duyne, App. Spect. **61**, 994 (2007)
41. C.L. Brosseau, A. Gambardella, F. Casadio, C.M. Grzywacz, J. Wouters, R.P. van Duyne, Anal. Chem. **81**, 3056 (2009)
42. S. Malynych, G. Chumanov, J. Am. Chem. Soc. **125**, 2896 (2003)
43. A.S. Kumbhar, M.K. Kinnan, G. Chumanov, J. Am. Chem. Soc. **127**, 12444 (2005)
44. M.K. Kinnan, G. Chumanov, J. Phys. Chem. C **111**, 18010 (2007)
45. Y.D. Suh, G.K. Schenter, L. Zhu, H.P. Lu, Ultramicroscopy **97**, 89 (2003)
46. L. Zhu, G.K. Schenter, M. Micic, Y.D. Suh, N. Klymyshyn, H.P. Lu, Proc. SPIE **4962**, 70 (2003)
47. P.I. Geshev, K. Dickmann, J. Opt. A: Pure Appl. Opt. **8**, S161 (2006)
48. T. Shegai, Z. Li, T. Dadosh, Z. Zhang, H. Xu, G. Haran, Proc. Natl. Acad. Sci. U.S.A. **105**, 16448 (2008)
49. D. Pristinski, E.C. la Ru, S. Tan, S. Sukhishvili, H. Du, Opt. Express **16**, 20117 (2008)
50. C.F. Bohren, D.R. Huffman, *Absorption and Scattering of Light by Small Particles* (Wiley-VCH, Weinheim, 2004)
51. D.W. Mackowski, Proc. R. Soc. London Ser. A **433**, 599 (1991)
52. D.W. Mackowski, J. Opt. Soc. Am. A **11**, 2851 (1994)
53. D.W. Mackowski, M.I. Mishchenko, J. Opt. Soc. Am. A **13**, 2266 (1996)
54. L.K. Ausman, G.C. Schatz, J. Chem. Phys. **129**, 054704 (2008)
55. C.T. Tai, *Dyadic Green's Functions in Electromagnetic Theory* (Intext Educational, Scranton, San Francisco, Toronto, London, 1971)
56. P.B. Johnson, R.W. Christy, Phys. Rev. B **6**, 4370 (1972)
57. D.J. Masiello, G.C. Schatz, Phys. Rev. A **78**, 042505 (2008)

Chapter 6
Optical Force and Torque on Single and Aggregated Spheres: The Trapping Issue

Rosalba Saija, Paolo Denti and Ferdinando Borghese

Abstract Radiation force and radiation torque stem from the conservation theorems governing the interaction between radiation and particles. As a result, even simple plane waves may yield force and torque that are nonnegligible when studying the dynamics of particles. We show how these forces and torques can be put into formulas of practical use starting with the expansion of the electromagnetic field in a series of vector multipole fields, and adapt the formalism for particles that are fairly well modeled as single or aggregated spheres. Then, we extend the formalism to deal with the case the field is a highly focalized laser beam, that, as is well known, may trap particles within the focal region. Finally, we present our calculations performed by the theory we exposed, finding a convincing concordance as long as comparison can be made with available experimental data.

6.1 Introduction

The interaction of radiation field with particles obeys the conservation of energy as well as the conservation both of the linear and of the angular momentum. As a result, particles illuminated by a radiation field may experience both a force $\mathbf{F}_{\mathrm{Rad}}$ and a torque $\boldsymbol{\Gamma}_{\mathrm{Rad}}$ that contribute in determining their dynamical behavior. For instance, in an astrophysical context, the radiation force may determine the space distribution of the cosmic dust grains [1], whereas the radiation torque is considered as a non-

R. Saija (✉) · P. Denti · F. Borghese
Dipartimento di Fisica della Materia e Ingegneria Elettronica,
Università di Messina Messina, Italy
e-mail: rosalba.saija@unime.it

P. Denti
e-mail: paolo.denti@unime.it

F. Borghese
e-mail: ferdinando.borghese@unime.it

W. Hergert and T. Wriedt (eds.), *The Mie Theory*, Springer Series in Optical Sciences 169, DOI: 10.1007/978-3-642-28738-1_6,

negligible agent of their partial alignment [2]. In laboratory applications both $\mathbf{F}_{\mathrm{Rad}}$ and $\boldsymbol{\Gamma}_{\mathrm{Rad}}$ proved useful in manipulating microsized and nanosized particles under controlled conditions.

The actual calculation of $\mathbf{F}_{\mathrm{Rad}}$ and $\boldsymbol{\Gamma}_{\mathrm{Rad}}$ requires the knowledge of the electromagnetic field in the presence of the particles. This amounts to say that besides the incident radiation fields $\mathbf{E}_{\mathrm{I}}$ and $\mathbf{B}_{\mathrm{I}}$ one has to know the fields $\mathbf{E}_{\mathrm{S}}$ and $\mathbf{B}_{\mathrm{S}}$ scattered by the particles. The latter fields can be calculated by several methods among which we prefer the one based on the expansion of the electromagnetic field in a series of vector spherical multipole fields in the framework of the transition matrix approach [3]. Actually, this is the approach on which Mie theory is based [4], although in 1908 the concept itself of transition matrix was still to be devised. The advantages of resorting to the transition matrix stem from its independence of the direction of incidence and of the polarization, even for nonspherical particles, and from its well-defined transformation properties under rotation of the coordinate frame. As a consequence, the elements of the transition matrix need to be calculated only once for an arbitrarily chosen orientation of the particle with respect to the incident field.

In this contribution we will describe how the multipole fields expansion can be applied to calculate the force and torque exerted by a radiation field on particles that either are spherical or can be modeled as aggregates of spheres that may be radially inhomogeneous and need not be identical to each other. In any case, most of the theory can be formulated without referring to the shape of the particles or to their dielectric properties. It is only when one has to perform the calculation of $\mathbf{E}_{\mathrm{S}}$ that the shape and properties of the particles enter into play.

In particular, the theory we are going to describe will prove suitable to deal with the problem of trapping of microsized and nanosized particles and to study the orientation they assume under the effect of the radiation torque.

Anyway, we do not present here a review of the theories of trapping and manipulation of small particles but only a description of our own approach and some of its applications.

6.2 Radiation Force and Radiation Torque

The geometry of the problem we deal with is depicted in Fig. 6.1, which also applies when the incident radiation is a focalized laser beam. We choose a reference frame Σ fixed in the laboratory, with respect to which the particle, i.e., its center of mass, has the vector position $\mathbf{R}_{O'}$. Of course, the coincidence of O' with the center of mass is non-necessary, but it is useful when one is interested both in the translational and in the rotational dynamics of the particle. We also attach to the particle a local frame of reference Σ' with origin at O' and whose axes are parallel to the axes of Σ. Then, the radiation force and the radiation torque on the particle are given by the integrals [5]

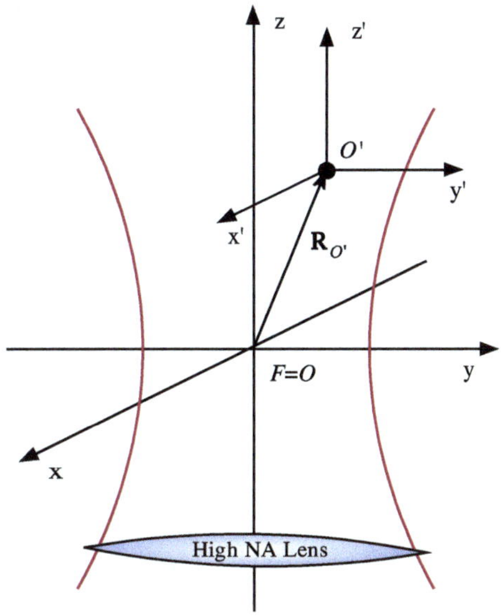

Fig. 6.1 Coordinate system we adopt to describe the radiation force and the radiation torque on a particle located at $\mathbf{R}_{O'}$. In case the radiation field is a focalized laser beam, the origin of the coordinates O and the z axis coincide with the nominal focus F of the lens and with the optical axis, respectively

$$\mathbf{F}_{\mathrm{Rad}} = r'^2 \int_{\Omega'} \hat{\mathbf{r}}' \cdot \langle \mathsf{T}_{\mathrm{M}} \rangle \, \mathrm{d}\Omega' \tag{6.1}$$

and

$$\boldsymbol{\Gamma}_{\mathrm{Rad}} = -r'^3 \int_{\Omega'} \hat{\mathbf{r}}' \cdot \langle \mathsf{T}_{\mathrm{M}} \rangle \times \hat{\mathbf{r}}' \, \mathrm{d}\Omega' \,, \tag{6.2}$$

respectively. In (6.1) and (6.2) the integration is over the full solid angle, r' is the radius of a large (possibly infinite) sphere with center at $\mathbf{R}_{O'}$ surrounding the particle, and

$$\langle \mathsf{T}_{\mathrm{M}} \rangle = \frac{1}{8\pi} \mathrm{Re} \left[n^2 \mathbf{E}' \otimes \mathbf{E}'^* + \mathbf{B}' \otimes \mathbf{B}'^* - \frac{1}{2} \left(n^2 |\mathbf{E}'|^2 + |\mathbf{B}'|^2 \right) \mathsf{I} \right] \tag{6.3}$$

is the time averaged Maxwell stress tensor in the form of Minkowski [5]. In (6.3), the fields $\mathbf{E}'$ and $\mathbf{B}'$ are considered in the frame Σ', $\otimes$ denotes dyadic product, the asterisk denotes complex conjugation, and I is the unit dyadic. In turn, n is the real refractive index of the homogeneous, non absorbing, isotropic medium surrounding the particle. Of course, choosing for $\langle \mathsf{T}_{\mathrm{M}} \rangle$ the form of Minkowski rather than the form of Becker is somewhat arbitrary as discussed by Pfeiffer et al. [6], who also stress that no experiment has been able, till date, to encourage the choice of the Minkowski form over the Becker form. Actually, the form of Minkowski is nonsymmetric in an anisotropic medium, but since the particles we deal with are embedded in an isotropic medium, no inconvenience arises because of our choice.

6.2.1 Radiation Force by Plane Waves

According to our statement in Sect. 6.1, the fields $\mathbf{E}'$ and $\mathbf{B}'$ that enter (6.3) are the superposition of the incident field and of the field scattered by the particle. We, for now, assume that the incident radiation is the plane wave

$$\mathbf{E}_{\mathrm{I}} = E_0 \hat{\mathbf{u}}_{\mathrm{I}} e^{i\mathbf{k}_{\mathrm{I}}\cdot\mathbf{r}} = E_0 \hat{\mathbf{u}}_{\mathrm{I}} e^{i\mathbf{k}_{\mathrm{I}}\cdot(\mathbf{r}'+\mathbf{R}_{O'})} = E_0' \hat{\mathbf{u}}_{\mathrm{I}} e^{i\mathbf{k}_{\mathrm{I}}\cdot\mathbf{r}'} = \mathbf{E}_{\mathrm{I}}' \tag{6.4}$$

of wavevector $\mathbf{k}_{\mathrm{I}} = \hat{\mathbf{k}}_{\mathrm{I}} k = \hat{\mathbf{k}}_{\mathrm{I}} n k_{\mathrm{v}}$, where $\hat{\mathbf{k}}_{\mathrm{I}}$ is the unit vector in the direction of incidence, $k_{\mathrm{v}} = \omega/c$, and $\hat{\mathbf{u}}_{\mathrm{I}}$ is the (unit) polarization vector that may be either real (linear polarization) or complex (elliptic polarization). Obviously, in (6.4), $|E_0'|^2 = |E_0|^2 = I_{\mathrm{I}}$, the intensity of the incident field. In view of our choice of the incident field and of the large radius of the integration sphere, the integral in (6.1) can be calculated by resorting to the asymptotic expansion of a plane wave [7]

$$\mathbf{E}_{\mathrm{I}}' = E_0' \hat{\mathbf{u}}_{\mathrm{I}} \frac{2\pi i}{r'k} \left[\delta\left(\hat{\mathbf{k}}_{\mathrm{I}} + \hat{\mathbf{r}}'\right) \exp\left(-ikr'\right) - \delta\left(\hat{\mathbf{k}}_{\mathrm{I}} - \hat{\mathbf{r}}'\right) \exp\left(ikr'\right) \right] .$$

Then, a straightforward calculation leads to the conclusion that the first two terms, i.e., the dyadic products in the expression of $\langle \mathsf{T}_{\mathrm{M}} \rangle$ give a vanishing contribution to the radiation force [8]. In Appendix 1 we will prove this result through a different approach that applies even when the incident field is not a single plane wave. Anyway, since $\mathsf{I}\cdot\hat{\mathbf{r}}' = \hat{\mathbf{r}}'$, the component of the radiation force along the direction characterized by the unit vector $\hat{\mathbf{v}}_\zeta$ turns out to be

$$\begin{aligned} F_{\mathrm{Rad}\,\zeta} &= -\frac{r'^2}{16\pi} \mathrm{Re} \int_{\Omega'} (\hat{\mathbf{r}}' \cdot \hat{\mathbf{v}}_\zeta) \left[n^2 \left(|\mathbf{E}_{\mathrm{S}}'|^2 + 2\mathbf{E}_{\mathrm{I}}'^* \cdot \mathbf{E}_{\mathrm{S}}' \right) \right. \\ &\quad \left. + \left(|\mathbf{B}_{\mathrm{S}}'|^2 + 2\mathbf{B}_{\mathrm{I}}'^* \cdot \mathbf{B}_{\mathrm{S}}' \right) \right] \mathrm{d}\Omega' . \end{aligned} \tag{6.5}$$

Of course, choosing $\hat{\mathbf{v}}_\zeta$ to coincide in turn with the unit vectors that characterize the axes of Σ one gets the cartesian components of the radiation force, whereas, choosing $\hat{\mathbf{v}}_\zeta$ along $\hat{\mathbf{k}}_{\mathrm{I}}$, one gets the radiation pressure in the form derived by Mishchenko [8] in terms of the customary asymmetry parameter. The components of $\mathbf{F}_{\mathrm{Rad}}$ in a plane orthogonal to $\hat{\mathbf{k}}_{\mathrm{I}}$ can also be calculated and imply the definition of two new asymmetry parameters [9].

Obviously, since the incident field is a plane wave, the integral (6.5) gets no contribution from the terms $\mathbf{E}_{\mathrm{I}}' \cdot \mathbf{E}_{\mathrm{I}}'^*$, and $\mathbf{B}_{\mathrm{I}}' \cdot \mathbf{B}_{\mathrm{I}}'^*$ that, accordingly, have been omitted. However, it is less obvious, although conceptually expected, that the integral (6.5) gets no contribution from the same terms even when the incident field is a superposition of plane waves with different directions of propagation. This statement will be proved in Appendix 1.

We expand both the incident and the scattered field in a series of vector multipole fields [10]. Accordingly,

$$\mathbf{E}'_{\mathrm{I}} = E'_0 \sum_{plm} \mathbf{J}^{(p)}_{lm}\left(\mathbf{r}', k\right) W^{(p)}_{lm}\left(\hat{\mathbf{u}}_{\mathrm{I}}, \hat{\mathbf{k}}_{\mathrm{I}}\right), \tag{6.6}$$

$$\mathbf{E}'_{\mathrm{S}} = E'_0 \sum_{plm} \mathbf{H}^{(p)}_{lm}\left(\mathbf{r}', k\right) A^{(p)}_{lm}\left(\hat{\mathbf{u}}_{\mathrm{I}}, \hat{\mathbf{k}}_{\mathrm{I}}\right), \tag{6.7}$$

hence the multipole expansion of the magnetic field can be inferred through the equation

$$\mathbf{B} = -\frac{\mathrm{i}}{k_{\mathrm{V}}} \nabla \times \mathbf{E}.$$

In (6.7) the amplitudes of the scattered field bear the arguments $\hat{\mathbf{u}}_{\mathrm{I}}$ and $\hat{\mathbf{k}}_{\mathrm{I}}$ to recall that in general they depend both on the polarization and on the direction of propagation of the incident field. The multipole fields in (6.6) are

$$\mathbf{J}^{(1)}_{lm}\left(\mathbf{r}, k\right) = j_l(kr)\mathbf{X}_{lm}\left(\hat{\mathbf{r}}\right), \quad \mathbf{J}^{(2)}_{lm}\left(\mathbf{r}, k\right) = \frac{1}{k} \nabla \times \mathbf{J}^{(1)}_{lm}\left(\mathbf{r}, k\right),$$

where $\mathbf{X}_{lm}(\hat{\mathbf{r}})$ denotes vector spherical harmonics [5]. The multipole fields $\mathbf{H}^{(p)}_{lm}(\mathbf{r}, k)$ in (6.7) are identical to the $\mathbf{J}^{(p)}_{lm}(\mathbf{r}, k)$ fields except for the substitution of the spherical Hankel functions of the first kind $h_l(kr)$ in place of the spherical Bessel functions $j_l(kr)$. The superscript $p = 1, 2$ is a parity index that distinguishes the magnetic multipole fields from the electric ones, respectively. The $\mathbf{J}$ fields are everywhere regular, whereas the $\mathbf{H}$ fields satisfy the radiation condition at infinity. The multipole amplitudes of the incident field are defined as

$$W^{(p)}_{lm}\left(\hat{\mathbf{u}}_{\mathrm{I}}, \hat{\mathbf{k}}_{\mathrm{I}}\right) = 4\pi \mathrm{i}^{p+l-1} \hat{\mathbf{u}}_{\mathrm{I}} \cdot \mathbf{Z}^{(p)*}_{lm}\left(\hat{\mathbf{k}}_{\mathrm{I}}\right),$$

where

$$\mathbf{Z}^{(1)}_{lm}\left(\hat{\mathbf{k}}\right) = \mathbf{X}_{lm}\left(\hat{\mathbf{k}}\right), \quad \mathbf{Z}^{(2)}_{lm}\left(\hat{\mathbf{k}}\right) = \mathbf{X}_{lm}\left(\hat{\mathbf{k}}\right) \times \hat{\mathbf{k}}$$

are transverse vector harmonics [10, 11], and the amplitudes of the scattered field $A^{(p)}_{lm}$ are calculated by imposing the customary boundary conditions across the surface of the particle. The actual calculation of the A-amplitudes in the framework of the transition matrix method will be discussed in Appendix 2.

Once the scattered field has been calculated, we perform the integration in (6.5) by exploiting the asymptotic expansion of the multipole fields [10] up to terms which give a contribution of order $1/r$. Actually, these expansions for the incident and the scattered field give

$$\mathbf{E}'_{\mathrm{I}} \to E'_0 \sum_{plm} \mathbf{Z}^{(p)}_{lm}\left(\hat{\mathbf{r}}'\right) W^{(p)}_{\mathrm{I}lm} \frac{(-)^{p-1}}{kr'} \sin\left[kr' - (l+1-p)\pi/2\right], \tag{6.8a}$$

$$\mathbf{E}'_{\mathrm{S}} \to E'_0 \sum_{plm} \mathbf{Z}^{(p)}_{lm}\left(\hat{\mathbf{r}}'\right) A^{(p)}_{lm} \frac{\exp\left(ikr'\right)}{kr'} \mathrm{i}^{-l-p} \tag{6.8b}$$

which, when substituted into (6.5), yield the result

$$\begin{aligned} F_{\mathrm{Rad}\,\zeta} = & -\frac{I_{\mathrm{I}}}{16\pi k_{\mathrm{v}}^2} \mathrm{Re}\left[\sum_{plm}\sum_{p'l'm'} \left(A^{(p)*}_{lm} A^{(p')}_{l'm'} + A^{(p'')*}_{lm} A^{(p''')}_{l'm'}\right) \mathrm{i}^{l-l'} I^{(pp')}_{\zeta\, lml'm'}\right] \\ & - \frac{2I_{\mathrm{I}}}{16\pi k_{\mathrm{v}}^2} \mathrm{Re}\left[\sum_{plm}\sum_{p'l'm'} \left(W^{(p)*}_{\mathrm{I}lm} A^{(p')}_{l'm'} + W^{(p'')*}_{\mathrm{I}lm} A^{(p''')}_{l'm'}\right)\right. \\ & \left. \times \sin\left[kr - (l-1+p)\pi/2\right] e^{ikr} (-\mathrm{i})^{l+p} \mathrm{i}^{l-l'} I^{(pp')}_{\zeta\, lml'm'}\right] \\ = & -F^{(\mathrm{Sca})}_{\mathrm{Rad}\,\zeta} + F^{(\mathrm{Ext})}_{\mathrm{Rad}\,\zeta}, \end{aligned} \tag{6.9}$$

with $p'' \neq p$ and $p''' \neq p'$. In (6.9) we define [12, 13]

$$\begin{aligned} I^{(pp')}_{\zeta\, lml'm'} &= \int_{\Omega'} \left(\hat{\mathbf{r}}' \cdot \hat{\mathbf{v}}_\zeta\right) i^{p-p'} \mathbf{Z}^{(p)*}_{lm}\left(\hat{\mathbf{r}}'\right) \cdot \mathbf{Z}^{(p')}_{l'm'}\left(\hat{\mathbf{r}}'\right) \mathrm{d}\Omega' \\ &= \frac{4\pi}{3} \sum_{\mu} Y^*_{1\mu}\left(\hat{\mathbf{v}}_\zeta\right) \frac{\mathrm{i}^{l'-l}}{16\pi^2} K^{(pp')}_{\mu;lml'm'}, \end{aligned}$$

where $Y_{1\mu}(\hat{\mathbf{v}}_\zeta)$ denotes spherical harmonics whose arguments are the polar angles of $\hat{\mathbf{v}}_\zeta$, and

$$K^{(pp')}_{\mu;lml'm'} = 16\pi^2 \mathrm{i}^{l-l'} \int_{\Omega'} Y_{1\mu}\left(\hat{\mathbf{r}}'\right) \mathrm{i}^{p-p'} \mathbf{Z}^{(p)*}_{lm}\left(\hat{\mathbf{r}}'\right) \cdot \mathbf{Z}^{(p')}_{l'm'}\left(\hat{\mathbf{r}}'\right) \mathrm{d}\Omega' .$$

The integrals $K^{(pp')}_{\mu;lml'm'}$ can be performed in closed form [9, 10] with the result

$$K^{(pp')}_{\mu;lml'm'} = 16\pi^2 \sqrt{\frac{3}{4\pi}} C\left(1, l', l; \mu, m', m\right) i^{l-l'} O^{(pp')}_{ll'}, \tag{6.10}$$

where $C\left(1, l', l; \mu, m', m\right)$ are Clebsch-Gordan coefficients [14] and

$$O_{ll}^{(pp')} = -\frac{1}{\sqrt{l(l+1)}}\left(1-\delta_{pp'}\right), \quad O_{l,l-1}^{(pp')} = \sqrt{\frac{(l-1)(l+1)}{l(2l+1)}}\delta_{pp'},$$

$$O_{l,l+1}^{(pp')} = -\sqrt{\frac{l(l+2)}{(l+1)(2l+1)}}\delta_{pp'}.$$

The K-integrals have the symmetry properties

$$K_{\zeta\, lml'm'}^{(11)} = K_{\zeta\, lml'm'}^{(22)}, \quad K_{\zeta\, lml'm'}^{(12)} = K_{\zeta\, lml'm'}^{(21)},$$

which help us to get a more compact expression for $F_{\mathrm{Rad}\,\zeta}^{(\mathrm{Sca})}$ and $F_{\mathrm{Rad}\,\zeta}^{(\mathrm{Ext})}$. In fact, from (6.9) we get

$$F_{\mathrm{Rad}\,\zeta}^{(\mathrm{Sca})} = \frac{I_{\mathrm{I}}}{8\pi k_{\mathrm{v}}^2}\mathrm{Re}\sum_{plm}\sum_{p'l'm'} A_{lm}^{(p)*} A_{l'm'}^{(p')} i^{l-l'} I_{\zeta\, lml'm'}^{(pp')}, \tag{6.11a}$$

$$F_{\mathrm{Rad}\,\zeta}^{(\mathrm{Ext})} = -\frac{I_{\mathrm{I}}}{8\pi k_{\mathrm{v}}^2}\mathrm{Re}\sum_{plm}\sum_{p'l'm'} W_{\mathrm{I}\,lm}^{(p)*} A_{l'm'}^{(p')} i^{l-l'} I_{\zeta\, lml'm'}^{(pp')}. \tag{6.11b}$$

Note that in (6.11b) above the apparent undue dependence on r remaining in (6.9) disappeared. The meaning of the superscripts (Ext) and (Sca) will be explained in Sect. 6.2.3.

6.2.2 Radiation Torque by Plane Waves

In order to calculate the radiation torque, we remark that, since $\hat{\mathbf{r}}' \cdot \mathsf{I} \times \hat{\mathbf{r}}' = 0$, the last two terms in $\langle \mathsf{T}_{\mathrm{M}} \rangle$ give no contribution to the integral in (6.2), which, therefore, reduces to

$$\boldsymbol{\Gamma}_{\mathrm{Rad}} = -r'^3 \int_{\Omega'} \left[n^2\left(\hat{\mathbf{r}}' \cdot \mathbf{E}'\right)\mathbf{E}'^* \times \hat{\mathbf{r}}' + \left(\hat{\mathbf{r}}' \cdot \mathbf{B}'\right)\mathbf{B}'^* \times \hat{\mathbf{r}}'\right] \mathrm{d}\Omega'. \tag{6.12}$$

Since the radius of the integration sphere is large, possibly infinite, we can resort to the asymptotic expression of the fields. Nevertheless, the reader is warned that, as regards the scattered field, the customary far zone expression in terms of the scattering amplitude

$$\mathbf{E}_{\mathrm{S}} = E_0 \frac{\exp(ikr)}{r}\mathbf{f}\left(\hat{\mathbf{k}}_{\mathrm{S}}, \hat{\mathbf{k}}_{\mathrm{I}}\right)$$

cannot be used lest to get a vanishing result [8] because $\hat{\mathbf{r}} \cdot \mathbf{f} = 0$. The correct result is obtained by solving the problem of scattering and then by expanding the fields for large r and retaining all terms that give contributions of order $1/r^3$ to the integrand in (6.2).

Assuming the scattered field has been calculated, let us search for its expression in the far zone. For the **H**-multipole fields we get [15]

$$\begin{aligned}\mathbf{H}_{lm}^{(1)} &\to (-\mathrm{i})^{l+1}\frac{e^{\mathrm{i}kr'}}{kr'}\mathbf{Z}_{lm}^{(1)}\,,\\ \mathbf{H}_{lm}^{(2)} &\to \frac{(-\mathrm{i})^{l}}{k^2r'^2}\sqrt{l(l+1)}e^{\mathrm{i}kr'}Y_{lm}\hat{\mathbf{r}}' - \frac{(-\mathrm{i})^{l+1}}{k^2r'^2}e^{\mathrm{i}kr'}\mathbf{Z}_{lm}^{(2)} - \frac{(-\mathrm{i})^{l}}{kr'}e^{\mathrm{i}kr'}\mathbf{Z}_{lm}^{(2)}\,.\end{aligned} \tag{6.13}$$

Analogously, for the **J**-multipole fields we get [15]

$$\begin{aligned}\mathbf{J}_{lm}^{(1)} &\to \frac{1}{kr'}\sin\left(kr'-l\pi/2\right)\mathbf{Z}_{lm}^{(1)}\,,\\ \mathbf{J}_{lm}^{(2)} &\to \frac{\mathrm{i}}{k^2r'^2}\sqrt{l(l+1)}\sin\left(kr'-l\pi/2\right)Y_{lm}\hat{\mathbf{r}}' - \frac{1}{k^2r'^2}\sin(kr'-l\pi/2)\mathbf{Z}_{lm}^{(2)}\\ &\quad - \frac{1}{kr'}\sin\left[kr'-(l-1)\pi/2\right]\mathbf{Z}_{lm}^{(2)}\,.\end{aligned} \tag{6.14}$$

Now, the **H**-fields and **J**-fields enter the integrand in (6.2) through the dot products

$$\begin{aligned}\hat{\mathbf{r}}'\cdot\mathbf{H}_{lm}^{(2)} &= \sqrt{l(l+1)}(-\mathrm{i})^{l}\frac{e^{\mathrm{i}kr'}}{k^2r'^2}Y_{lm} = c_{\mathrm{r}2l}Y_{lm}\,,\\ \hat{\mathbf{r}}'\cdot\mathbf{J}_{lm}^{(2)} &= \mathrm{i}\sqrt{l(l+1)}\frac{\sin\left(kr'-l\pi/2\right)}{k^2r'^2}Y_{lm} = c_{\mathrm{r}1l}Y_{lm}\,,\end{aligned}$$

and through the cross products

$$\begin{aligned}\mathbf{H}_{\bar{l}\bar{m}}^{(\bar{p})*}\times\hat{\mathbf{r}}' &= \mathrm{i}^{\bar{l}+\bar{p}}\frac{e^{-\mathrm{i}kr'}}{kr'}\mathbf{Z}_{\bar{l}\bar{m}}^{(\bar{p})*}\times\hat{\mathbf{r}}' = c_{\mathrm{t}2\bar{l}}^{(\bar{p})}\mathbf{Z}_{\bar{l}\bar{m}}^{(\bar{p})*}\times\hat{\mathbf{r}}'\,,\\ \mathbf{J}_{\bar{l}\bar{m}}^{(\bar{p})*}\times\hat{\mathbf{r}}' &= \frac{\sin\left[kr'-(\bar{l}+\bar{p}-1)\pi/2\right]}{kr'}\mathbf{Z}_{\bar{l}\bar{m}}^{(\bar{p})*}\times\hat{\mathbf{r}}' = c_{\mathrm{t}1\bar{l}}^{(\bar{p})}\mathbf{Z}_{\bar{l}\bar{m}}^{(\bar{p})*}\times\hat{\mathbf{r}}'\,,\end{aligned}$$

where we neglected the further terms that would come from (6.13) and (6.14) because they vanish at infinity to an order higher than the order we have to retain.

At this stage we found it convenient to focus on the spherical components of the radiation torque Γ_μ, that are related to the rectangular components of $\boldsymbol{\Gamma}_{\mathrm{Rad}}$ through [14]

$$\begin{aligned}&\Gamma_{\mathrm{Rad}\,x} = \mathrm{Re}\left[\frac{1}{\sqrt{2}}(\Gamma_{-1}-\Gamma_{1})\right],\quad \Gamma_{\mathrm{Rad}\,y} = \mathrm{Re}\left[\frac{\mathrm{i}}{\sqrt{2}}(\Gamma_{-1}+\Gamma_{1})\right],\\ &\Gamma_{\mathrm{Rad}\,z} = \mathrm{Re}\left(\Gamma_0\right)\,.\end{aligned} \tag{6.15}$$

Accordingly, we note that the integrand in (6.12) gets contributions that can be written as

$$K^{(p\bar{p})}_{\mu;\alpha\bar{\alpha}lm\bar{l}\bar{m}} = \delta_{p2}\, d_{\alpha lm}\, v^{(\bar{p})}_{\mu;\bar{\alpha}\bar{l}\bar{m}}, \tag{6.16}$$

where

$$d_{1lm} = \hat{\mathbf{r}}' \cdot \mathbf{J}^{(2)}_{lm}, \quad d_{2lm} = \hat{\mathbf{r}}' \cdot \mathbf{H}^{(2)}_{lm},$$

$$v^{(\bar{p})}_{\mu;1\bar{l}\bar{m}} = \left(\mathbf{J}^{(\bar{p})*}_{\bar{l}\bar{m}} \times \hat{\mathbf{r}}'\right)_{\mu}, \quad v^{(\bar{p})}_{\mu;2\bar{l}\bar{m}} = \left(\mathbf{H}^{(\bar{p})*}_{\bar{l}\bar{m}} \times \hat{\mathbf{r}}'\right)_{\mu}.$$

The subscripts α and $\bar{\alpha}$ in (6.16) denote terms coming either from the incident field ($\alpha,\ \bar{\alpha} = 1$) or from the scattered field ($\alpha,\ \bar{\alpha} = 2$). Thus, performing in (6.16) the appropriate substitutions and integrating according to (6.2), we get

$$\begin{aligned}
I^{(21)}_{1;\alpha\bar{\alpha}lm\bar{l}\bar{m}} &= \mathrm{i}c_{\mathrm{r}\alpha l}c^{(1)}_{\mathrm{t}\bar{\alpha}\bar{l}}\sqrt{\frac{2l+1}{2\bar{l}+1}}C(1,l,\bar{l};0,0)\delta_{m,\bar{m}-1} \\
&\quad\times\big[C(1,\bar{l},\bar{l};1,\bar{m}-1)C(1,l,\bar{l};0,m,\bar{m}-1) \\
&\quad-C(1,\bar{l},\bar{l};0,\bar{m})C(1,l,\bar{l};1,m,\bar{m})\big], \\
I^{(21)}_{0;\alpha\bar{\alpha}lm\bar{l}\bar{m}} &= \mathrm{i}c_{\mathrm{r}\alpha l}c^{(1)}_{\mathrm{t}\bar{\alpha}\bar{l}}\sqrt{\frac{2l+1}{2\bar{l}+1}}C(1,l,\bar{l};0,0)\delta_{m,\bar{m}} \\
&\quad\times\big[C(1,\bar{l},\bar{l};1,\bar{m}-1)C(1,l,\bar{l};-1,m,\bar{m}-1) \\
&\quad-C(1,\bar{l},\bar{l};-1,\bar{m}+1)C(1,l,\bar{l};1,m,\bar{m}+1)\big], \\
I^{(21)}_{-1;\alpha\bar{\alpha}lm\bar{l}\bar{m}} &= \mathrm{i}c_{\mathrm{r}\alpha l}c^{(1)}_{\mathrm{t}\bar{\alpha}\bar{l}}\sqrt{\frac{2l+1}{2\bar{l}+1}}C(1,l,\bar{l};0,0)\delta_{m,\bar{m}+1} \\
&\quad\times\big[C(1,\bar{l},\bar{l};0,\bar{m})C(1,l,\bar{l};-1,m,\bar{m}) \\
&\quad-C(1,\bar{l},\bar{l};-1,\bar{m}+1)C(1,l,\bar{l};0,m,\bar{m}+1)\big],
\end{aligned}$$

and

$$I^{(22)}_{\mu;\alpha\bar{\alpha}lm\bar{l}\bar{m}} = c_{\mathrm{r}\alpha l}c^{(2)}_{\mathrm{t}\bar{\alpha}\bar{l}}C(1,l,\bar{l};\mu,\bar{m}-\mu)\delta_{l\bar{l}}\delta_{m,\bar{m}-\mu}, \tag{6.17}$$

where the C's denote Clebsch-Gordan coefficients [14]. Finally, collecting all the terms one obtains

$$\Gamma_{\mu} = -\frac{n^2r^3}{8\pi}\sum_{\alpha\bar{\alpha}}\left[\sum_{\bar{p}}\sum_{lm}\sum_{\bar{l}\bar{m}} I^{(2\bar{p})}_{\mu;\alpha\bar{\alpha}lm\bar{l}\bar{m}}\left(a^{(2)}_{\alpha lm}a^{(\bar{p})*}_{\bar{\alpha}\bar{l}\bar{m}} + a^{(1)}_{\alpha lm}a^{(\bar{p}')*}_{\bar{\alpha}\bar{l}\bar{m}}\right)\right], \tag{6.18}$$

where $\bar{p}' \neq \bar{p}$, $a^{(p)}_{1lm} = W^{(p)}_{\mathrm{I}lm}$, and $a^{(p)}_{2lm} = A^{(p)}_{lm}$.

At this stage we remark that (6.18) could be used for a brute force calculation of the spherical components of the radiation torque. However, we can obtain a more efficient formula through a detailed inspection of the terms present. In fact, (6.18) is built as a sum of contributions with different α, $\bar{\alpha}$, and $\bar{p}$ or $\bar{p}'$.

- The contributions with $\alpha = \bar{\alpha} = 1$ are just vanishing both for $\bar{p} = 2$ or $\bar{p} = 1$ and for $\bar{p}' = 1$ or $\bar{p}' = 2$. This result, that is identical to the one of Marston and Crichton [16], comes from the fact that these terms describe the flux of the momentum of the Maxwell stress tensor through a closed surface in the absence of any scatterer.
- The contribution with $\alpha = \bar{\alpha} = 2$ vanishes both for $\bar{p} = 1$ and for $\bar{p}' = 2$.
- The contributions for $\alpha \neq \bar{\alpha}$ depend on r, i.e., on the radius of the spherical surface of integration. Nevertheless, those with $\bar{p} = 1$ are identical but of opposite sign and cancel each other; on the other hand, the sum of those with $\bar{p} = 2$ turns out to be independent of r. Analogous considerations hold true for the contributions with $\bar{p}' = 2$ and $\bar{p}' = 1$, respectively. This result is a consequence of the fact that the torque cannot depend on the choice of the surface of integration.

Now, separating in (6.18) the sum over $\alpha \neq \bar{\alpha}$ from the one over $\alpha = \bar{\alpha}$, we can write

$$\Gamma_\mu = \Gamma_\mu^{(\mathrm{Ext})} - \Gamma_\mu^{(\mathrm{Sca})}, \tag{6.19}$$

with

$$\Gamma_\mu^{(\mathrm{Sca})} = -\frac{I_\mathrm{I}}{8\pi n k_\mathrm{V}^3} \sum_{plm} s_{\mu;lm} A_{l,m-\mu}^{(p)} A_{lm}^{(p)*}, \tag{6.20a}$$

$$\Gamma_\mu^{(\mathrm{Ext})} = \frac{I_\mathrm{I}}{8\pi n k_\mathrm{V}^3} \sum_{plm} s_{\mu;lm} W_{\mathrm{I}l,m-\mu}^{(p)} A_{lm}^{(p)*}, \tag{6.20b}$$

where

$$s_{-1;lm} = -\sqrt{\frac{(l-m)(l+1+m)}{2}}, \quad s_{1;lm} = \sqrt{\frac{(l+m)(l+1-m)}{2}},$$

$$s_{0;lm} = -m\ .$$

We want to stress that the radiation torque does not vanish for a homogeneous sphere provided the polarization is elliptical and the sphere is absorbing. This result has been first proved by Marston and Chricton [16] just in terms of multipole fields expansion. The radiation torque due to an elliptically polarized plane wave does not vanish either for an absorbing scatterer with axial symmetry provided the incidence is along the cylindrical axis.

6.2.3 Contributions to Radiation Force and Torque

Although in our approach the radiation force and torque have been separated into two contributions labeled (Sca) and (Ext), there is no similarity with the customary separation into a field gradient contribution and a scattering contribution. The latter contributions, in fact, arise when exploiting the dipole approximation [17], whereas the separation effected in (6.11a) and (6.11b) as well as in (6.20a) and (6.20b) can be tracked back to (6.1) and (6.2).

Equation (6.1), due to the structure of the Maxwell stress tensor, includes $|\mathbf{E}'_{\mathrm{S}}|^2$ and $\mathbf{E}'^{*}_{\mathrm{I}} \cdot \mathbf{E}'_{\mathrm{S}}$ as well as the corresponding terms from the magnetic field. When these terms are expanded as a series of multipole fields we get $F^{(\mathrm{Sca})}_{\mathrm{Rad}\,\zeta}$, that depends on the multipole amplitudes of the scattered field only, and $F^{(\mathrm{Ext})}_{\mathrm{Rad}\,\zeta}$ that depends on the multipole amplitudes both of the incident and of the scattered fields, in analogy to what one finds for the scattering cross section and for the extinction cross section of a particle, respectively [10]. As a consequence, $F_{\mathrm{Rad}\,\zeta}$ can loosely be related to the absorptivity of a particle.

In turn, (6.2) gets from the Maxwell tensor the terms $(\mathbf{r}' \cdot \mathbf{E}'_{\mathrm{S}})(\mathbf{E}'^{*}_{\mathrm{S}} \times \mathbf{r}')$ and $(\mathbf{r}' \cdot \mathbf{E}'_{\mathrm{I}})(\mathbf{E}'^{*}_{\mathrm{S}} \times \mathbf{r}')$, and the corresponding terms from the magnetic field. Again, the multipole expansion of these terms yields just $\Gamma^{\mathrm{Sca}}_{\mathrm{Rad}\,\mu}$ and $\Gamma^{\mathrm{Ext}}_{\mathrm{Rad}\,\mu}$ which depend on the multipole amplitudes of the scattered field only and by the multipole amplitudes of the incident and of the scattered field, respectively. Thus, considerations analogous to those made above for the radiation force hold true also for the radiation torque [18, 19]: in particular, the torque exerted by an elliptically polarized plane wave on a spherical scatterer can be explicitly written in terms of the difference of the extinction and of the scattering cross section [16].

6.3 Radiation Force and Torque on Aggregated Spheres

The theory that we summarized in Sect. 6.2 can be applied to particles of any shape, provided we are able to calculate their transition matrix and the incident field is well represented by simple plane waves. This is often the case in an astrophysical environment, because the cosmic dust grains are well represented either by single spheres or by aggregates of spheres [20]. The component of the radiation force along the direction of propagation is the well-known *radiation pressure*, whereas the transverse components, although less well-known, produce dynamical effects whose relevance has been recognized in the last few years [21]. In turn, the radiation torque may produce the so-called *superthermal spin-up* whose relevance for the alignment of the cosmic dust grains has been investigated, e.g., by Purcell [22] and by Draine and Weingartner [23]. Although we dealt with the calculation of the radiation force and torque on reliable models of cosmic dust grains [24], in the present context we prefer to expound a more illuminating study on clusters of selected geometry. Our aim is, indeed, to highlight the dependence of radiation force and torque on the structure and refractive index of particles of known geometry and dielectric properties.

6.3.1 Radiation Pressure on Aggregated Spheres

We apply the theory to binary aggregates of latex spheres whose dielectric function, according to Ma et al. [25] has the value $\varepsilon = 2.490087 + \mathrm{i}0.001578$ at $\lambda = 600$ nm. More precisely, we considered a reference sphere with radius 200 nm and the aggregate composed by two mutually contacting spheres whose radius is 158.77 nm so that their total volume equals that of the reference sphere. The frame Σ' attached to the aggregate is chosen to coincide to the laboratory frame Σ and the point of contact of the spheres always coincide with the origin, whereas the axis of the aggregate lies in the z'-x' plane. We considered four configurations each characterized by the angle ϑ between the axis of the aggregate and the z' axis: (i) $\vartheta = 90°$, (ii) $\vartheta = 60°$, (iii) $\vartheta = 30°$, and (iv) $\vartheta = 0°$. We assume the incident field to be linearly polarized and that the plane of reference is the meridional plane through the $z = z'$ axis and the incident wavevector $\mathbf{k}_\mathrm{I}$ whose direction is individuated by the polar angles ϑ_I and φ_I. We decompose the radiation force along the triplet formed by $\hat{\mathbf{k}}_\mathrm{I} = \hat{\mathbf{v}}_3$, $\hat{\boldsymbol{\vartheta}}_\mathrm{I} = \hat{\mathbf{v}}_1$, and $\hat{\boldsymbol{\varphi}}_\mathrm{I} = \hat{\mathbf{v}}_2$. The component $F_k = \mathbf{F}_\mathrm{Rad} \cdot \hat{\mathbf{k}}_\mathrm{I}$ is the well-known radiation pressure, whereas the transverse components are $F_{\mathrm{T}\eta} = \mathbf{F}_\mathrm{Rad} \cdot \hat{\mathbf{v}}_\eta$, $\eta = 1, 2$. Actually, the quantities that we report in the following figures are $\mathscr{F}_k = 8\pi F_k / I_\mathrm{I}$ and $\mathscr{F}_\mathrm{T} = 8\pi \left(F_{\mathrm{T}1}^2 + F_{\mathrm{T}2}^2\right)^{1/2} / I_\mathrm{I}$, as well as their *axial averages* around the z axis of Σ. All these quantities are given in $\mu\mathrm{m}^2$ and, although they depend on the polarization of the incident field, are reported only for polarization parallel to the plane of scattering. The interested reader can find a more complete set of figures in [9, 10]. As our calculations are based on the multipole expansion of the fields, their convergence must carefully be checked. Actually the size parameter of the smallest mathematical sphere that includes a whole binary aggregate is $x = 3.1$, so that convergence to 4 significant digits required $l_\mathrm{M} = 8$.

We start by stating that the radiation pressure on the reference sphere is $0.125027\,\mu\mathrm{m}^2$, whereas the transverse components vanish due to the spherical symmetry. On the contrary, the results reported in Fig. 6.2 show that, as expected, the transverse components on the binary aggregate do not vanish for any value of φ_I and their value is a significant fraction of the radiation pressure. These results are confirmed by the axial averages we report in Fig. 6.3 for all the configurations we considered. Perhaps the most interesting results are those reported in Fig. 6.4 where we study the dependence of the axial averages on the absorptivity of the aggregates. In other words, we considered particles whose real part of the dielectric function is that of Ma et al. [25], whereas the imaginary part changes from 0 to 0.1. In fact, the transverse components of the radiation force show a weak dependence on $\mathrm{Im}(\varepsilon)$, whereas the radiation pressure has a comparatively stronger dependence. This behav-

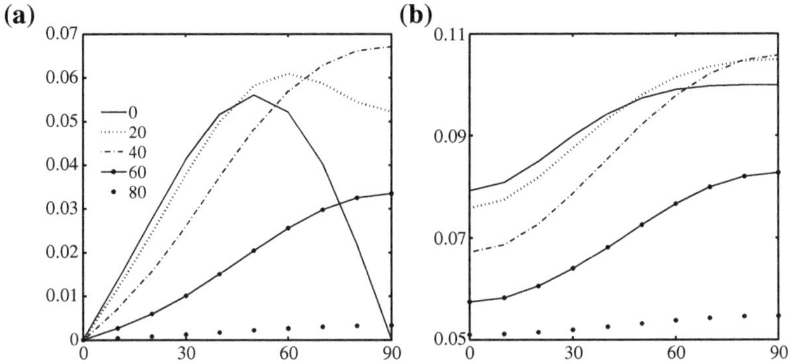

Fig. 6.2 $\mathscr{F}_{\eta\mathrm{T}}$ in **a**, and $\mathscr{F}_{\eta k}$, in **b**, in $\mu\mathrm{m}^2$ for the binary cluster in configuration (i), for polarization of the incident wave parallel to the plane of reference ($\eta = 1$), as a function of ϑ_{I} for the values of φ_{I} indicated in the inset in **a**

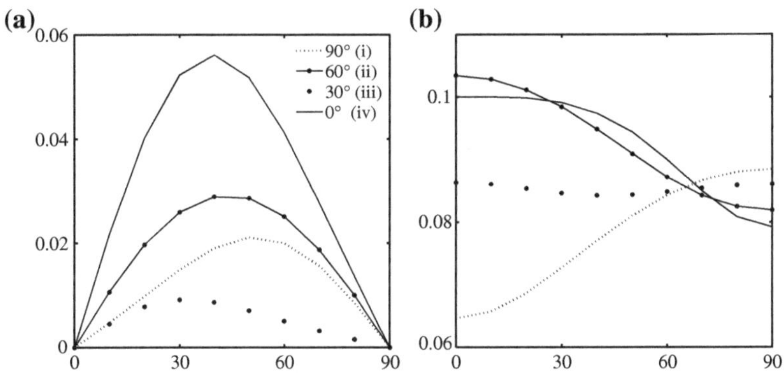

Fig. 6.3 Axial averages of $\mathscr{F}_{\eta\mathrm{T}}$ in **a**, and of $\mathscr{F}_{\eta k}$ in **b** in $\mu\mathrm{m}^2$ for the binary cluster in configurations (i), (ii), (iii), and (iv), for polarization of the incident wave parallel to the plane of reference ($\eta = 1$), as a function of ϑ_{I}. The averages over random orientation give $\mathscr{F}_{\eta k} = 0.08543\,\mu\mathrm{m}^2$ both for $\eta = 1$ and for $\eta = 2$

ior may be important for understanding the possible expulsion of cosmic dust grains from galaxies and protoplanetary disks [26].

6.3.2 Radiation Torque on Aggregated Spheres

The absorptivity of the particles is important also for understanding their rotational behavior when driven by the radiation torque. To show this we report in the following figures the adimensional vector

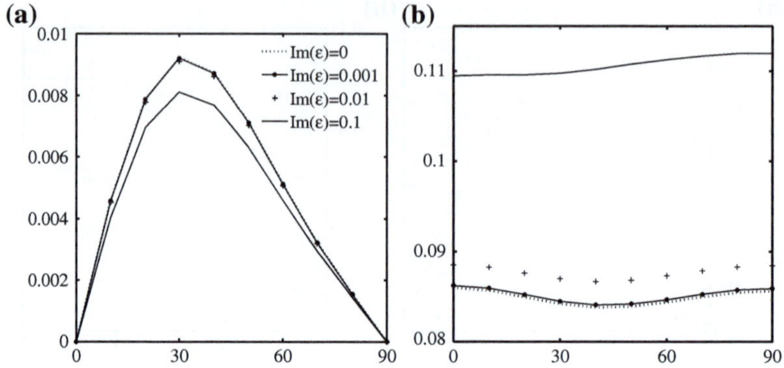

Fig. 6.4 Dependence on $\mathrm{Im}(\varepsilon)$ of the axial averages of $\mathscr{F}_{\eta\mathrm{T}}$ in **a**, and of $\mathscr{F}_{\eta k}$ in **b** in $\mu\mathrm{m}^2$ for the binary cluster in configuration (iii) for polarization parallel to the plane of reference ($\eta = 1$) as a function of ϑ_I. The averages over random orientation of $\mathscr{F}_{\eta k}$ (in $\mu\mathrm{m}^2$) are $0.08595,\ 0.08621,\ 0.08499$, and 0.1095 for $\mathrm{Im}(\varepsilon) = 0,\ 0.001,\ 0.01$, and 0.1, respectively, both for $\eta = 1$ and for $\eta = 2$. Note that in **a** the curves for $\mathrm{Im}(\varepsilon) = 0.0,\ 0.001$, and 0.01 are almost coincident on the scale of the figure

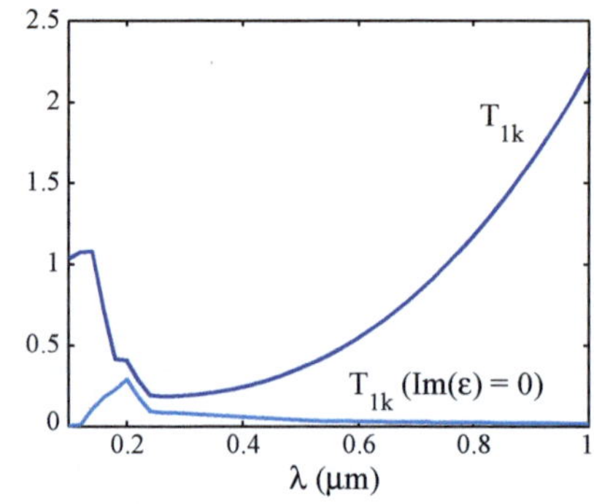

Fig. 6.5 Component of $\mathbf{T}_1$ along $\mathbf{k}_\mathrm{I}$ for a binary cluster composed of astronomical silicates both when the imaginary part of their dielectric function has its actual value [27] and when it is arbitrarily set to zero. The axis of the cluster is orthogonal to $\mathbf{k}_\mathrm{I}$

$$\mathbf{T}_\eta = \frac{8\pi k}{n^2 I_\mathrm{I} \sigma_{\mathrm{T}\eta}} \boldsymbol{\Gamma}_{\mathrm{Rad}\,\eta}, \tag{6.21}$$

where $\boldsymbol{\Gamma}_{\mathrm{Rad}\,\eta}$ and $\sigma_{\mathrm{T}\eta}$ are the radiation torque and the extinction cross section calculated for incident light with polarization η, and $n = 1$. The incident field is circularly polarized and we show the results for $\eta = 1$ only, because, on account of the symmetry of the scatterer, the results for $\eta = 2$ are identical, except for the sign.

First, we consider a binary aggregate of identical spheres with radius 50 nm, composed of astronomical silicates [27]. The origin of the frame of reference lies at the point of contact of the spheres. Since, according to Sect. 6.2.2, the torque depends on the absorptivity of the particles, we report in Fig. 6.5 the quantity $T_{\eta k} = \mathbf{T}_\eta \cdot \hat{\mathbf{k}}_\mathrm{I}$ as a function of the wavelength for the binary cluster with its axis perpendicular to $\mathbf{k}_\mathrm{I}$ both in the case in which the imaginary part of the dielectric function is set to zero or assumes its actual value. As expected, the transverse components of $\boldsymbol{\Gamma}_{\mathrm{Rad}\,\eta}$,

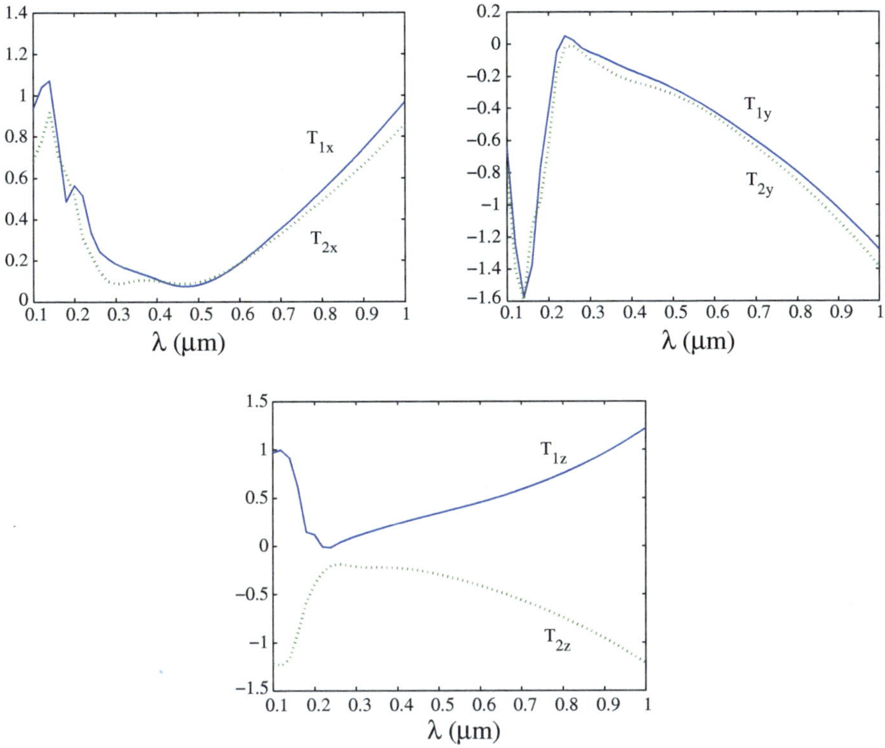

Fig. 6.6 Cartesian components of $\mathbf{T}_1$ and $\mathbf{T}_2$ for the five-spheres cluster whose geometry is given in Table 6.1. The direction of the incident wavevector is along the z axis

Table 6.1 Coordinates of the centers of the spheres in nm

x	y	z
0.000	0.000	0.000
70.710	−70.710	0.000
−5.982	50.218	−86.851
70.855	81.028	−143.837
−61.734	7.789	78.923

i.e., those in the plane orthogonal to $\mathbf{k}_\mathrm{I}$, were found to be zero. We do not report the results for the case in which the axis of the aggregate is parallel to $\mathbf{k}_\mathrm{I}$, because, in this case, $\boldsymbol{\Gamma}_{\mathrm{Rad}\,\eta} \cdot \hat{\mathbf{k}}_\mathrm{I}$ is nonvanishing only for complex refractive index, and changes its sign with the change of polarization. Even in this case the transverse components are rigorously zero for symmetry reasons. Anyway, the results in Fig. 6.5 show the great importance of the absorptivity of the particles on the value of the torque, and in our opinion do not deserve further comments.

As a second example we now consider a more complex cluster composed of five spheres identical to each other and to the spheres we used above to compose the binary cluster. The geometry is chosen so that no symmetry is present, and the coordinates of the centers of the spheres are listed in Table 6.1. The incident wavevector $\mathbf{k}_\mathrm{I}$ is parallel to the z axis ($\hat{\mathbf{k}}_\mathrm{I} \equiv \hat{\mathbf{e}}_z$) and circular polarization is assumed. In Fig. 6.6 we report the cartesian components of $\mathbf{T}_\eta$ both for $\eta = 1$ and for $\eta = 2$. We assume here that the refractive index is just that of astronomical silicates [27]. We first notice that the x and y components of $\boldsymbol{\Gamma}_{\mathrm{Rad}}$ do not change their sign with changing polarization, unlike the z component which does change sign. Of course, the lack of symmetry prevents the curves of each component to coincide with changing polarization. Nevertheless, the most striking result is the coincidence of the axial average around the z axis, $\langle \boldsymbol{\Gamma}_{\mathrm{Rad}\,\eta} \cdot \hat{\mathbf{k}}_\mathrm{I} \rangle$, with $\boldsymbol{\Gamma}_{\mathrm{Rad}\,\eta} \cdot \hat{\mathbf{k}}_\mathrm{I}$. This result would be quite evident for the axial average of the binary cluster with its axis along $\mathbf{k}_\mathrm{I}$. In the present case, because of the lack of symmetry, the occurrence of this coincidence deserves a few additional comments. It is not easy to extract any conclusion from the results on the basis of the formulas of Sect. 6.2.2. Nevertheless, a few heuristic remarks can be drawn with the help of a complete set of calculations, in the sense that they were also performed using linearly polarized incident radiation. Both when we assume circular and when we assume linear polarization, the components of $\boldsymbol{\Gamma}_{\mathrm{Rad}\,\eta}$ in a plane orthogonal to $\mathbf{k}_\mathrm{I}$ are found, in general, nonzero and different from each other, whereas their axial averages around $\mathbf{k}_\mathrm{I}$ do vanish. Actually, averaging around $\mathbf{k}_\mathrm{I}$ makes the average particle akin to an axially symmetric particle, for which the mentioned result is to be expected. Furthermore, assuming circular polarization, we found $\langle \boldsymbol{\Gamma}_{\mathrm{Rad}\,\eta} \cdot \hat{\mathbf{k}}_\mathrm{I} \rangle = \boldsymbol{\Gamma}_{\mathrm{Rad}\,\eta} \cdot \hat{\mathbf{k}}_\mathrm{I}$ that follows at once, if one considers the behavior of the field of a circularly polarized wave. On the other hand, in case linear polarization is assumed, the axial average of $\boldsymbol{\Gamma}_{\mathrm{Rad}\,\eta}$ around $\mathbf{k}_\mathrm{I}$ turns out to be independent of η. In fact, as regards the axial averaging, the different polarization is seen only as a different orientation of the particle around z'. Of course, in general, the axial average $\langle \boldsymbol{\Gamma}_{\mathrm{Rad}\,\eta} \cdot \hat{\mathbf{k}}_\mathrm{I} \rangle \neq \boldsymbol{\Gamma}_{\mathrm{Rad}\,\eta} \cdot \hat{\mathbf{k}}_\mathrm{I}$ for linear polarization.

At this stage, since $\boldsymbol{\Gamma}_{\mathrm{Rad}}$ depends on the orientation of the particle, the problem arises out of the stability of the rotational motion driven by the electromagnetic torque. Actually, we performed a study of this problem using as a model just the aggregate whose geometry is given in Table 6.1. The interested reader is referred to [28] for a detailed discussion of the problem.

6.4 Laser Beams as a Superposition of Plane Waves

The theory of Sect. 6.2 can be extended to deal with the issue of trapping and manipulation of microsized and nanosized particles. The extension is necessary because the so-called optical trapping occurs when a laser beam is focalized through a high numerical aperture lens. The radiation force may then trap a particle within the focal region. At the same time, the torque exerted by the same laser beam may force the particle to assume particular orientations or to perform rotational motions. Now, a laser beam is quite different from a plane wave, nevertheless, whatever its transverse profile, it can be represented as a superposition of plane waves with the same frequency but with a different direction of propagation. This is the so-called *angular spectrum representation* of an optical field. Moreover, when the angular spectrum representation is considered in the far field zone, one may exploit the geometrical optics to describe the focalization in terms of rays.

6.4.1 The Angular Spectrum Representation

Let us consider a monochromatic field $\mathbf{E}(\mathbf{r})$ that propagates within a region of space that contains no sources and is filled by a homogeneous isotropic medium of real refractive index n. The propagation of such a field is described by the Helmholtz equation. Let us now choose a rectangular system of axes, the direction of the z axis being quite arbitrary, and perform the bidimensional space Fourier transform of $\mathbf{E}(\mathbf{r})$ in a plane orthogonal to z,

$$\hat{\mathbf{E}}(k_x, k_y; z) = \frac{1}{4\pi^2} \int_{-\infty}^{\infty} \mathbf{E}(x, y, z) \exp[-\mathrm{i}(k_x x + k_y y)] \, \mathrm{d}x \mathrm{d}y$$

with inverse

$$\mathbf{E}(x, y, z) = \int_{-\infty}^{\infty} \hat{\mathbf{E}}(k_x, k_y; z) \exp[\mathrm{i}(k_x x + k_y y)] \, \mathrm{d}k_x \mathrm{d}k_y \; . \tag{6.22}$$

Substituting (6.22) into the Helmholtz equation and putting

$$k_z = (k^2 - k_x^2 - k_y^2)^{1/2}, \quad \mathrm{Im}(k_z) \geq 0, \tag{6.23}$$

it is an easy matter to see that the space Fourier transform in the plane $z = 0$ is related to the one in the plane at $z \neq 0$ by

$$\hat{\mathbf{E}}(k_x, k_y; z) = \hat{\mathbf{E}}(k_x, k_y; 0) \exp(\pm \mathrm{i} k_z z) \; .$$

As a result, we can write

$$\mathbf{E}(x, y, z) = \int_{-\infty}^{\infty} \hat{\mathbf{E}}(k_x, k_y; 0) \exp[\mathrm{i}(k_x x + k_y y \pm k_z z)] \, \mathrm{d}k_x \, \mathrm{d}k_y, \tag{6.24}$$

that is, by definition, the *angular spectrum representation* of the field $\mathbf{E}(\mathbf{r})$. Note that the $\pm$ sign in front of k_z is due to the fact that the field can propagate either in the positive or in the negative z direction. However, henceforth we consider only beams propagating along the positive z axis, so that only the positive sign will be retained in front of k_z. We stress that, although n is real, k_z may be imaginary, so that (6.24) describes $\mathbf{E}(\mathbf{r})$ as a superposition of plane waves (with real k_z) and of evanescent waves (those with $\mathrm{Im}(k_z) \neq 0$).

6.4.2 Far Field in the Angular Spectrum Representation

We will now show that the geometrical optics stems from the angular spectrum representation when we consider the field in the far zone, i.e., at a large distance from its source, that is assumed to be of finite extent. Let us thus consider an optical field in the plane $z = 0$ and let us use the angular spectrum representation to describe how the field transforms when it is considered in a plane at $z = z_\infty$ at large distance from the chosen origin. To this end we need to find the limiting expression of (6.24) for $\mathbf{r} \to \mathbf{r}_\infty$. Let us introduce the unit vector

$$\hat{\mathbf{s}} = s_x\hat{\mathbf{e}}_x + s_y\hat{\mathbf{e}}_y + s_z\hat{\mathbf{e}}_z = x/r\hat{\mathbf{e}}_x + y/r\hat{\mathbf{e}}_y + z/r\hat{\mathbf{e}}_z$$

in the direction of $\mathbf{r}_\infty$. Of course $r = (x^2 + y^2 + z^2)^{1/2}$ is the distance of $\mathbf{r}_\infty$ from the origin. Then the field in the far zone in the direction of $\hat{\mathbf{s}}$ is given by

$$\mathbf{E}_\infty(s_x, s_y, s_z) = \lim_{kr\to\infty} \int_{(k_x^2+k_y^2)\le k^2} \hat{\mathbf{E}}(k_x, k_y; 0) \mathrm{e}^{ikr(k_x s_x/k + k_y s_y/k + k_z s_z/k)}\, \mathrm{d}k_x\, \mathrm{d}k_y \,. \tag{6.25}$$

We remark that in (6.25) the evanescent waves, due to their fast decay, do not contribute to the field at infinity so that the limits of integration have accordingly been set. The integral in (6.25) can be performed through the stationary phase method [29] with the result

$$\mathbf{E}_\infty(s_x, s_y, s_z) = -2\pi \mathrm{i} k s_z \hat{\mathbf{E}}(k s_x, k s_y; 0)\frac{\mathrm{e}^{ikr}}{r} \,. \tag{6.26}$$

The field (6.26) has the typical form of the scattered field in the far zone with $\hat{\mathbf{E}}$ playing the role of the scattering amplitude. Note, however, that the argument of $\hat{\mathbf{E}}$ is parallel to $\mathbf{r}_\infty$, so that a single plane wave, the one that propagates along $\hat{\mathbf{s}}$, enters the description of the far field. This is enough to justify the use of the geometrical optics for any optical field. As a consequence, on account that $\hat{\mathbf{s}}$ is fully determined by k_x and k_y only, we can write

$$\hat{\mathbf{E}}(k_x, k_y; 0) = \frac{\mathrm{i} r \mathrm{e}^{-ikr}}{2\pi k_z}\mathbf{E}_\infty(k_x, k_y)$$

and the angular spectrum representation becomes

$$\mathbf{E}(x, y, z) = \frac{\mathrm{i}r\mathrm{e}^{-\mathrm{i}kr}}{2\pi k_z} \int_{(k_x^2+k_y^2)\leq k^2} \mathbf{E}_\infty(k_x, k_y)\mathrm{e}^{\mathrm{i}(k_x x+k_y y+k_z z)}\, \mathrm{d}k_x\, \mathrm{d}k_y\, . \tag{6.27}$$

Equation (6.27) can now be interpreted as a description of the field in terms of rays.

6.5 Radiation Force from a Highly Focalized Laser Beam

Exploiting the geometrical optics simplifies the description of the focalization of a paraxial optical beam, i.e., of an optical field whose angular spectrum representation includes only **k**-vectors that are almost parallel to a fixed direction. For our purposes we choose this direction to coincide with the axis of an aplanatic optical system, i.e., a stigmatic optical system that satisfies the Abbe sine condition. Let f be the focal length of the lens, whose exit pupil has the radius $f \sin\vartheta_{\mathrm{Max}}$; of course, ϑ_{Max} is the angle under which the radius of the exit pupil is seen from the focus and is thus related to the numerical aperture of the lens by

$$\mathrm{NA} = n \sin\vartheta_{\mathrm{Max}},$$

where n is the refractive index of the medium that fills the image space. We consider, as an example, the focalization of a gaussian TEM_{00} beam, whose waist, of radius w_0, is taken to coincide with the entrance pupil of the lens. Then, according to Novotny and Hecht [17], the field at any point within the focal region can be written in the angular spectrum representation as

$$\mathbf{E}(\mathbf{r}) = \int\limits_{k_x^2+k_y^2\leq k_\perp^2} E_{\mathrm{PW}}(\hat{\mathbf{k}})\hat{\mathbf{u}}_{\hat{\mathbf{k}}}\mathrm{e}^{\mathrm{i}\mathbf{k}\cdot\mathbf{r}}\, \mathrm{d}k_x\, \mathrm{d}k_y, \tag{6.28}$$

where $k_\perp = k\sin\vartheta_{\mathrm{Max}}$, $\mathbf{k}$ has polar angles $\vartheta_{\mathbf{k}}$ and $\varphi_{\mathbf{k}}$, $\hat{\mathbf{u}}_{\hat{\mathbf{k}}} = \hat{\mathbf{u}}(\vartheta_{\mathbf{k}}, \varphi_{\mathbf{k}})$; the limits of integration ensure that only the rays that actually traverse the exit pupil of the optical system are considered. In (6.28)

$$E_{\mathrm{PW}}(\hat{\mathbf{k}}) = E_0\, i f \frac{\mathrm{e}^{ikf}}{2\pi k}\sqrt{\frac{n_{\mathrm{I}}}{n}}\, (\cos\vartheta_{\mathbf{k}})^{-1/2} f_w,$$

where n_{I} is the refractive index of the object space and f_w is the apodization function

$$f_w = \exp\left[-\frac{1}{f_0^2}\frac{\sin^2\vartheta_{\mathbf{k}}}{\sin^2\vartheta_{\mathrm{Max}}}\right], \tag{6.29}$$

in which f_0 is the filling factor

$$f_0 = \frac{w_0}{f \sin \vartheta_{\mathrm{Max}}} .$$

The apodization function (6.29) is common also to the higher gaussian modes TEM_{10} and TEM_{01}, and when $f_w = 1$, i.e., for $w_0 \to \infty$, one recovers the description of the field of Richards and Wolf [30].

Often, the image space is not filled by a single homogeneous medium but rather by two homogeneous media, of refractive indexes n and n_{F}, separated by a plane interface orthogonal to the optical axis. We assume the interface to be located at $z_{\mathrm{s}} = -D$ between the exit pupil and the nominal focus. Hereafter, the quantities considered in the region $z > z_{\mathrm{s}}$ will be characterized by the index F, even when, strictly speaking, this notation would not be necessary. For instance, since k_x and k_y are unaffected by the refraction, we have $k_x = k_{\mathrm{F}x}$ and $k_y = k_{\mathrm{F}y}$, and thus also $k_\perp = k_{\mathrm{F}\perp}$. On the contrary, k_z is affected by the refraction according to

$$k_{\mathrm{F}z} = \left(k_{\mathrm{F}}^2 - k_x^2 - k_y^2\right)^{1/2} = \left[\left(\frac{n_{\mathrm{F}}}{n}\right)^2 k^2 - k_x^2 - k_y^2\right]^{1/2} .$$

The refraction of the rays through the interface introduces a spherical aberration and a polarization-dependent transmission that can be taken into account by the Fresnel coefficients T_η [5]; $\eta = 1$ stands for polarization parallel and $\eta = 2$ for polarization perpendicular to the plane of incidence that, for each ray, coincides with the meridional plane defined by the $\mathbf{k}$ or by the $\mathbf{k}_{\mathrm{F}}$ vector, and the z axis (optical axis). The decomposition of the polarization vectors of each of the plane waves in (6.28) into their components parallel and perpendicular to the plane of incidence can be effected by introducing for each plane of incidence a pair of unit vectors $\hat{\mathbf{u}}_{\eta\hat{\mathbf{k}}}$. Thus, we have

$$\hat{\mathbf{u}}_{\hat{\mathbf{k}}} = \sum_\eta \left(\hat{\mathbf{u}}_{\hat{\mathbf{k}}} \cdot \hat{\mathbf{u}}_{\eta\mathbf{k}}\right) \hat{\mathbf{u}}_{\eta\mathbf{k}} = \sum_\eta c_\eta \hat{\mathbf{u}}_{\eta\hat{\mathbf{k}}} \tag{6.30}$$

and, the refraction through the interface yields

$$\mathbf{E}(\mathbf{r}) = \int\limits_{k_{\mathrm{F}x}^2 + k_{\mathrm{F}y}^2 \le k_{\mathrm{F}\perp}^2} E_{\mathrm{FPW}}\left(\hat{\mathbf{k}}_{\mathrm{F}}\right) \sum_\eta c_\eta T_\eta(\vartheta_{\mathbf{k}}) \hat{\mathbf{u}}_{\eta\hat{\mathbf{k}}_{\mathrm{F}}} \mathrm{e}^{\mathrm{i}\mathbf{k}_{\mathrm{F}}\cdot\mathbf{r}} \, \mathrm{d}k_{\mathrm{F}x} \, \mathrm{d}k_{\mathrm{F}y}, \tag{6.31}$$

where

$$E_{\mathrm{FPW}}\left(\hat{\mathbf{k}}_{\mathrm{F}}\right) = \exp[\mathrm{i}z_{\mathrm{s}}(k_z - k_{\mathrm{F}z})] E_{\mathrm{PW}}(\hat{\mathbf{k}}) .$$

We stress that, although $\vartheta_{\mathbf{k}}$ appears in place of $\vartheta_{\mathbf{k}\mathrm{F}}$ in (6.31), Snell's law grants an immediate relation between these angles. Moreover, the same law makes the apodization function (6.29) insensitive to the use of $\vartheta_{\mathbf{k}}$ or of $\vartheta_{\mathbf{k}\mathrm{F}}$ (Fig. 6.7).

We are now able to calculate the radiation force that the field exerts on a particle at O'. To this end we rewrite (6.31) as

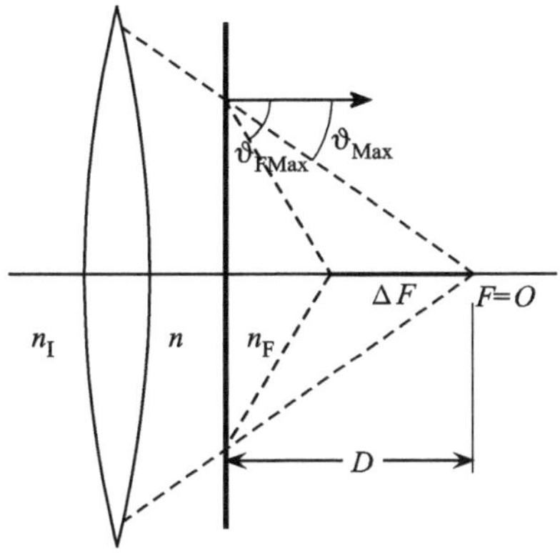

Fig. 6.7 Sketch of the focalization of an aberrated laser beam

$$\mathbf{E}'(\mathbf{r}') = \int_{k_{\mathrm{F}x}^2+k_{\mathrm{F}y}^2 \le k_{\mathrm{F}\perp}^2} E_{\mathrm{FPW}}\left(\hat{\mathbf{k}}_{\mathrm{F}}\right) \mathrm{e}^{\mathrm{i}\mathbf{k}_{\mathrm{F}}\cdot\mathbf{R}_{O'}} \times \sum_{\eta} c_\eta T_\eta(\vartheta_{\mathbf{k}}) \hat{\mathbf{u}}_{\eta\hat{\mathbf{k}}_{\mathrm{F}}} \mathrm{e}^{\mathrm{i}\mathbf{k}_{\mathrm{F}}\cdot\mathbf{r}'} \, \mathrm{d}k_{\mathrm{F}x} \, \mathrm{d}k_{\mathrm{F}y}, \tag{6.32}$$

and perform the multipole expansion

$$\hat{\mathbf{u}}_{\eta\hat{\mathbf{k}}\mathrm{F}} \mathrm{e}^{\mathrm{i}\mathbf{k}_{\mathrm{F}}\cdot\mathbf{r}'} = \sum_{plm} \mathbf{J}_{lm}^{(p)}(\mathbf{r}', k_{\mathrm{F}}) W_{lm}^{(p)}\left(\hat{\mathbf{u}}_{\eta\hat{\mathbf{k}}_{\mathrm{F}}}, \hat{\mathbf{k}}_{\mathrm{F}}\right) .$$

Since the $\mathbf{J}$ multipole fields depend on the magnitude of $\mathbf{k}_{\mathrm{F}}$ only, they can be carried outside the integral with the result

$$\mathbf{E}' = \sum_{plm} \mathbf{J}_{lm}^{(p)}(\mathbf{r}', k_{\mathrm{F}}) \mathscr{W}_{lm}^{(p)}(\mathbf{R}_{O'}), \tag{6.33}$$

where

$$\mathscr{W}_{lm}^{(p)}(\mathbf{R}_{O'}) = \sum_{\eta} c_\eta \int_{k_{\mathrm{F}x}^2+k_{\mathrm{F}y}^2 \le k_{\mathrm{F}\perp}^2} E_{\mathrm{FPW}}\left(\hat{\mathbf{k}}_{\mathrm{F}}\right) \mathrm{e}^{\mathrm{i}\mathbf{k}_{\mathrm{F}}\cdot\mathbf{R}_{O'}} \times T_\eta(\vartheta_{\mathbf{k}}) W_{lm}^{(p)}\left(\hat{\mathbf{u}}_{\eta\hat{\mathbf{k}}\mathrm{F}}, \hat{\mathbf{k}}_{\mathrm{F}}\right) \mathrm{d}k_{\mathrm{F}x} \mathrm{d}k_{\mathrm{F}y} \, . \tag{6.34}$$

Of course, (6.33) refers to the case in which there is a plane of separation between two media of different refractive index, so that the consequent refraction must be taken into account. It is an easy matter to show that, when $n_{\mathrm{F}} \to n$, (6.33) simplifies into

$$\mathbf{E}' = \sum_{plm} \mathbf{J}_{lm}^{(p)}(\mathbf{r}', k)\mathcal{W}_{lm}^{(p)}(\mathbf{R}_{O'}), \tag{6.35}$$

where

$$\mathcal{W}_{lm}^{(p)}(\mathbf{R}_{O'}) = \int_{k_x^2+k_y^2 \le k_\perp^2} E_{\rm PW}\left(\hat{\mathbf{k}}\right) \mathrm{e}^{\mathrm{i}\mathbf{k}\cdot\mathbf{R}_{O'}} W_{lm}^{(p)}\left(\hat{\mathbf{u}}_{\hat{\mathbf{k}}}, \hat{\mathbf{k}}\right) \mathrm{d}k_x \mathrm{d}k_y \,. \tag{6.36}$$

According to the case we deal with, (6.33) and (6.35) show that the multipole expansion of the field in the focal region resembles the expansion of a plane wave, whose amplitudes $\mathcal{W}_{lm}^{(p)}(\mathbf{R}_{O'})$, according to (6.34) or (6.36) depend on the position of the particle. Thus, when calculating the radiation force or the radiation torque exerted on a particle, we only have to substitute the newly defined amplitudes $\mathcal{W}_{lm}^{(p)}$ into (6.38) of Appendix 2 to get the amplitudes of the scattered field $\mathcal{A}_{lm}^{(p)}$, and into (6.11a) and (6.11b) to get $F_{\mathrm{Rad}\,\zeta}$, and into (6.20a) and (6.20b) to get Γ_μ. In practice, the required result is still given by (6.11a) and (6.11b) and by (6.20a) and (6.20b) provided that

$$E_0' W_{\mathrm{I}lm}^{(p)} \to \mathcal{W}_{lm}^{(p)}(\mathbf{R}_{O'}) \,, \quad E_0' A_{lm}^{(p)} \to \mathcal{A}_{lm}^{(p)} .$$

The preceding considerations highlight the importance of the quantities $\mathcal{W}_{lm}^{(p)}(\mathbf{R}_{O'})$ that describe the lens, as they depend on the characteristics of the latter. Then, the integrals (6.34) or (6.36) can be calculated numerically once for all at the nodes of a suitably chosen grid for a given lens and stored for further use.

6.6 Trapping and Manipulation of Single and Aggregated Spheres

The theory we expounded in Sect. 6.2 is, in a sense, nonconventional. In fact, although both the radiation force and torque were separated into a scattering contribution and an extinction contribution, these parts have no relation with the customary gradient contribution and scattering contributions, that, as stated in Sect. 6.2.3, come from the application of the dipole approximation [17]. Our theory is exact, as it goes beyond the dipole approximation, its accuracy depending on the use of a sufficiently large l_{M} to ensure the convergence of the multipole field expansions.

6.6.1 Trapping of Dielectric Spheres

We calculate the trapping position and the stiffness constants (see later) of homogeneous spheres of latex ($n_{\mathrm{p}} = 1.57$) of various sizes and compare our results with the experimental data existing in the literature. To this end, we assumed an aplanatic optical system with numerical aperture NA $= 1.2$ both in case the image space is filled of water with refractive index $n = 1.33$ and in case the exit pupil is immersed in oil with refractive index $n = 1.52$, whereas the trapping particle is immersed in water. These media are separated by a cover slip orthogonal to the optical axis located at $z_\mathrm{S} = -20\,\mu\mathrm{m}$ with respect to the origin at the nominal focus of the lens. Since NA < 1.33 no plane wave is totally reflected at the separation interface. As the presence of two media of different refractive indexes introduces a spherical aberration, the maximum field intensity occurs at a point F' with $z_{\mathrm{F}'} \approx -4.0\,\mu\mathrm{m}$. The incident field is assumed to be a TEM_{00} Gaussian beam, linearly polarized along the x axis, with filling factor $f_0 = 2$ and wavelength $\lambda_0 = 1064\,\mathrm{nm}$ in vacuo, i.e., $\lambda = 800\,\mathrm{nm}$ in water. The contour plot of the focalized field [28] shows that the effect of the aberration manifests as a large shift of the field maximum and a more accentuated lack of cylindrical symmetry around the optical axis due to the linear polarization of the incident beam. These results are in substantial agreement with those reported by Rohrbach and Stelzer [31].

We now go to consider the trapping of single spheres both in the unaberrated and in the aberrated field. According to the experimentalists we define the so-called trapping efficiency as

$$\mathbf{Q}(\mathbf{r}) = \mathbf{F}_{\mathrm{Rad}}(\mathbf{r})c/(Pn),$$

where P is the power of the trapping beam and n is the refractive index of the medium surrounding the particle. Thus, $n = 1.33$ in all the cases we deal with. The argument $\mathbf{r} = \mathbf{R}_{O'}$ (see Fig. 6.1) denotes the position of the particles that is assumed to coincide with their center. In Fig. 6.8 we report, as a representative example of spheres to which, according to our experience, the Rayleigh or the Born approximations do not apply, the components of the trapping efficiency for single spheres with diameter of $d = 850\,\mathrm{nm}$ both for the case of an unaberrated field and in the case spherical aberration is present. The trapping occurs where the components of the trapping efficiency vanish with a negative derivative. Figure 6.8 shows that the spheres we considered undergo trapping on the optical axis. In fact, both $Q_x(x, 0, 0)$ and $Q_y(0, y, 0)$ vanish on the optical axis for evident symmetry reasons. As regards $Q_z(0, 0, z)$, we see that there exist at least one value of z that satisfies the trapping condition. The values of z at which the trapping occurs both for unaberrated (z_0) and for the aberrated field ($z_{0\mathrm{a}}$) in Table 6.2 for spheres of diameter $d = 220$–$1900\,\mathrm{nm}$.

Since the behavior of the components of the trapping efficiency is almost linear in the vicinity of the trapping point, one is able to define the stiffness of the optical trap by introducing the constants κ_x, κ_y, and κ_z such that the components of the radiation force, in the absence of aberration, can be written as

$$F_x = -\kappa_x x, \quad F_y = -\kappa_y y, \quad F_z = -\kappa_z(z - z_0) \tag{6.37}$$

in the neighborhood of the point where they vanish. In case aberration is present we must put

$$F_z = -\kappa_z(z - z_{0\mathrm{a}})\,.$$

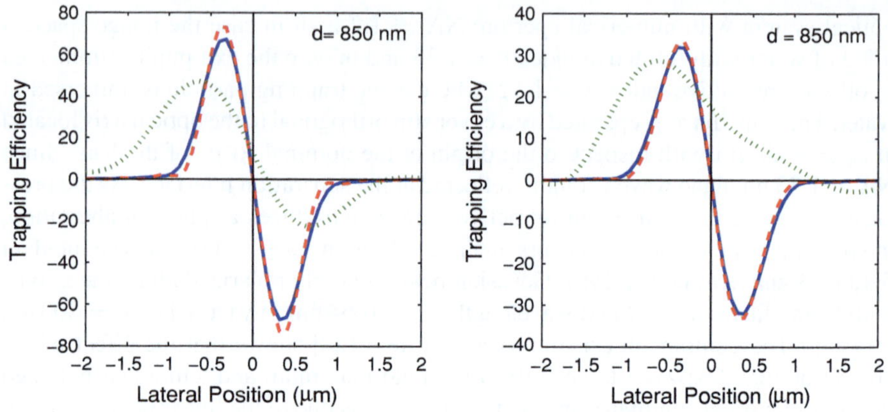

Fig. 6.8 $Q_x(x, 0, 0)$ (*blue solid line*), $Q_y(0, y, 0)$ (*red dashed line*), and $Q_z(0, 0, z)$ (*green dotted line*) for single spheres with diameter $d = 850$ nm both in unaberrated gaussian (*left panel*) and aberrated gaussian beam (*right panel*). Note that the origin of the z axis in the *right panels* is shifted by $-z_{F'} = 4.0\,\mu$m

Table 6.2 Calculated and experimental values of $s_l = 1 - \kappa_x/\kappa_y$ as well as the trapping position for the spheres we deal with in this contribution

d (nm)	z_0 (nm)	s_l calc.	z_{0a} (nm)	s_{la} calc.	s_l exp.	s_{la} exp.
220	120	0.38	800	0.19	0.38	
530	200	0.27	1280	0.16	0.31	
690	240	0.11	1520	0.09	0.08	
850	120	−0.07	1200	0.05	0.03	
1030	80	−0.13	1040	−0.01	−0.12	
1280	0	−0.02	480	−0.30		−0.33
1500	40	0.20	560	−0.20		−0.18
1660	40	0.02	400	−0.10	−0.10	
1900	160	0.06	480	0.12		0.08

A good parameter to be compared with the experimental data is the so-called stiffness asymmetry factor $s_l = 1 - \kappa_x/\kappa_y$ that we report, for all the spheres we dealt with, in Table 6.2 both for the unaberrated field (s_l) and for the aberrated field (s_{la}). The experimental values of the asymmetry factor were taken from the paper of Rohrbach [32] for the unaberrated field and from the paper of Zakharian et al. [33] for the aberrated field. We note that the size parameter $x = \pi n d/\lambda_0$ for the spheres considered in Table 6.2 goes from 0.6 to 5.61 in vacuo, i.e., from 0.80 to 7.46 in water, so that we had to use up to $l_M = 12$ to get full convergence of all the calculated values [34]. Table 6.2 seems to show a fair agreement between theory and experiment. This agreement seems confirmed by Fig. 6.9, where we report the curves of the asymmetry factor calculated both for unaberrated and for the aberrated field, together with the experimental values for the spheres used by Rohrbach (circles) and by Zakharian et al. (dots). Nevertheless, we call the attention on the values of the asymmetry parameters for the spheres with $d = 850$ and 1660 nm. Actually, these

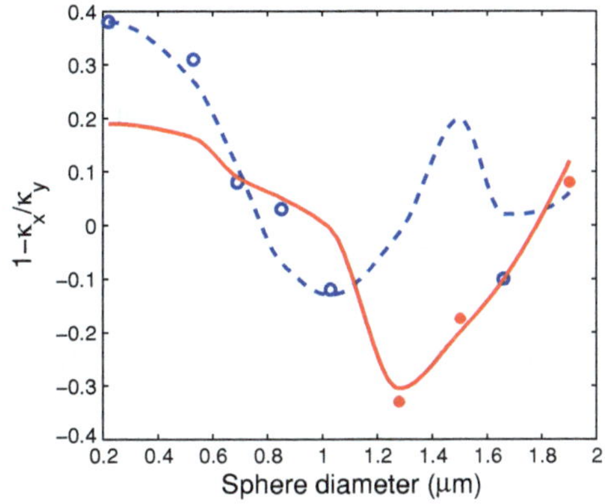

Fig. 6.9 Comparison of calculated asymmetry factors s_l (*dashed blue line*) and s_{la} (*solid red line*) with experimental data of Rohrbach [32] (*blue circles*) and of Zakharian et al. [33] (*red dots*)

values are favorably located on at least one of the two theoretical curves we draw in Fig. 6.9. We stress, however, that Rohrbach declares the experimental values at $d = 850$ and 1660 nm as obtained in the absence of aberration [32], whereas in Fig. 6.9 they appear to be on the curve for the aberrated field. On the other hand, the experimental values of Zakharian et al. appear very close to our curve for the aberrated field, that has been used in [33]. In this respect, it may be interesting to notice that the values of s_{la} measured by Zakharian et al. are not located on their theoretical curve (see Fig. 6.12 of [33]). This discrepancy, in our opinion, may be due to an inadequate consideration of the effect of aberration, and/or to the fact that with so large a diameter any approximation used to calculate the Lorenz force density does not apply. Moreover, the diameter of these spheres is larger than the size of the trap, that, according to our calculations, spans $\approx\lambda = 800$ nm so that one must be very careful both in computations and in the interpretation of the experimental data when dealing with such large objects.

6.6.2 Trapping of Gold Spheres

In this section we calculate the optical forces on gold nanospheres with the purpose of checking up to what size our theoretical approach foresees the trapping of spherical particles of noble metals. To this end, we adopt even in this section the geometry that is sketched in Fig. 6.1.

A critical point in the calculation of the radiation force, whatever the approach one chooses to use, is the knowledge of the dielectric properties of the particles. Here, we use the dielectric function tabulated by Johnson and Christy [35], which yields a good agreement between calculated and experimental extinction spectra of such nanoparticles. The main feature for gold spheres is the structure of the plasmon resonances whose position and complexity depends on the radius of the particles. In this respect, we notice that at $\lambda = 1064$ nm, the wavelength used by Hansen et al.

[36] in their experiments, particles with $r = 77$ nm still show a large extinction in comparison with particles of smaller radius.

All the parameters that we adopt in our calculations are those of [36]. Thus, we assume an optical system with numerical aperture NA=1.2 in water illuminated by an x-polarized TEM_{00} laser beam. Since in [36] the objective has been only slightly overfilled, we chose the filling factor $f_0 = 2$ according to the common choice reported in the literature [17, 28, 32]. We also consider the case of an oil immersion lens, and consequently change the numerical aperture to NA=1.32 (see [36]). We did not compensate for the spherical aberration by using oils of different refractive indexes [37]. We preferred to use $n_{\text{oil}} = 1.54$ and to consider several values of $D = -z_{\text{s}}$, the distance from the nominal focus of the cover slip that separates oil from water, as in all cases the particles are embedded in water. In fact, the effect of the aberration produced by the presence of the cover slip can be minimized by locating the latter as near as possible to the trapping volume [36]. On the other hand, in [36] it is stressed that the effect of the aberration is to reduce the maximum size of the trappable particles and that the best results are obtained using a water immersion lens. Since trapping particles of different sizes require different powers of the laser beam, in [36] the experimental stiffnesses of the trap are normalized to the laser power P. Our calculated κ_z/P and κ_x/P versus the size of the spheres are reported in Fig. 6.10 together with the experimental data of [36]. Figure 6.10 also includes some points that are marked by an arrow: these points will be discussed later. At present, we see at once that we do not succeed in the trapping of spheres with $r > 77$ nm, whereas Hansen et al. were able to trap spheres with r up to 127 nm. Nevertheless, in [36] the authors stress that they were unable to get stable trapping of spheres with $r = 77$ nm. In a sense, this is not surprising, because the experimental measurements include all effects, linear and nonlinear, yielded by the radiation field with scarce or no possibility of discrimination. On the other hand, our calculations are performed for fixed parameters of the model, i.e., fixed radius and refractive index of the particles, and fixed refractive index of the surrounding medium.

Hansen et al. report their measured κ_z/P for the nonaberrated setup only, but our calculated curves for nonaberrated and aberrated setup, provided $-6\,\mu\text{m} \leq z_{\text{s}} \leq -2\,\mu\text{m}$, are almost indistinguishable from each other and show a good agreement with the experimental data up to radii at which trapping is predicted by the theory. Our results for κ_x/P in Fig. 6.10 show that the calculated curves for nonaberrated and aberrated setup are quite superposable, and agree rather well with the experimental data for the nonaberrated setup. In fact, the experimental data for aberrated and non-aberrated setup show strong differences. We believe that this is due to experimental difficulties in the power normalization when using oil immersion objectives.

In Fig. 6.10 a clear size scaling is evident for small radii. The optical trapping stiffnesses lie on straight lines with slope 3 on a log–log scale, thus showing the existence of a scaling law with the volume of the particle. The fact that our calculated results change their slope for radii larger that 50 nm suggests the transition from a proportionality to the volume to a dependence on some other feature. Indeed, in the range of radii between 70 and 80 nm occurs a change in the trapping regime. In order to understand the mechanism of this change we first studied the behavior of the

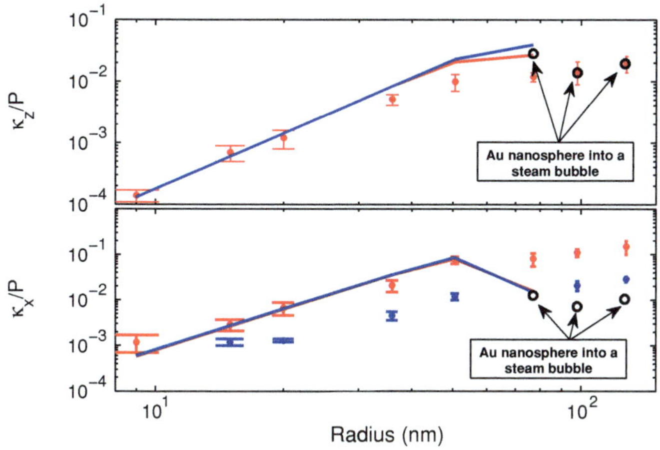

Fig. 6.10 Experimental (from [36]) and calculated stiffness κ_z/P and κ_x/P in pN/(nm · W) for gold spheres, both for water immersion lens (*red points* and *lines*) and for oil immersion lens (*blue points* and *lines*). The points marked by an arrow refer to our calculations for unaberrated field on particles embedded into a steam bubble

radiation force both in the case of unaberrated field and, in the case of an aberrated field, for different values of D, in order to see to what extent the aberration plays a significant role in degrading the trapping in the axial direction. In agreement with [36], we found that the more the field is aberrated the more its ability to trap relatively large particles weakens, perhaps by reduction of the trapping region. As regards the behavior of the radiation force within that region, it strongly suggests that a given model with parameters chosen once for all cannot explain the trapping of the larger particles, i.e., of spheres with $r > 77$ nm: the theory is linear in power, indeed. We must thus consider the possibility that the parameters of the model, essentially the refractive index, may change with increasing power. In this respect, it is quite natural to suspect the heating both of the medium and of the particles yielded by the laser beam. Therefore, we first considered the study of Liz-Marzán and Mulvaney [38] who measured the change of the dielectric properties of Au colloids, more precisely of the surface plasmon absorption, in the temperature range of 14–70° C. These authors, after carefully considering all the physical effects that stem from the increase in temperature, were led to the conclusion that the most important factor is the change of resistivity of gold with a consequent change of its refractive index. Nevertheless, the experiments of Liz-Marzán and Mulvaney were performed on Au spheres with an average diameter $d = 15$ nm. When the corrections that they suggest on the basis of the findings of Doremus [39, 40] and of Kreibig [41] (these corrections depend on the size of the particles) are calculated for spheres with a radius $r \geq 77$ nm, the change of the refractive index turns out to be negligible and leads to no increase in the radius of the trappable particles.

Table 6.3 Temperature change with respect to room temperature for the Au nanospheres used in [36] at their trapping position. r is the radius (in nm) of the spheres used in [36], σ_{abs} (in μm^2) is the absorption cross section, and $\Delta T/P$ is the change of temperature (in K/W) due to heating by the trapping beam. ΔT has been estimated using the formulas in [42] at 5 nm from the surface of the nanospheres

r	σ_{abs}	$\Delta T/P$
15	3.48×10^{-6}	36.43
25	1.87×10^{-5}	130.46
35	6.01×10^{-5}	313.75
45	1.47×10^{-5}	614.78
50	2.15×10^{-4}	817.74
77	1.05×10^{-3}	1679.90
98	2.50×10^{-3}	5072.50
127	4.61×10^{-3}	7303.20

Next, we considered that, according to Hansen et al., the power needed to trap the larger Au spheres was of 135 mW, and that Seol et al. [42] in their study of the heat developed by the trapping beam, estimate an increase in temperature at the surface of Au nanospheres as high as $260°$ C/W. Much lower temperatures were estimated by Peterman et al. [43] at the surface of dielectric particles, but in a recent paper Lapotko [44] demonstrated that the laser-induced heating around plasmonic nanoparticles, in particular Au nanospheres, may excite detectable vapor bubbles. These studies suggested us to use the formulas reported by Seol et al. to estimate the change in temperature, with respect to room temperature, at 5 nm from the surface of the Au nanospheres considered in [36]. The results of our estimate are reported in Table 6.3 where we also reported the absorption cross sections that, unlike [42], we calculated without resorting to the dipole approximation. The other parameters were taken as reported by Seol et al. because they are compatible even with our focalized beam.

The increases in temperature reported in Table 6.3 should be taken with some caution as they are due to a simplified model of the thermal equilibrium of the naked metal spheres and the surrounding water, yielding perhaps too large values for the largest Au spheres. Anyway, these increases turn out to be large enough to justify the working hypothesis that the particles, especially the largest ones, may become embedded into a steam bubble. In turn, the presence of a steam bubble, provided it is stable enough, will attenuate the effect of the metallic nature of the particles. Of course, we kept the thickness of the steam layer, which we assumed to have the refractive index $n_{steam} = 1$, as small as possible, for the sake of stability, and made also the assumption that the particle remains steadily at the center of the bubble. The spheres with $r = 98$ nm are surrounded by a steam bubble with thickness 100 nm, and those with $r = 127$ nm are surrounded by a bubble with thickness 130 nm. We stress that the total diameters (particle plus bubble) are smaller than the width of the trapping spot and of the resolution of the imaging system used in the experiments. Thus we calculated the field scattered by the stratified particle that includes the metal

Table 6.4 Trapping positions of Au nanospheres (the largest embedded into a steam bubble with thickness of the steam layer t). r is the radius of the spheres used in [36], z_0 is the calculated trapping position, and z_E the trapping position determined by the extinction contribution alone. r, t, z_0, and z_E are all in nm

r	t	z_0	z_E
15		40	40
25		65	64
35		118	113
45		213	200
50		283	262
77		916	841
77	5	864	781
98	100	−1492	−1470
127	130	−1379	−1362

nanosphere surrounded by the steam layer and embedded into water. This has been done by exploiting the procedure that we developed to study the resonances of layered metal spheres [45]. The resulting scattered field, when introduced into the Maxwell stress tensor, yields through (6.1) the radiation force acting on the center of mass of the particle. We also calculated the trapping position for spheres with $r = 77$ nm surrounded by a steam layer with thickness 5 nm in order to see whether the presence of the bubble improves the trapping. All these calculations were performed for the unaberrated field, the one able to trap the largest particles we are considering. In fact, in our calculations the steam layer grants stable trapping (at negative values of z) also for the $r = 98$ and $r = 127$ nm spheres. We collect in Table 6.4 all the trapping positions for the unaberrated field together with the thickness of the steam layer enveloping the largest spheres. As regards the entries in Table 6.4 one may wonder about the relevance of z_E. For the Au spheres we considered here that absorption and extinction prevail over scattering. Thus, according to Sect. 6.2.3 we found that for the spheres with $r \leq 77$ nm $F^{(\mathrm{Sca})}_{\mathrm{Rad}\,z}$ as a function of z within the focal region is, in general, two orders of magnitude smaller than $F^{(\mathrm{Ext})}_{\mathrm{Rad}\,z}$. As a consequence, it is mainly $F^{(\mathrm{Ext})}_{\mathrm{Rad}\,z}$ that determines the trapping near the point where it vanishes. However, for $r = 77$ nm at that point the zero is in no way a sharp one, and for $r > 77$ nm for $F^{(\mathrm{Ext})}_{\mathrm{Rad}\,z}$ there is no zero point at all. Adding the steam layer results in a slightly less stable trapping of the $r = 77$ nm spheres but makes a reshaping of $F^{(\mathrm{Ext})}_{\mathrm{Rad}\,z}$ for the $r = 98$ and for the $r = 127$ nm spheres, so that $F^{(\mathrm{Ext})}_{\mathrm{Rad}\,z}$ has again a zero point for stable trapping. In conclusion, the z_E entry shows that for the Au spheres the trapping is due to the behavior of $F^{(\mathrm{Ext})}_{\mathrm{Rad}\,z}$.

The stiffnesses of the trap for spheres into a steam bubble are reported and marked by arrows in Fig. 6.10. We see at once that our calculated values of κ_z/P, rather surprisingly, coincide with the experimental values. In particular, the addition of the bubble to the 77 nm spheres produces little change of κ_z/P. As for κ_x/P, we see that for 77 nm spheres it lies on the calculated curve, whereas, for the largest spheres, it lies on a line parallel to the experimental data.

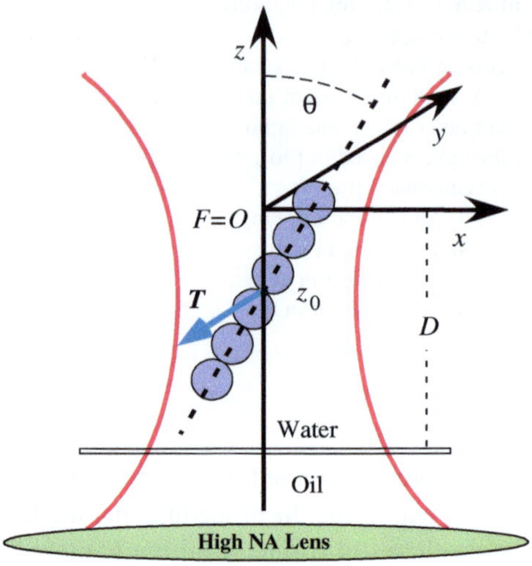

Fig. 6.11 Sketch of the geometry of the optical trap for a x-polarized TEM_{00} laser beam propagating in the positive direction of the z axis. The vector **T** represents the radiation torque exerted on the chain trapped at z_0 with its axis in the xz plane

6.7 Radiation Torque on Elongated Nanostructures

In the last few years there has been a growth of interest in trapping and manipulating elongated nanostructures such as nanowires and fibers. Indeed, these structures are characterized by a large aspect ratio and a subwavelength diameter that makes them suitable for several applications such as nanoprobe microscopy and spectroscopy [46]. However, for these applications to take hold it is of fundamental importance to understand the optical trapping properties of quasi-1D nanostructures, to evaluate quantitatively radiation force and torque acting on them and eventually to shed light on their dynamics within the optical trap.

In this section we study the torque exerted on optically trapped linear nanostructures that we model as linear aggregates of identical spheres of diameter d and length L. The geometry of the problem is sketched in Fig. 6.11 that is a specialized version of Fig. 6.1.

The setup we assume is the aberrated one that we considered in Sect. 6.6.1. The L and d values of the particles we deal with were chosen so as to highlight the orientational behavior of the chains, as we will show below. We also investigate long chains (up to 20 spheres) with $d = 50$ nm in order to compare our calculated results with some experimental data available for semiconductor nanowires [46, 47]. In fact, the $d = 50$ nm spheres are much smaller than the wavelength, so that these chains are a good model for quasi-1D nanostructures [10, 48].

We consider linear chains composed of latex spheres (refractive index $n_{\rm p} = 1.57$), with diameter 220, 100, and 50 nm, and for each chain we calculate the radiation force $\mathbf{F}_{\rm Rad}(\mathbf{r})$, the argument $\mathbf{r}$ denoting the position of O', the center of mass of the chain. The trapping occurs on the optical axis [31] where all the components of $\mathbf{F}_{\rm Rad}(\mathbf{r})$ vanish with a negative derivative, and in the vicinity of the trapping point

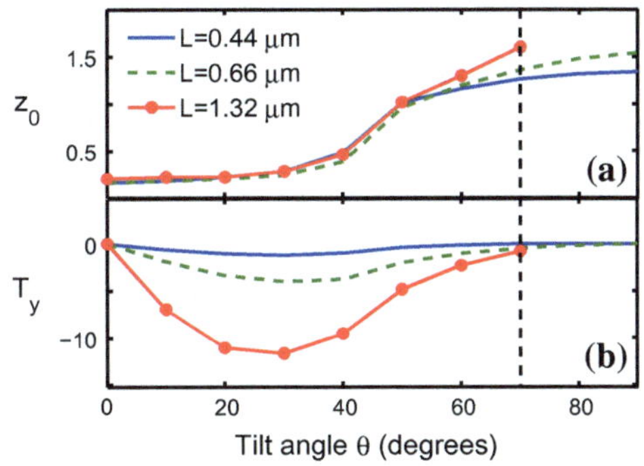

Fig. 6.12 Calculated trapping position z_0 (μm) **a** and T_y **b** as a function of the tilt angle ϑ for chains of spheres of latex with diameter $d = 220$ nm and lengths $L = 0.44$ (2 spheres), 0.66 (3 spheres), and 1.32 μm (6 spheres). In **a** the origin of the z axis is shifted by $-z_s$ (see Sect. 6.6.1). The vertical dashed line marks the tilt beyond which the 6-spheres chains do not trap

$\mathbf{r}_0 \equiv (0, 0, z_0)$ the components of $\mathbf{F}_{\text{Rad}}(\mathbf{r})$ can be approximated as in (6.37). The orientation of the aggregates is individuated by the polar angles ϑ and φ of the axis of the linear chain. For each orientation we determined first the trapping position of the center of mass of the chain, then calculated the torque $\boldsymbol{\Gamma}_{\text{Rad}}$ at that position. As regards the torque, we actually calculate the adimensional vector $\mathbf{T}_\eta$ defined in (6.21). The orientational stability occurs when the cartesian components of $\mathbf{T}$ vanish with a negative derivative with respect to both ϑ and φ. In this respect we notice that, in all the cases we dealt with the components T_x and T_z vanish (with negative derivatives) when the chain lies in the xz plane, i.e., when $\varphi = 0$. In other words, the xz plane is a plane of orientational stability with respect to φ, as expected on account of the symmetry of the chains and of the polarization of the incident beam. Accordingly, in all the following figures we assume $\varphi = 0$. On this assumption, when $T_y < 0$ the chains are forced to align to the optical axis, whereas when $T_y > 0$ they tend to align along the polarization axis. Let us first consider the linear chains composed of 2, 3, and 6 latex spheres of diameter 220 nm. In Fig. 6.12 (a) we report the trapping position as a function of ϑ for all the chains. Even a cursory examination shows that the trapping position is strongly dependent on the tilt angle and that the 6 spheres chains are not trapped for $\vartheta > 70°$. It is tempting to attribute this result to the length of these chains ($L = 1.32\,\mu$m) that exceeds the width of the trapping spot ($\approx$800 nm). Nevertheless, we will show below that chains with a length of 1 μm and diameter of 50 nm do trap even when orthogonal to the optical axis ($\vartheta = 90°$). In this respect, we remark that σ_{T}, the extinction cross section of the particles, depends on their orientation and that the dominant term in $F_{\text{Rad}\,z}$, the component of the radiation force along the optical axis, is proportional just to σ_{T} [8]. Now, our calculations show that, at any tilt angle, the extinction cross section of the 220 nm chains is about 1000 times larger than that of the 50 nm chains. Therefore, $F_{\text{Rad}\,z}$ may grow so large as to prevent the trapping of the 220 nm chains for $\vartheta > 70°$. However, the curves in Fig. 6.12 (b), show that $T_y < 0$ so that the torque forces the chains to align their axis to the optical axis, the strongest torque being exerted just on the 6 spheres chains.

The chains of diameter 100 nm include up to 6 spheres and thus their maximum length is 600 nm, whereas those with a diameter of 50 nm include up to 20 spheres with a maximum length of 1 μm. We first state that all the chains of 100 nm spheres

Fig. 6.13 Calculated T_y as a function of the tilt angle ϑ for several values of the length L of the chains of latex spheres with diameter $d = 100$ nm in **a** and 50 nm in **b**

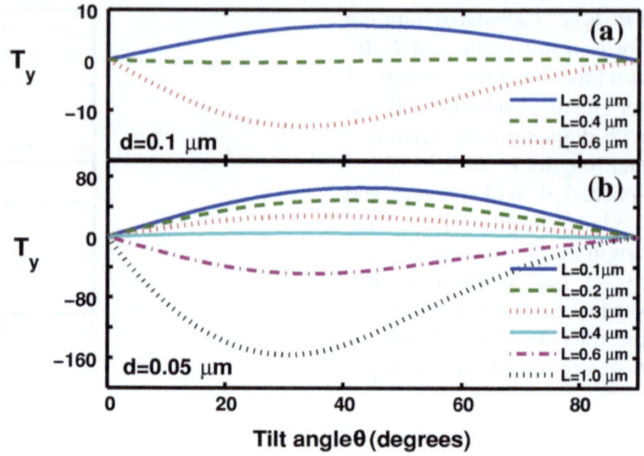

undergo trapping, whatever their tilt angle, and that the trapping position is almost independent of ϑ. Moreover, Fig. 6.13a, in which we report T_y as a function of ϑ for these chains, shows that for the 2-spheres chains $T_y > 0$, whereas $T_y < 0$ for the 6-spheres chains. Thus, the shorter chains tend to orient along the polarization axis, whereas the longer ones tend to align with the optical axis. The chains with $L = 400$ nm mark a change in the sign of T_y, hence a change in the orientational stability.

The behavior of the chains with diameter 50 nm is quite analogous, i.e., whatever their length they undergo trapping. According to Fig. 6.13b, in which we report T_y as a function of ϑ, for the shorter chains $T_y > 0$, whereas for the longer ones $T_y < 0$. The point of inversion lies somewhere between the 8 and the 9-spheres chains, i.e., at $L \approx 400$ nm.

The results discussed above show that the trapping of our model linear chains presents features that are akin to those observed for trapped nanowires. In fact, force constants and torque we calculated for several materials for long chains of spheres modeling nanowires are in qualitative agreement with the few available and comparable experimental results. For instance, calculations confirmed trapping, alignments along the optical axis, and linear dependence of κ_z on laser power, as expected according to [47].

6.8 Appendix 1

As anticipated in Sect. 6.2, with the help of the asymptotic multipole expansions it is easily proved that the dyadic terms

$$\hat{\mathbf{r}}' \cdot \Big[n^2(\mathbf{E}' \otimes \mathbf{E}'^*) + (\mathbf{B}' \otimes \mathbf{B}'^*)\Big] = n^2\big(\hat{\mathbf{r}}' \cdot \mathbf{E}'\big)\mathbf{E}' + \big(\hat{\mathbf{r}}' \cdot \mathbf{B}'\big)\mathbf{B}'$$

give a vanishing contribution at the radiation force. In fact, a look to (6.8a) and (6.8b) shows that the asymptotic multipole expansions of $\mathbf{E}'_{\mathrm{I}}$ and $\mathbf{E}'_{\mathrm{S}}$ contain the transverse harmonics $\mathbf{Z}^{(p)}_{lm}(\hat{\mathbf{r}}')$ which, according to their definition are orthogonal, in the ordinary vector sense, to $\hat{\mathbf{r}}'$. This decrees the vanishing of the contribution of the dyadic terms for whatever form of the incident amplitudes $W^{(p)}_{\mathrm{I}lm}$, even when the latter are substituted by the $\mathscr{W}^{(p)}_{lm}(\mathbf{R}_{O'})$.

Now, we show how it happens that the terms $\mathbf{E}'_{\mathrm{I}} \cdot \mathbf{E}'^{*}_{\mathrm{I}}$ and $\mathbf{B}'_{\mathrm{I}} \cdot \mathbf{B}'^{*}_{\mathrm{I}}$ give a vanishing contribution to the radiation force even when the incident field is not a single plane wave but rather a superposition of plane waves with the same magnitude of k but different direction of propagation, i.e., different $\hat{\mathbf{k}}$.

Let us thus assume that the incident electric field is a superposition of plane waves of the kind of (6.28). Thus a typical term that would enter (6.1) is

$$I_E = \mathrm{Re}\left[r'^2 n^2 \int_{\Omega'} \mathbf{E}'_{\mathrm{I}}(\hat{\mathbf{k}}) \cdot \mathbf{E}'^{*}_{\mathrm{I}}\left(\hat{\mathbf{k}}'\right) \hat{\mathbf{r}}' \, \mathrm{d}\Omega' \right],$$

where $\mathbf{E}'_{\mathrm{I}}\left(\hat{\mathbf{k}}\right) = E'_{\mathrm{PW}}\left(\hat{\mathbf{k}}\right) \hat{\mathbf{u}}_{\hat{\mathbf{k}}} \exp(\mathrm{i}\mathbf{k} \cdot \mathbf{r}') = \mathbf{E}'_{\mathrm{PW}}\left(\hat{\mathbf{k}}\right) \exp(\mathrm{i}\mathbf{k} \cdot \mathbf{r}')$. An analogous term I_B comes from $\mathbf{B}'_{\mathrm{I}}$. Since r' is large, we can use the asymptotic form of a plane wave [7, 8] so that I_E becomes

$$\begin{aligned} I_E = \mathrm{Re}\Big\{ \frac{4\pi n^2}{k^2} \int_{\Omega'} & \mathbf{E}'_{\mathrm{PW}}\left(\hat{\mathbf{k}}\right) \cdot \mathbf{E}'^{*}_{\mathrm{PW}}\left(\hat{\mathbf{k}}'\right) \\ & \times \left[\delta\left(\hat{\mathbf{k}} + \hat{\mathbf{r}}'\right) \exp(-\mathrm{i}kr') - \delta\left(\hat{\mathbf{k}} - \hat{\mathbf{r}}'\right) \exp(\mathrm{i}kr') \right] \\ & \times \left[\delta\left(\hat{\mathbf{k}}' + \hat{\mathbf{r}}'\right) \exp(\mathrm{i}kr') - \delta\left(\hat{\mathbf{k}}' - \hat{\mathbf{r}}'\right) \exp(-\mathrm{i}kr') \right] \hat{\mathbf{r}}' \, \mathrm{d}\Omega' \Big\} . \end{aligned}$$

Due to the properties of the δ-function, the result of the integration is

$$\begin{aligned} I_E = \mathrm{Re}\Big\{ \frac{4\pi n^2}{k^2} & \mathbf{E}'_{\mathrm{PW}}(\hat{\mathbf{k}}) \cdot \mathbf{E}'^{*}_{\mathrm{PW}}\left(\hat{\mathbf{k}}'\right) \Big[-\hat{\mathbf{k}}\delta\left(\hat{\mathbf{k}} - \hat{\mathbf{k}}'\right) + \hat{\mathbf{k}}'\delta\left(\hat{\mathbf{k}} - \hat{\mathbf{k}}'\right) \\ & + \hat{\mathbf{k}}'\delta\left(\hat{\mathbf{k}} + \hat{\mathbf{k}}'\right) \exp(2\mathrm{i}kr') - \hat{\mathbf{k}}'\delta\left(\hat{\mathbf{k}} + \hat{\mathbf{k}}'\right) \exp(-2\mathrm{i}kr') \Big] \Big\} . \end{aligned}$$

A quite similar expression, except for the absence of n, is obtained for I_B, so that collecting all the terms we get

$$\begin{aligned} I &= I_E + I_B \\ &= \mathrm{Re}\Big\{ \frac{4\pi}{k^2} \left[n^2 \mathbf{E}'_{\mathrm{PW}}\left(\hat{\mathbf{k}}\right) \cdot \mathbf{E}'^{*}_{\mathrm{PW}}\left(\hat{\mathbf{k}}'\right) + \mathbf{B}'_{\mathrm{PW}}\left(\hat{\mathbf{k}}\right) \cdot \mathbf{B}'^{*}_{\mathrm{PW}}\left(\hat{\mathbf{k}}'\right) \right] \\ &\qquad \times \hat{\mathbf{k}}'\delta\left(\hat{\mathbf{k}} + \hat{\mathbf{k}}'\right) \left[\exp(2\mathrm{i}kr') - \exp(-2\mathrm{i}kr') \right] \Big\} \end{aligned}$$

which is easily seen to vanish, even when $\hat{\mathbf{k}}' = -\hat{\mathbf{k}}$, on account that

$$\mathbf{B}'_{\mathrm{PW}}(\hat{\mathbf{k}}) = -in\,\hat{\mathbf{k}} \times \mathbf{E}'_{\mathrm{PW}}\left(\hat{\mathbf{k}}\right).$$

Similar conclusions can be reached, starting from (6.31), for the case of an aberrated laser beam.

Finally we note that, on account of the definition of the amplitudes $\mathscr{W}_{lm}^{(p)}(\mathbf{R}_{O'})$ and of the related amplitudes $\mathscr{A}_{lm}^{(p)}$, in analogy to Sect. 6.2.2, the radiation torque $\boldsymbol{\Gamma}_{\mathrm{Rad}}$ does not get contributions from terms of the form $\mathscr{W}_{lm}^{(p)}(\mathbf{R}_{O'})\mathscr{W}_{l'm'}^{(p')*}(\mathbf{R}_{O'})$.

In conclusion, neither a plane wave field nor a focal field gives direct contributions to the radiation force or torque, as they are just the fields in the absence of particles.

6.9 Appendix 2

In Sect. 6.1 we stated that the calculation of $\mathbf{F}_{\mathrm{Rad}}$ and $\boldsymbol{\Gamma}_{\mathrm{Rad}}$ requires the knowledge of the field scattered by the particles. We solve the scattering problem for the particles concerned by expanding both the incident and the scattered field in a series of vector multipole fields (see (6.6) and (6.7)). Then, according to Waterman [3] we relate the multipole amplitudes, the incident field $W_{\mathrm{I}lm}^{(p)}$, to those of the scattered field $A_{lm}^{(p)}$, by the equation [10]

$$A_{lm}^{(p)}\left(\hat{\mathbf{u}}_{\mathrm{I}}, \hat{\mathbf{k}}_{\mathrm{I}}\right) = \sum_{p'l'm'} \mathscr{S}_{lml'm'}^{(pp')} W_{\mathrm{I}l'm'}^{(p')}\left(\hat{\mathbf{u}}_{\mathrm{I}}, \hat{\mathbf{k}}_{\mathrm{I}}\right), \tag{6.38}$$

that defines the elements of the *transition matrix* of the particle. Let us stress that, in principle, the sums in (6.6) and (6.7) and thus also (6.38) include an infinite number of terms. In practice, for computational reasons, these sums must be truncated to some suitable l_{M} chosen so as to ensure a fair description of the fields. Therefore, all sums over the multipole order l should be understood to extend up to $l = l_{\mathrm{M}}$. Since the particles we deal with either are, or can be modeled as aggregates of spheres, we calculate $\mathscr{S}_{lml'm'}^{(pp')}$ for such aggregates by inverting the matrix of the linear system that is obtained by imposing to the fields the boundary conditions across each of the spherical surfaces [10, 49]. The order of the matrix to be inverted is $2Nl_{\mathrm{M}}(l_{\mathrm{M}}+2)$, where N is the number of the spheres of the aggregate. The convergence of such kinds of calculations, i.e., the choice of the appropriate l_{M} is studied in Refs. [34]. A comprehensive treatment of all the topics mentioned above related to the calculation of the transition matrix can be found in Ref. [10]. It may be useful to note that for a homogeneous sphere

$$\mathscr{S}_{lml'm'}^{(pp')} = \mathscr{S}_{l}^{(p)}\delta_{pp'}\delta_{ll'}\delta_{mm'},$$

and $\mathscr{S}_l^{(1)} = b_l$, $\mathscr{S}_l^{(2)} = a_l$, a_l and b_l being the well-known Mie coefficients [4, 10]. The transition matrix is related to the scattering amplitude of the particle $\mathbf{f}(\hat{\mathbf{k}}_\mathrm{S}, \hat{\mathbf{k}}_\mathrm{I}; \hat{\mathbf{u}}_\mathrm{I})$ through the equation

$$\mathbf{f}\left(\hat{\mathbf{k}}_\mathrm{S}, \hat{\mathbf{k}}_\mathrm{I}; \hat{\mathbf{u}}_\mathrm{I}\right) = \frac{1}{k} \sum_{plm} (-\mathrm{i})^{l+p} \mathbf{Z}_{lm}^{(p)} \left(\hat{\mathbf{k}}_\mathrm{S}\right) A_{lm}^{(p)} \left(\hat{\mathbf{u}}_\mathrm{I}, \hat{\mathbf{k}}_\mathrm{I}\right) \tag{6.39}$$

where the amplitudes of the scattered field are given by (6.38). Equation (6.39) thus allows to express the scattering cross section and the extinction cross section in terms of the elements of the transition matrix.

The usefulness of the transition matrix stems from the its transformation properties under rotation of the coordinate frame. In fact, if the frame Σ' attached to the particle rotates to an orientation characterized by the Eulerian angles α, β, and γ, Θ for short, the elements of the transition matrix become

$$\mathscr{S}_{lml'm'}^{(pp')} = \sum_{\mu\mu'} D_{m\mu}^{(l)*}(\Theta) \bar{\mathscr{S}}_{lml'm'}^{(pp')} D_{m'\mu'}^{(l')}(\Theta) \tag{6.40}$$

where $\bar{\mathscr{S}}_{lml'm'}^{(pp')}$ denotes the elements of the transition matrix calculated in the frame Σ' so that they are independent of the orientation of the particle. As a result, the elements of the transition matrix depend on the orientation through the rotation matrices $D^{(l)}$ only. Then, the averages over the orientation of the particles for all the physically significant quantities result in integrals of products of up to four elements of rotation matrices further multiplied by $P(\Theta)$, the latter being the appropriate function that describes the distribution of orientations. There are relevant cases, e.g., the case of random orientations (random average), or the case of random orientation around a fixed axis (axial average), in which the integration above can be performed analytically. These averaging procedures, although straightforward, yield rather complex formulas so that we refer the interested reader to the details that are fully expounded elsewhere [9, 10, 50].

References

1. R.Y. Chiao, N.C. Wickramasinghe, Mon. Not. R. Astr. Soc. **159**, 361 (1972)
2. J. Cho, A. Lazarian, Astrophys. J. **631**, 361 (2005)
3. P.C. Waterman, Phys. Rev. D **3**, 825 (1971)
4. G. Mie, Ann. Phys. **25**, 377 (1908)
5. J.D. Jackson, *Classical electrodynamics*, 2nd edn. (Wiley, New York, 1975)
6. R.N.C. Pfeifer, T.A. Nieminen, N.R. Heckenberg, H. Rubinsztein-Dunlop, Rev. Mod. Phys. **79**, 1197 (2007)
7. D.S. Saxon, Phys. Rev. **100**, 1771 (1955)
8. M.I. Mishchenko, J. Quant. Spectrosc. Radiat. Transfer **70**, 811 (2001)
9. R. Saija, M.A. Iatì, A. Giusto, P. Denti, F. Borghese, J. Quant. Spectrosc. Radiat. Transf. **94**, 163 (2005)

10. F. Borghese, P. Denti, R. Saija, *Scattering from model nonspherical particles*, 2nd edn. (Springer, Heidelberg, 2007)
11. E. Fucile, F. Borghese, P. Denti, R. Saija, O.I. Sindoni, IEEE Trans. Antennas Propag. AP **45**, 868 (1997)
12. F. Borghese, P. Denti, R. Saija, M.A. Iatì, Opt. Express **15**, 11984 (2007)
13. F. Borghese, P. Denti, R. Saija, M.A. Iatì, Opt. Express **15**, 14618 (2007)
14. M.E. Rose, *Elementary theory of angular momentum* (Wiley, New York, 1957)
15. M. Abramowitz, I.A. Stegun, *Handbook of mathematical functions* (Dover Publications, New York, 1972)
16. P.L. Marston, J.H. Crichton, Phys. Rev. A **30**, 2508 (1984)
17. L. Novotny, B. Hecht, *Principles of nano-optics* (Cambridge University Press, New York, 2007)
18. F. Borghese, P. Denti, R. Saija, M.A. Iatì, Opt. Express **14**, 9508 (2006)
19. F. Borghese, P. Denti, R. Saija, M.A. Iatì, Opt. Express **15**, 6946 (2007)
20. G. Wurm, M. Schnaiter, Astrophys. J. **567**, 370 (2002)
21. J. Klačka, M. Kocifai, J. Quant. Spectrosc. Radiat. Transf. **70**, 595 (2001)
22. E.M. Purcell, Astrophys. J. **231**, 404 (1979)
23. B.T. Draine, J.C. Weingartner, Astrophys. J. **470**, 551 (1996)
24. R. Saija, M.A. Iatì, A. Giusto, F. Borghese, P. Denti, S. Aiello, C. Cecchi-Pestellini, Mon. Not. R. Astr. Soc. **341**, 1239 (2003)
25. X. Ma, J.Q. Lu, R.S. Brock, K.M. Jacobs, P. Yang, X.H. Hu, Phys. Med. Biol. **48**, 4165 (2003)
26. M.J. Greenberg, F. Ferrini, B. Barsella, S. Aiello, Nature **327**, 214 (1987)
27. B.T. Draine, H.M. Lee, Astrophys. J. **285**, 89 (1984)
28. F. Borghese, P. Denti, R. Saija, M.A. Iatì, Opt. Express **15**(14), 8960 (2007)
29. L. Mandel, E. Wolf, *Optical coherence and quantum optics* (Cambridge University Press, New York, 1995)
30. B. Richards, E. Wolf, Proc. Roy. Soc. (London) **253**, 358 (1959)
31. A. Rohrbach, E.H.K. Stelzer, Appl. Opt. **41**, 2494 (2005)
32. A. Rohrbach, Phys. Rev. Letters **95**, 168102 (2005)
33. A.R. Zakharian, P. Polynkin, M. Mansuripur, J.V. Moloney, Opt. Express **14**(8), 3660 (2005)
34. R. Saija, M.A. Iatì, P. Denti, F. Borghese, A. Giusto, O.I. Sindoni, Appl. Opt. **42**, 2785 (2003)
35. P.B. Johnson, R.W. Christy, Phys. Rev. B **6**(12), 4370 (1972)
36. P.M. Hansen, V.K. Bhatia, N. Harrit, L. Oddershede, Nano Letters **5**, 1937 (2005)
37. S.N.S. Reihani, L.B. Oddershede, Opt. Lett. **32**, 1998 (2007)
38. L.M. Liz-Marzán, P. Mulvaney, New J. Chem. **22**, 1285 (1998)
39. R.H. Doremus, J. Chem. Phys. **40**, 2389 (1964)
40. R.H. Doremus, J. Chem. Phys. **41**, 3259 (1964)
41. U. Kreibig, J. Phys. (Paris) **C2**, 97 (1977)
42. Y. Seol, A.E. Carpenter, T. Perkins, Opt. Lett. **31**, 2429 (2006)
43. E.J.G. Peterman, F. Gittes, C.F. Schmidt, Biophys. J. **1984**, 1308 (2003)
44. D. Lapotko, Opt. Express **17**, 2538 (2009)
45. F. Borghese, P. Denti, R. Saija, G. Toscano, O.I. Sindoni, J. Opt. Soc. Am. A **4**, 1984 (1987)
46. Y. Nakayama, P.J. Pauzauskie, A. Radenovic, R.M. Onorato, R.J. Saykally, J. Liphardt, P. Yang, Nature **447**, 1098 (2007)
47. P.J. Pauzauskie, A. Radenovic, E. Trepagnier, H. Shroff, P. Yang, J. Liphardt, Nature Materials **5**, 97 (2006)
48. R. Saija, M.A. Iatì, P. Denti, F. Borghese, O. Sindoni, Appl. Opt. **40**, 5337 (2001)
49. R. Saija, M.A. Iatì, F. Borghese, P. Denti, S. Aiello, C. Cecchi-Pestellini, Astrophys. J. **559**, 993 (2001)
50. F. Borghese, P. Denti, R. Saija, M.A. Iatì, O.I. Sindoni, J. Quant. Spectrosc. Radiat. Transf. **70**, 237 (2001)

Chapter 7
Rainbows, Coronas and Glories

Philip Laven

Abstract Rainbows, coronas and glories are examples of atmospheric optical phenomena caused by the scattering of sunlight from spherical drops of water. It is surprising that the apparently simple process of scattering of light by spherical drops of water can result in this wide range of colourful effects. However, the scattering mechanisms are very complicated. Eminent scientists (such as Descartes, Newton, Young, Airy and many others) offered various explanations for the formation of rainbows—thus making major contributions to our understanding of the nature of light. The basic features of rainbows can be explained by geometrical optics but, in the early 1800s, supernumerary arcs on rainbows provided crucial supporting evidence for the wave theory of light. In 1908, Mie provided a rigorous (but very complicated) solution to the problem of scattering of light by spherical particles. More than 100 years later, Mie's solution can now be used to produce excellent full-colour simulations. Examples of such simulations show how the appearance of these phenomena vary with the size of the water drops, as well as describing the scattering mechanisms that are responsible for their formation.

7.1 Introduction

The rainbow, corona and glory are examples of atmospheric optical phenomena caused by the scattering of sunlight from spherical drops of water. Eminent scientists (such as Descartes, Newton, Young, Airy and many others) offered various explanations for the formation of rainbows—thus making major contributions to our understanding of the nature of light [1, 3, 7, 37, 61].

Mie's rigorous solution [46] for the scattering of plane waves from homogeneous spheres has been available since 1908 but, for many years, its computational

P. Laven (✉)
9 Russells Crescent, Horley RH6 7DJ, United Kingdom
e-mail: philip@philiplaven.com

W. Hergert and T. Wriedt (eds.), *The Mie Theory*, Springer Series in Optical Sciences 169, DOI: 10.1007/978-3-642-28738-1_7,

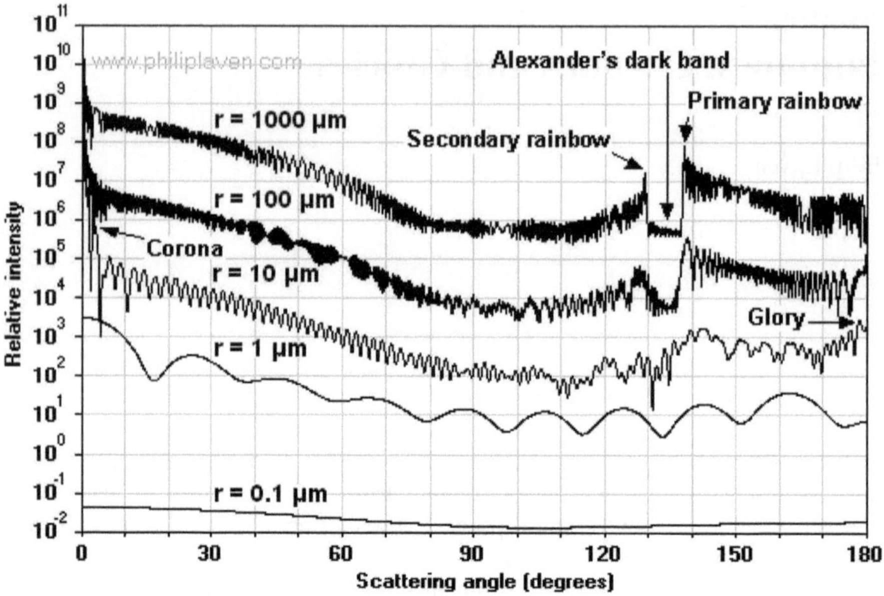

Fig. 7.1 Mie calculations for scattering of unpolarised monochromatic *red* light ($\lambda = 650\,\text{nm}$) by a spherical water drop with radius $r = 0.1\,\mu\text{m}$ to $r = 1{,}000\,\mu\text{m}$ for refractive index $n = 1.33257$

complexity was a formidable obstacle to its practical application, especially for large spheres. For scattering of light by water drops in the atmosphere, we are concerned with water drops of radius $r < 10\,\mu\text{m}$ in fog to $r > 1{,}000\,\mu\text{m}$ in heavy rain. Rapid advances in computer power now allow Mie's solution to be used for such calculations [15–18, 29, 30, 32, 36, 58–60] and, indeed, computer software for full-colour simulations of rainbows, coronas and glories is now freely available [11, 35].

Figure 7.1 shows the results of Mie calculations of scattered intensity for light of wavelength λ as a function of scattering angle θ for monochromatic red light for spherical water drops of different radius r. Note that $\theta = 0°$ implies forward scattering (i.e. in the original direction) and $\theta = 180°$ implies backscattering (i.e. back towards the source of the light). The primary and secondary rainbows are fairly obvious in Fig. 7.1 for large values of r, whereas the corona and the glory appear on the curve for $r = 10\,\mu\text{m}$ as a series of maxima and minima at $\theta \to 0°$ and $\theta \to 180°$ respectively.

Given the availability since 1908 of Mie's rigorous solution for scattering of light by homogeneous spherical particles, it might be assumed that everything is now known about the rainbow, corona and glory. However, as Mie's solution is based on the summation of a large number of terms (which converges when the number of terms is slightly greater than the size parameter $x = 2\pi r/\lambda$), it provides little information about the scattering mechanisms causing the rainbow, corona and glory. At this point, it is appropriate to refer to the Debye series (also published in 1908)

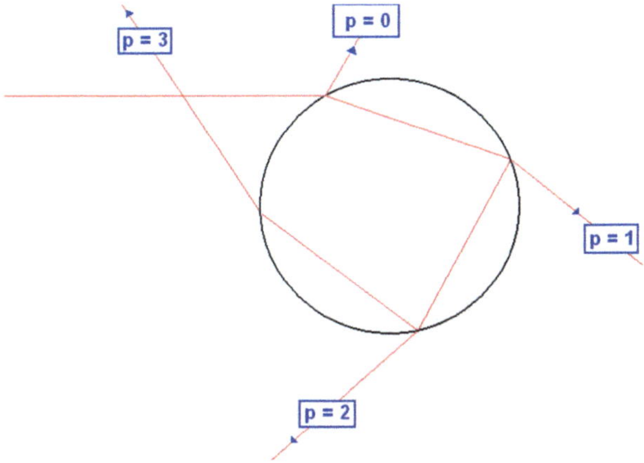

Fig. 7.2 Definitions of ray paths scattered from a sphere

[13, 23] which is essentially a reformulation of Mie's solution separating the contributions made by different scattering mechanisms of order p, where:

- $p = 0$ corresponds to external reflection plus diffraction;
- $p = 1$ corresponds to direct transmission through the sphere;
- $p = 2$ corresponds to one internal reflection;
- $p = 3$ corresponds to two internal reflections;
- and so on.

Figure 7.2 shows an example of the various ray paths, as indicated by different values of p. Unlike calculations based on geometrical optics, the Debye series is not an approximation: the summation of the Debye series for all integer values of p from zero to infinity gives the same result as Mie's solution. Figure 7.3 compares the results of Mie and Debye series calculations for scattering of red light from a water drop of radius $r = 100\,\mu$m. The results of Mie calculations are not easy to understand, whereas calculations using the Debye series reveal the intricacy of the scattering processes. For example, the Debye series calculations in Fig. 7.3 show that $p = 2$ rays are responsible for the primary rainbow at $\theta \approx 139°$ and that $p = 3$ rays are responsible for the secondary rainbow at $\theta \approx 128°$. Although such results are well known from geometrical optics, geometrical optics fails to predict that the appearance of rainbows varies with the radius r of the water drops (as is evident from the Mie results shown in Fig. 7.1). Similarly, Fig. 7.3 shows that the large variations in intensity in the Mie results at $\theta \approx 68°$ coincide with the crossing point of the $p = 0$ and $p = 1$ contributions in the Debye series calculations. In this area, the intensities of the two components are similar, but their relative phases vary rapidly as a function of θ thus producing the complicated interference pattern at $\theta \approx 68°$.

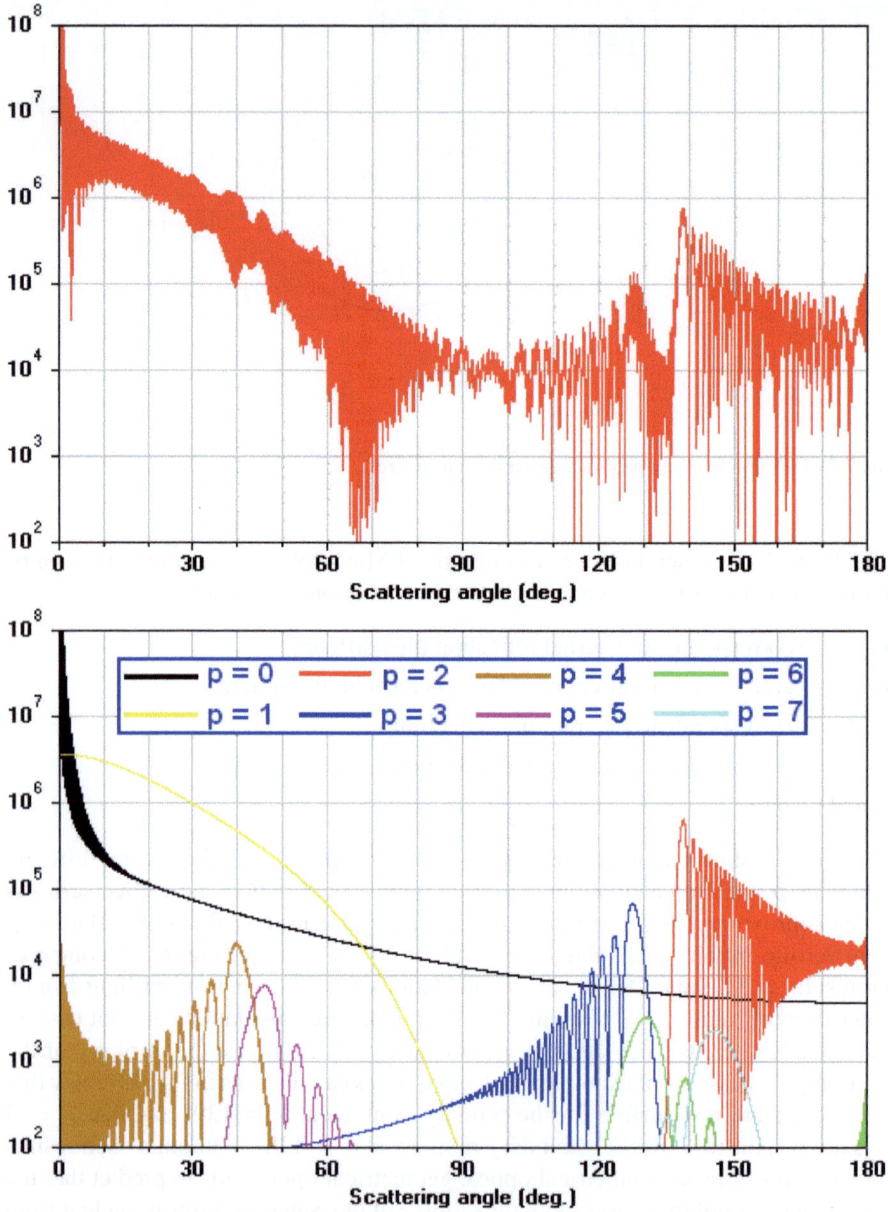

Fig. 7.3 Comparison of Mie calculations (*top*) and Debye series calculations (*bottom*) for scattering of monochromatic red light ($\lambda = 650\,\text{nm}$) from a spherical water drop with radius $r = 100\,\mu\text{m}$ for refractive index $n = 1.33257$ (perpendicular polarisation)

7.2 Rainbows

Rainbows have long played an important role in the development of scientific knowledge. Descartes and Newton used geometrical optics to explain the formation of rainbows. In discussing Figs. 7.1 and 7.3, the term *rainbow* has been used to denote the concentration of light in a narrow angular zone (such as $138° < \theta < 140°$) despite the fact that these graphs are concerned with the scattering of monochromatic light. To produce simulations of natural rainbows caused by the scattering of sunlight from water drops, Mie calculations must be performed at many wavelengths across the visible spectrum, taking account of the spectrum of sunlight and of the variations in the refractive index n of water.

According to geometrical optics, the rainbow angles θ_{rainbow} for $p = 2$ and $p = 3$ can be calculated according to the following formulae [2]:

Primary rainbow ($p = 2$):

$$\theta_{\text{rainbow}} = 2 \arccos \left[\frac{1}{n^2} \left(\frac{4 - n^2}{3} \right)^{\frac{3}{2}} \right] \tag{7.1}$$

Secondary rainbow ($p = 3$):

$$\theta_{\text{rainbow}} = 2 \arcsin \left[\sqrt{n^2 - 1} \left(\frac{\sqrt{9 - n^2}}{2n} \right)^3 \right] \tag{7.2}$$

As θ_{rainbow} is dependent on refractive index n, θ_{rainbow} varies with wavelength as shown in Table 7.1 which indicates that the colours of the secondary rainbow are reversed in order compared with the colours of the primary rainbow. Table 7.1 also shows that the width of the primary rainbow (from violet to red) is about 1.9°, whilst the width of the secondary rainbow is about 3.4°. However, it must be stressed that calculations using geometrical optics for rainbows do not entirely agree with the results of Mie calculations: first, geometric optics wrongly predicts infinite intensity at the rainbow angle and, second, Mie calculations show that the peak intensity is offset slightly from the geometric rainbow angle θ_{rainbow}. For example, Mie and Debye series results show that for $\lambda = 650$ nm and a water drop of radius $r = 100\,\mu$m, the maximum intensity occurs at $\theta \approx 138.8°$, whereas geometrical optics suggests a primary rainbow angle of $\theta_{\text{rainbow}} = 137.9°$ irrespective of r.

Another failing of geometric optics is that it does not predict the series of maxima and minima associated with rainbows (known as supernumerary arcs), which are clearly shown in Fig. 7.3 on the $p = 2$ and $p = 3$ curves obtained with the Debye series. What causes these supernumerary arcs? The primary rainbow marks the transition between no geometrical rays when $\theta < \theta_{\text{rainbow}}$ and two geometrical rays when $\theta > \theta_{\text{rainbow}}$ (as shown in Fig. 7.4). As the two rays corresponding to a given value of

Table 7.1 Geometric rainbow angles for water

Wavelength λ (nm)	n	Primary θ_{rainbow}	Secondary θ_{rainbow}
400	1.34451	139.57°	126.14°
450	1.34055	139.01°	127.15°
500	1.33772	138.60°	127.88°
550	1.33560	138.30°	128.43°
600	1.33393	138.06°	128.87°
650	1.33257	137.86°	129.22°
700	1.33141	137.69°	129.53°

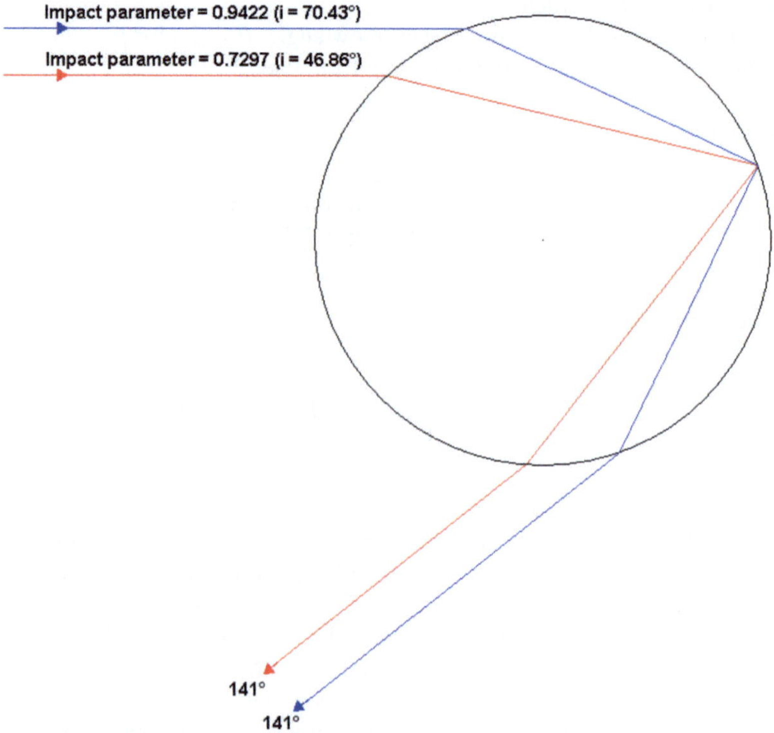

Fig. 7.4 Two $p = 2$ geometric rays contributing to scattering at $\theta = 141°$ by a water drop with refractive index $n = 1.33257$—in this case, the geometric rainbow angle $\theta_{\text{rainbow}} = 137.9°$ corresponds to an impact parameter $b = \sqrt{\arccos[(n^2 - 1)/(p^2 - 1)]} = 0.861$ (i.e. between the two rays shown)

$\theta > \theta_{\text{rainbow}}$ have different optical path lengths, there is a phase difference between the two rays. Constructive interference occurs when this phase difference is a multiple of 360° (leading to a maximum of intensity), whilst destructive interference occurs when the phase difference is an odd multiple of 180° (leading to a minimum of intensity). The maxima and minima in the scattered intensity as a function of θ

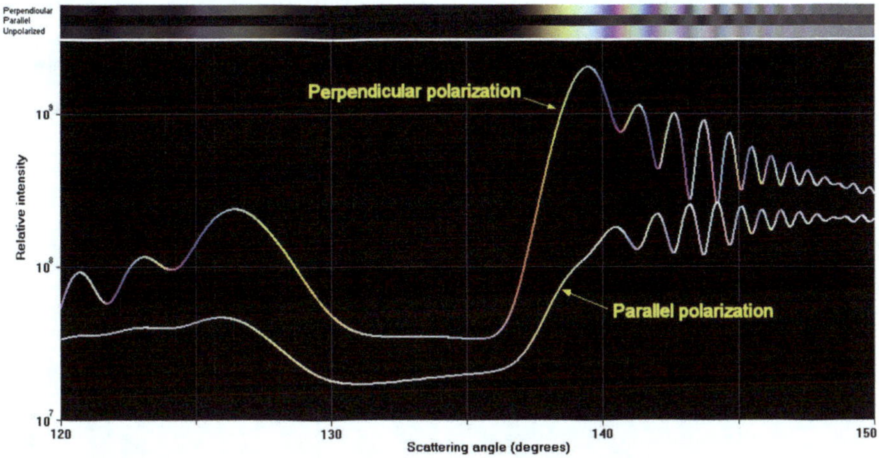

Fig. 7.5 Scattering of sunlight by a spherical drop of water with radius $r = 100\,\mu\text{m}$

occur because the phase difference varies with θ. From a historical perspective, it is noteworthy that the appearance of supernumerary arcs on natural rainbows was used by Young to support the wave theory of light [61].

Figure 7.5 shows Mie calculations for scattering of sunlight by a spherical drop of water with radius $r = 100\,\mu\text{m}$. The graph shows that the intensity of the primary rainbow is about one order of magnitude greater than that of the secondary rainbow, whilst the horizontal coloured bars immediately above the graph show simulations of the rainbows—with the top bar showing perpendicular polarisation, the middle bar showing parallel polarisation and the bottom bar showing unpolarised light. Both rainbows are strongly polarised, with perpendicular polarisation being dominant in both cases. Note that the maxima of the primary rainbow's supernumerary arcs for parallel polarisation coincide with the minima for perpendicular polarisation (and vice versa) [27]. The reason for this extraordinary behaviour is that, for parallel polarisation, an abrupt phase change of 180° (due to the Brewster angle) occurs when the impact parameter $b = \sin[\arctan[n]]$. For $n = 1.33257$, this critical value occurs when $b = 0.7998$. Figure 7.4 shows that, for $\theta = 141°$, one of the rays is at $b < 0.7998$, whilst the other is at $b > 0.7998$. This means that the phase difference between the two rays for parallel polarisation is 180° greater than the phase difference for perpendicular polarisation—thus explaining the coincidence of the maxima and minima of the supernumerary arcs. In fact, the real *coincidence* is that critical impact parameter $b = \sin[\arctan[n]]$ and the impact parameter $b = \sqrt{\arccos[(n^2 - 1)/(p^2 - 1)]}$ corresponding to θ_{rainbow} are very similar for $p = 2$ and for the refractive index n of water.

Figure 7.6 shows a series of simulations of the rainbows caused by the scattering of sunlight by spherical water drops of various values of r between 10 and $500\,\mu\text{m}$. For very small water droplets (e.g. for $r = 10\,\mu\text{m}$), the rainbow (or fogbow) is almost white and not well-defined. For $r = 50\,\mu\text{m}$, the primary rainbow is predom-

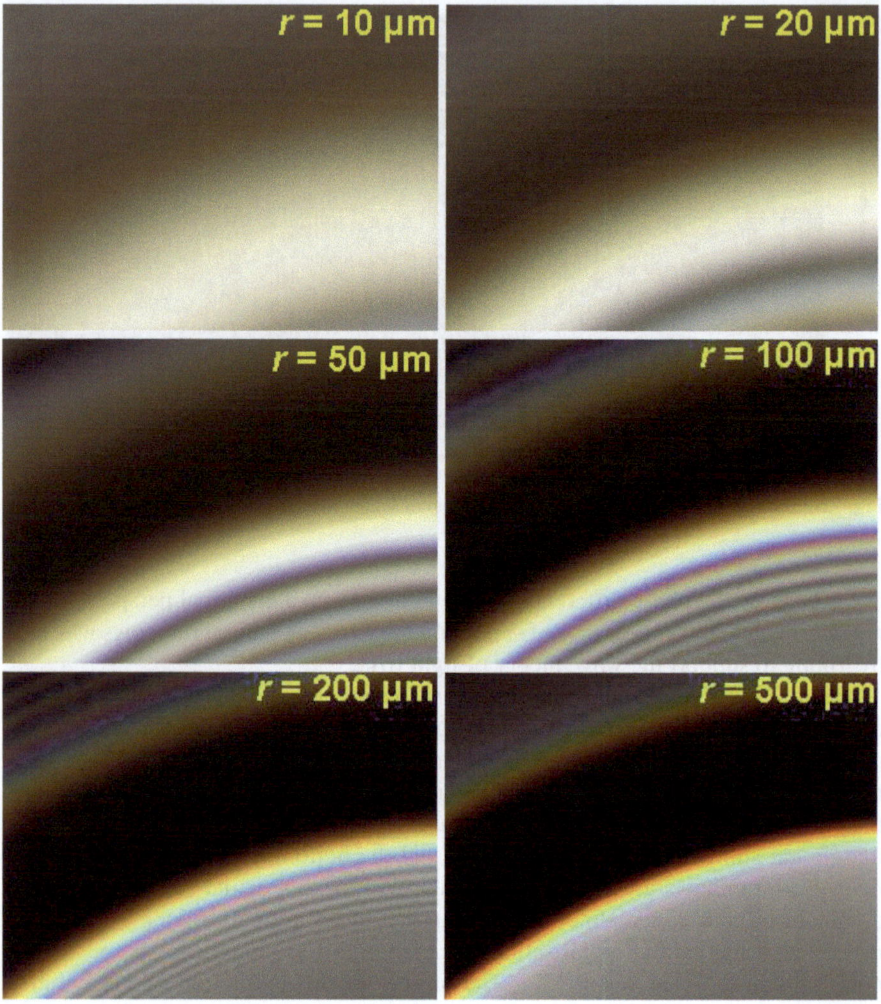

Fig. 7.6 Simulations of the primary and secondary rainbow caused by scattering of sunlight by spherical water drops of specified radius r

inantly white with a hint of red at its outer edge. The colours of the primary rainbow become more obvious for larger drops, together with the fainter secondary rainbow and Alexander's dark band. Supernumerary arcs can be seen inside the primary rainbow for $r = 50\,\mu\text{m}$, $r = 100\,\mu\text{m}$ and $r = 200\,\mu\text{m}$, as well as on the outside of the secondary rainbow for $r = 100\,\mu\text{m}$ and $r = 200\,\mu\text{m}$. However, the supernumerary arcs disappear for larger water drops (e.g. $r = 500\,\mu\text{m}$) because the angular separation between these arcs is less than the sun's apparent angular diameter of 0.5°.

The Lee diagram [36] in Fig. 7.7 illustrates more generally how the appearance of rainbows varies with the radius r of the water drops. Each coloured point in Fig. 7.7

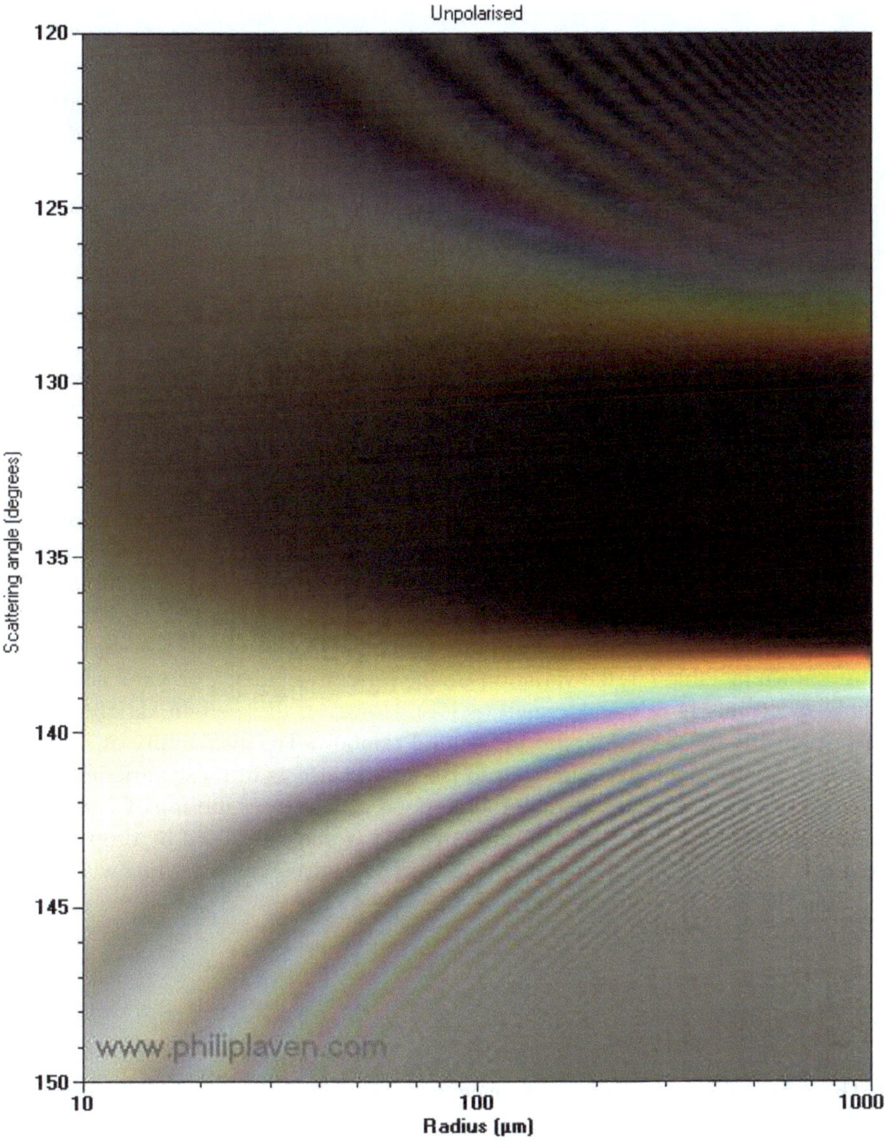

Fig. 7.7 Lee diagram showing variations in the appearance of primary and secondary rainbows caused by scattering of sunlight by spherical water drops as a function of drop radius from $r = 10\,\mu\text{m}$ to $r = 1{,}000\,\mu\text{m}$

represents the colour of light scattered in a specific direction θ by a drop of radius r. This diagram has been calculated for 400 values of r between $10\,\mu\text{m}$ and $1{,}000\,\mu\text{m}$. For each value of r, the primary and secondary rainbows are depicted by a vertical column of colours, in which the brightness of column has been normalised by the

Fig. 7.8 *Left* simulation of the primary and secondary rainbows caused by spherical drops of water with radius $r = 100\,\mu\text{m}$. *Right* simulation of primary and secondary rainbows caused by spherical drops of fixed refractive index $n = 1.33$ across the visible spectrum with radius $r = 100\,\mu\text{m}$

maximum luminance for that value of r. Figure 7.7 shows that the angle of maximum intensity for the primary rainbow decreases from $\theta \approx 142°$ when $r = 10\,\mu\text{m}$ to $\theta \approx 138°$ when $r = 1{,}000\,\mu\text{m}$. Similarly, the angle of maximum intensity for the secondary rainbow increases from $\theta \approx 124°$ when $r = 10\,\mu\text{m}$ to $\theta \approx 128°$ when $r = 1{,}000\,\mu\text{m}$.

As shown in Table 7.1, the variations in the refractive index n of water across the visible spectrum are relatively small ($\approx$1%). However, these slight variations in n are responsible for the colours of the natural rainbow. The availability of software for simulation of rainbows allows us to isolate the effects of these variations in n. For example, Fig. 7.8 compares natural rainbows caused by scattering of sunlight by water drops with the hypothetical rainbows that would be caused by drops with a fixed value of $n = 1.33$—the hypothetical rainbows are essentially white with tinges of colour on the supernumerary arcs.

7.3 Coronas

The corona appears as a series of concentric coloured rings around a cloud-covered sun or moon. As the brightness of the sun is generally overwhelming to the human eye, coronas are more frequently observed around the moon.

Figure 7.9 compares simulations of the corona and glory caused by scattering of sunlight by water droplets of radius $r = 10\,\mu\text{m}$. The brightness of the simulated corona in Fig. 7.9 has been increased by a factor of 20 so that the sun and its immediate surroundings are "over-exposed." The corona is obviously centred on the sun or moon at $\theta = 0$, whilst the glory is centred on the anti-solar point at $\theta = 180°$. The corona and the glory have different sizes, but the sequence of colors is almost identical. The key differences are that the glory for $r = 10\,\mu\text{m}$ has a dark ring around $\theta \approx 179.3°$

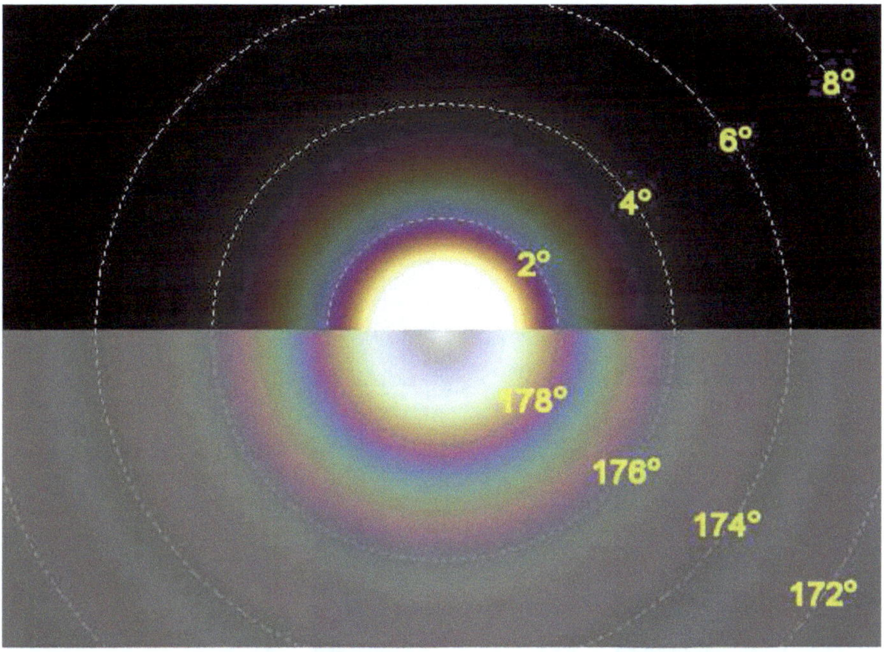

Fig. 7.9 Simulations of the corona (*top*) and glory (*bottom*) caused by scattering of sunlight by spherical water drops of radius $r = 10\,\mu\text{m}$

and that, in general, the outer rings of glories are relatively bright compared with those of coronas.

Although similar in appearance, these two phenomena are caused by completely different mechanisms. The corona is generally considered to be due to diffraction by small droplets of water in the clouds [10, 47, 54, 55, 58, 59]—and thus the intensity of the corona $I(\theta)$ is calculated using the Fraunhofer diffraction equation[1]:

$$I(\theta) \propto \left(x^2 \left[\frac{1+\cos\theta}{2}\right]\left[\frac{J_1(x\sin\theta)}{x\sin\theta}\right]\right)^2 \tag{7.3}$$

where:

- θ is the scattering angle;
- J_1 is the first-order Bessel function;
- $x = 2\pi r/\lambda$ where r is the radius of the scattering sphere and λ is the wavelength of the incident light.

[1] Such equations are frequently attributed to Fraunhofer, but Craig Bohren has pointed out in a private communication that, without diminishing the importance of Fraunhofer's pioneering work in experimental optics, there is no evidence to suggest that Fraunhofer developed any theoretical treatment of diffraction. Consequently, *Fresnel-Fraunhofer-Airy-Schwerd* might be a more appropriate designation. Schwerd's contribution is described in [22].

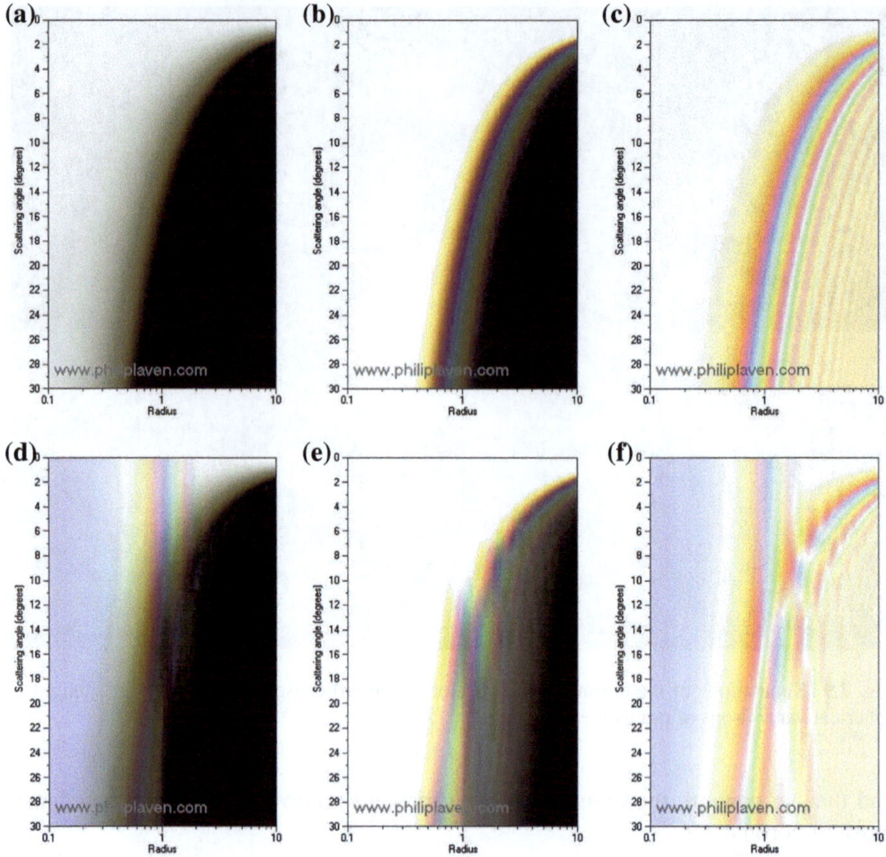

Fig. 7.10 Lee diagrams showing the corona caused by scattering of sunlight by spherical water drops as a function of drop radius from $r = 0.1\,\mu\text{m}$ to $r = 10\,\mu\text{m}$. **a–c** show diffraction calculations. **d–f** show Mie calculations. The brightness of (**b**) and (**e**) has been increased by a factor of ten to reveal the colours of the corona. The results shown in **a** and **b** have been processed to remove the brightness information, leaving only the saturated colours

The rings of the corona correspond to maxima in the last term of Eq. 7.3: the first 4 maxima occur when $x \sin\theta = 5.136$, 8.418, 11.62 and 14.796—corresponding to $\sin\theta = 0.8174\lambda/r$, $1.3398\lambda/r$, $1.8494\lambda/r$ and $2.3549\lambda/r$. If θ is measured in degrees and r is measured in μm, the inner rings of the corona for red light ($\lambda = 0.65\,\mu\text{m}$) are therefore defined by $\theta_1 = 30.5/r$, $\theta_2 = 50/r$, $\theta_3 = 69/r$ and $\theta_4 = 88/r$.

There are significant differences between the diffraction model and Mie's solution for the corona, as highlighted by van de Hulst [57] and by Lock and Yang [38]. The Lee diagrams for the corona in Fig. 7.10 compare the diffraction model with Mie calculations for scattering of sunlight from water droplets for $0.1 < r < 10\,\mu\text{m}$. Figure 7.10a shows the results of diffraction calculations. As the brightest scattering occurs at $\theta = 0°$, Fig. 7.10a contains little information because the corona is very much darker than the forward scattered light (i.e the sun or the moon) and computer displays cannot reproduce the necessary large dynamic range. To overcome this problem, the brightness of Fig. 7.10b has been increased by a factor of 10 so that the top part of is "over-exposed", allowing the colours of the corona to become visible. The data from Fig. 7.10a have been re-plotted in Fig. 7.10c so as to remove the brightness information, instead showing the saturated colour of each pixel. This form of presentation shows that, according to the diffraction model, the sequence of colours in the corona is independent of r. Figure 7.10c suggests that no more than two or three rings of the corona will be visible even under optimum viewing conditions. Although diagrams based on saturated colours (such as Figs. 7.10c and f) are useful for comparing colours, it must be emphasised that they do not represent the appearance of scattered light.

Looking now at the equivalent diagrams (Figs. 7.10d–f) produced using Mie calculations, the key difference between Fig. 7.10a and d is that, for $0.5\,\mu\text{m} < r < 2\,\mu\text{m}$, Mie calculations predict uniform bands of colour at scattering angles $\theta < 8°$: for example, red for $r = 0.8\,\mu\text{m}$ and for $r = 1.6\,\mu\text{m}$ and violet for $r = 1\,\mu\text{m}$. Comparison of Fig. 7.10b and e shows that Mie calculations produce very complex patterns for $r < 3\,\mu\text{m}$. Comparison of Fig. 7.10c and f indicates that Mie and diffraction calculations produce similar results for $r > 5\,\mu\text{m}$, but the diffraction model is totally inadequate for $r < 3\,\mu\text{m}$. Figure 7.10f also confirms the findings of Gedzelman and Lock [18] who reported:

> The sequence of corona colors changes rapidly for small droplets but becomes fixed once droplet radius exceeds about $6\,\mu\text{m}$.

The irregular patterns shown in Fig. 7.10 are caused by interference between $p = 0$ (diffraction) and $p = 1$ (transmission through the sphere) contributions. Obviously, the $p = 1$ contributions are not taken into account in diffraction calculations using Eq. 7.3.

More generally, the Lee diagrams in Fig. 7.10 for scattering of sunlight show that the size of the corona is inversely proportional to the radius r of the water droplets: Fig. 7.10c suggests that the inner 3 red rings occur at $\theta_1 \approx 16/r$, $\theta_2 \approx 31/r$ and $\theta_3 \approx 47/r$. These results for scattering of sunlight should be compared with those obtained above for monochromatic red light: the red ring at $\theta_1 \approx 16/r$ does not coincide with a maximum of the Fraunhofer equation. The "main beam" of the diffraction pattern is wider for red light than for the rest of the visible spectrum—thus producing a red ring that has no equivalent in the diffraction pattern of monochromatic red light. However, the next two red rings at $\theta_2 \approx 31/r$ and $\theta_3 \approx 47/r$ for sunlight are close to the values of $30.5/r$ and $50/r$ obtained for monochromatic red light.

The diffraction model is appealing because of its simplicity compared with Mie calculations, but the above results show that it can give misleading results for the corona—because the corona is not solely due to diffraction. In any event, the diffraction model relies on the evaluation of Bessel functions, thus requiring the user either to use published tables of Bessel functions or to adopt a numerical solution—thus negating some of its apparent simplicity. Given the ready availability of software for Mie calculations, it is suggested that the diffraction model should be avoided.

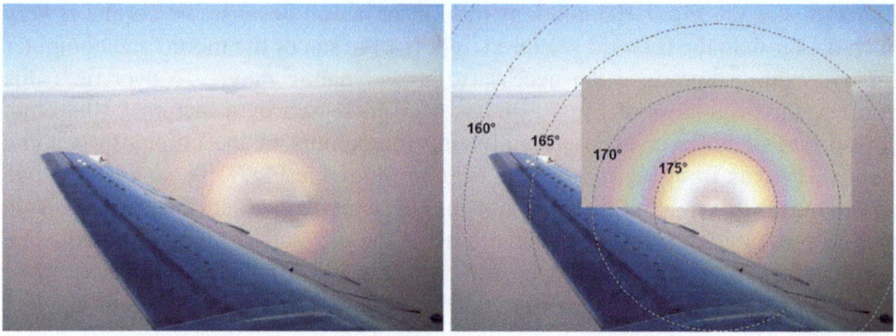

Fig. 7.11 *Left* A glory as seen from a commercial aircraft. *Right* Mie simulation of this glory assuming scattering of sunlight by spherical water droplets with radius $r = 4.3\,\mu\text{m}$

7.4 Glories

In the past, sightings of glories were rare and usually associated with the Brocken Spectre, which refers to the elongated shadows of mountaineers appearing on fog or clouds beneath them. In such circumstances, a mountaineer might observe that the shadow of his head was surrounded by a glory (consisting of a series of concentric coloured rings). Strangely, such coloured rings appear around the shadow of your head, but not apparently around the shadows of your companions! Nowadays, glories are seen much more frequently surrounding the shadow of an aircraft on clouds: the author has taken more than 4,000 pictures of glories whilst travelling on commercial aircraft within the last 10 years. Figure 7.11 shows one such picture, together with a matching simulation based on Mie calculations.

The effects of viewing a glory through a vertical polariser (equivalent to the view seen through polarised sunglasses worn in the usual orientation) are simulated in Fig. 7.12. The distinctive dark spots immediately above and below the antisolar point in the right-hand image in Fig. 7.12 were initially predicted by Können [28]. These theoretical simulations have been confirmed by the publication of several images of polarised glories [32]. Figure 7.13 offers an explanation of the formation of polarised glories.

Figure 7.14 is a Lee diagram showing how the appearance of the glory varies with the radius r of the water drop. The sequence of ring colors is essentially independent of r. Glories with large rings imply that they are caused by scattering from small droplets, whereas glories with small rings imply large droplets. To a first approximation, the radius R of a given ring from the anti-solar point $\theta = 180°$ is inversely proportional to r; for example, the four inner red rings have radii of $R_1 \approx 24/r$, $R_2 \approx 37/r$, $R_3 \approx 56/r$ and $R_4 \approx 75/r$, where R is measured in degrees and r is measured in μm. This very simple relationship allows the values of r of the clouds to be estimated very easily from images of glories [33].

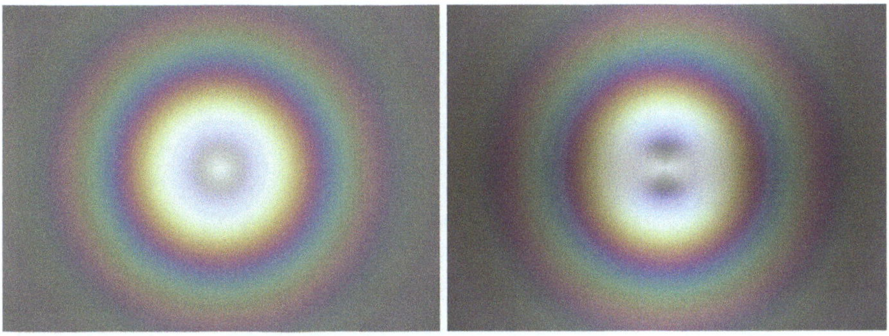

Fig. 7.12 The left image shows a simulation of a glory caused by scattering of sunlight from spherical water droplets with $r = 10\,\mu$m (the width of the image is about $\pm 5°$). The right image shows the same glory as viewed through a polariser with its transmission axis vertical

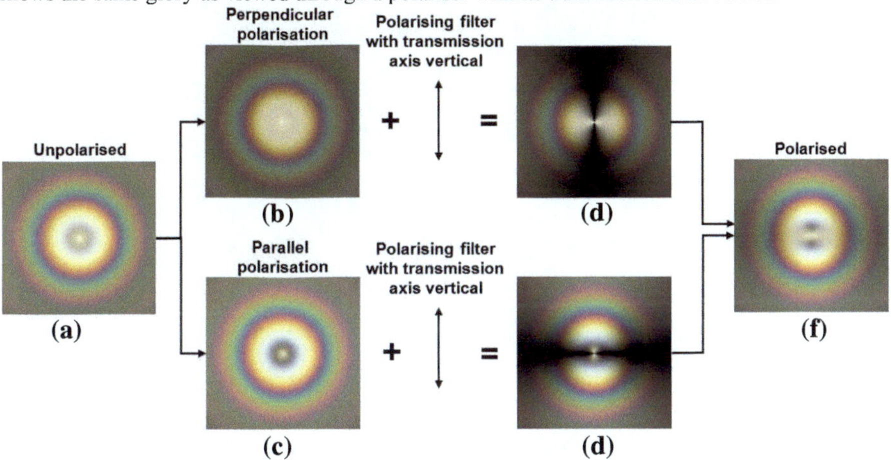

Fig. 7.13 An unpolarised glory, shown as image (**a**), can be separated into two components (**b**) and (**c**), which have polarisations perpendicular and parallel to the scattering plane, respectively. When viewed through a vertical polariser, (**b**) is transformed into (**d**), while (**c**) is transformed into (**e**). Image (**d**) shows that the polariser suppresses perpendicular polarisation along the vertical line through the antisolar point, but has no effect along the horizontal line. This is similar to the effect of viewing the primary rainbow through a polariser, because the primary rainbow is dominated by perpendicular polarisation. Image (**e**) shows that the polariser suppresses parallel polarisation along the horizontal line through the antisolar point. When (**d**) and (**e**) are combined, the resulting image (**f**) shows the *polarised* glory with its distinctive dark spots above and below the antisolar point. Note that the colors of the polarised glory along the horizontal line through the antisolar point correspond to perpendicular polarisation, whereas those along the vertical correspond to parallel polarisation. Other orientations obviously involve a mixture of the two polarisations

Most glories seem to be caused by water droplets with $4 < r < 25\,\mu$m, thus overlapping with the white fogbows described in Sect. 7.2. In fact, fogbows can usually be observed when glories are visible—and vice versa. In practice, it can be difficult to observe a glory and a fogbow simultaneously, especially as glories

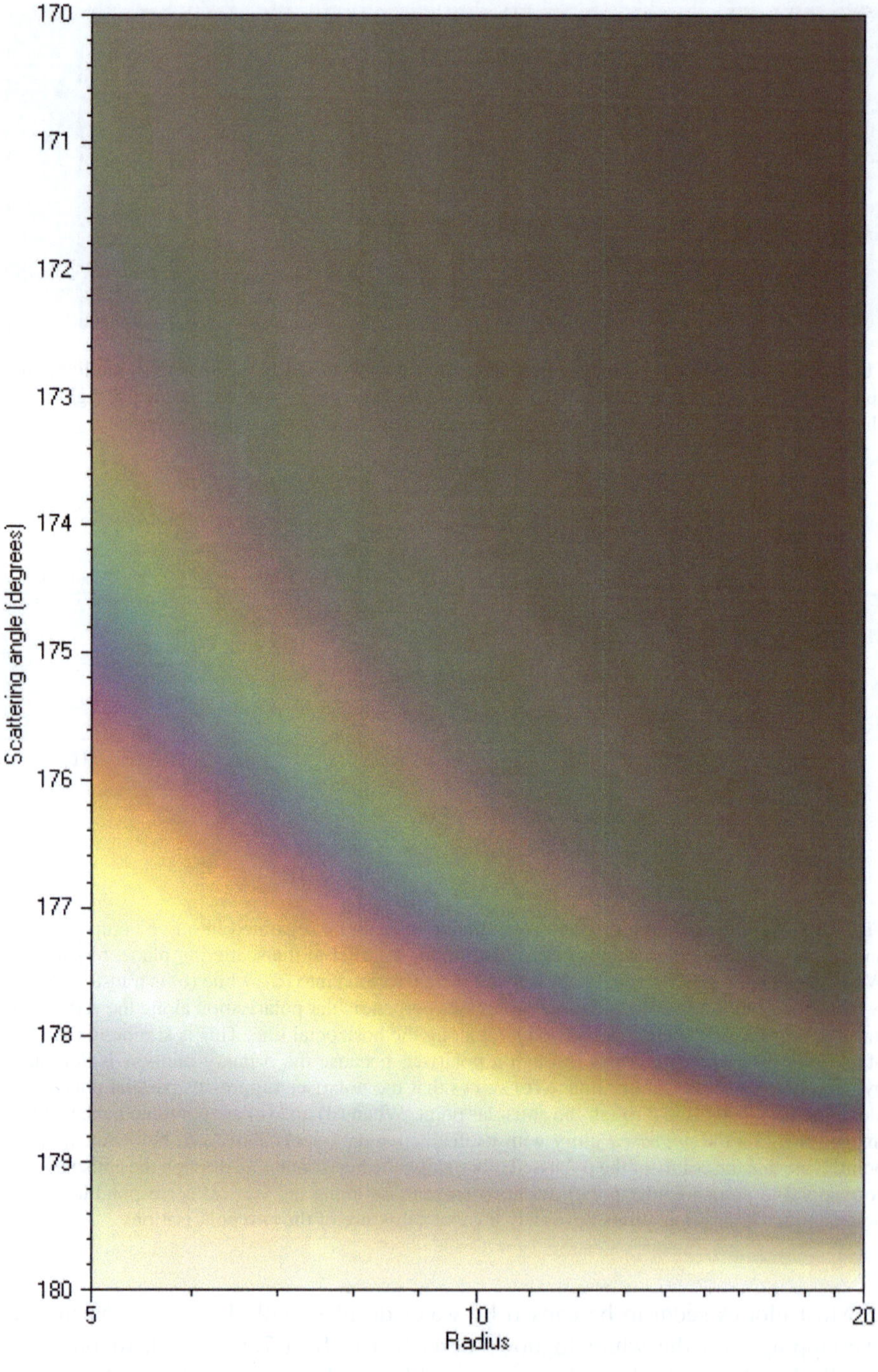

Fig. 7.14 Lee diagram showing variations in the appearance of glories caused by scattering of sunlight by spherical water drops as a function of drop radius r from $r = 5\,\mu\text{m}$ to $r = 20\,\mu\text{m}$

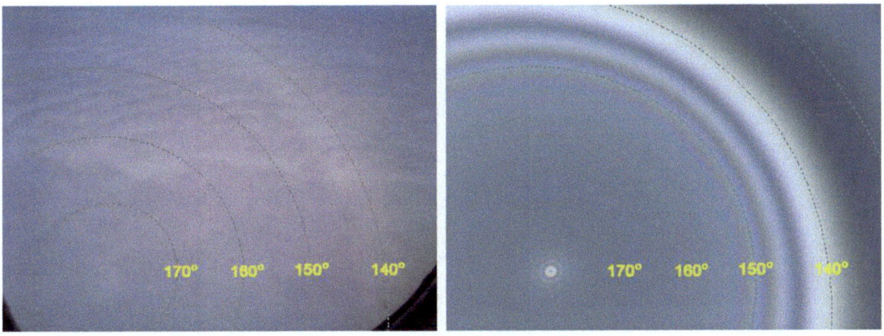

Fig. 7.15 *Left* image of a glory and a cloudbow. *Right* Mie simulation assuming scattering of sunlight by water drops of radius $r = 20\,\mu\text{m}$

are now most frequently seen from inside commercial aircraft where the field of view is generally very limited. Nevertheless, Fig. 7.15 shows a layer of cloud which produced a glory at $\theta > 176°$ and a primary rainbow at $\theta \approx 142°$—it would be more appropriate to call the latter a "cloudbow". In practice, such effects are not easy to observe because white clouds are typically very bright, but one useful tip is to look for the dark band just inside the fogbow/cloudbow (at $\theta \approx 144°$ in Fig. 7.15). Having found this dark band, it is much easier to identify the white band corresponding to the primary fogbow/cloudbow.

The Lee diagram in Fig. 7.16 shows that, although the angular size of the coloured rings of the glory decreases as r increases, the angular size of the fogbow or primary rainbow (as measured from the anti-solar point) increases with r. Although the rings of the glory and the fogbow (and its supernumerary arcs) are easily separable when seen in Fig. 7.16, the simulation in Fig. 7.17 shows a puzzling mixture of glory rings, supernumerary arcs and fogbows that can be observed when $r = 5\,\mu\text{m}$. Confronted with such a complex display, it is almost impossible for observers to identify which are glory rings and which are supernumerary arcs!

Mie's solution can be used to simulate glories, but it does not offer any explanation for their formation. In his seminal paper *A theory of the anti-coronae* [56], van de Hulst suggested that the glory could be explained as the diffraction pattern corresponding to a toroidal wave front and noted that back-scattering $\theta = 180°$ by geometrical $p = 2$ rays (apart from the trivial case of central rays with impact parameter $b = 0$) would require refractive index $n > \sqrt{2} \approx 1.414$. Given that $n \approx 1.333$ for water, this explanation left something to be desired! However, in his 1957 book [57], van de Hulst extended his treatment of the glory to propose that it is caused by $p = 2$ rays with $b = \pm 1$ and that the 14.4° gap between $\theta = 165.6°$ and $\theta = 180°$ could be bridged by surface waves, as illustrated in Fig. 7.18. This gap of 14.4° is depicted in Fig. 7.18 for simplicity as the last part of the ray path before it emerges from the sphere, but this gap could be covered by three separate segments of surface waves covering a total of 14.4°. Because of symmetry of the sphere, ray

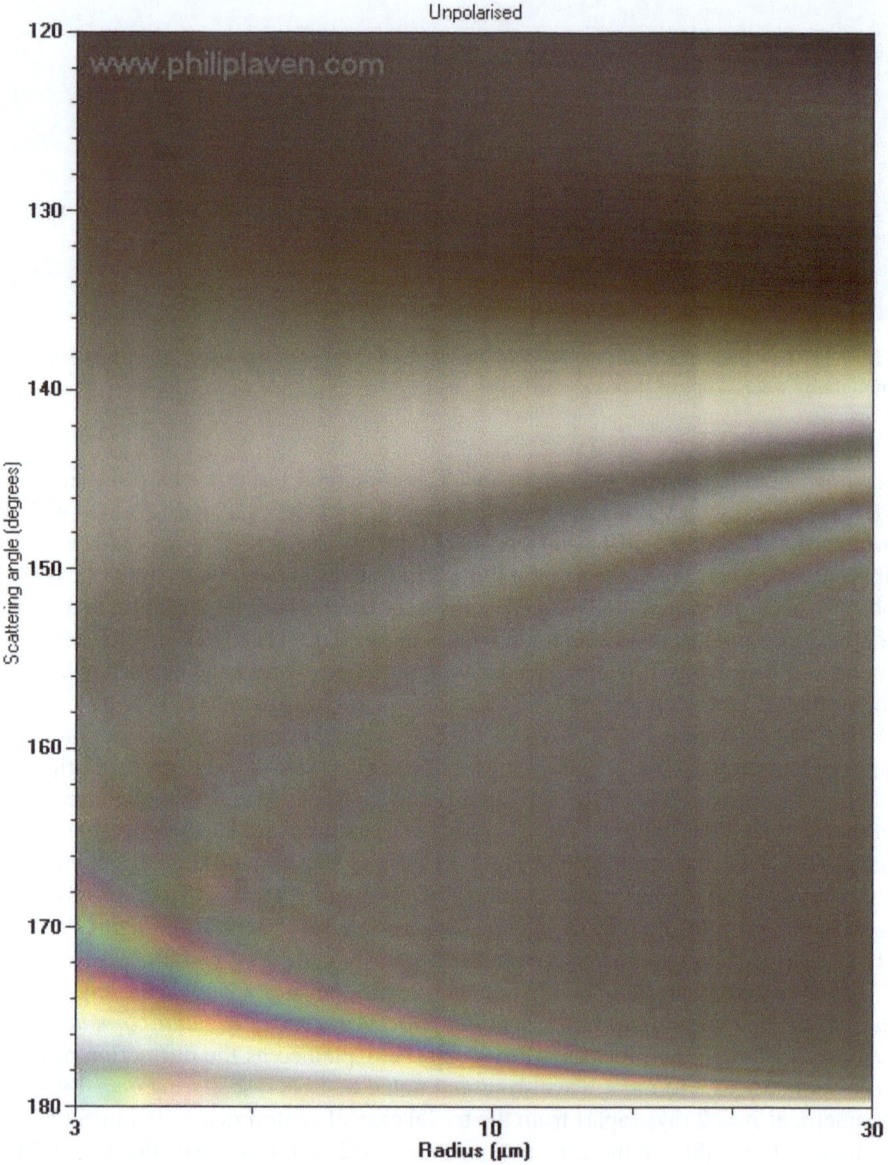

Fig. 7.16 Lee diagram showing variations in the appearance of glories and fogbows caused by scattering of sunlight by spherical water drops as a function of drop radius from $r = 3\,\mu\text{m}$ to $r = 30\,\mu\text{m}$

paths of the form shown in Fig. 7.18 generate a toroidal wavefront with a diameter nearly equal to that of the sphere, propagating in the direction $\theta = 180°$.

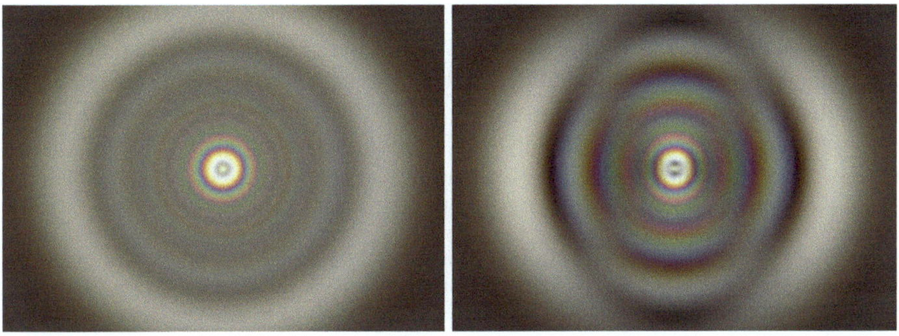

Fig. 7.17 The *left* image shows a Mie simulation of scattering of sunlight by spherical water drops of radius $r = 5\,\mu\text{m}$ as would be captured by a camera with a very wide-angle lens: the width of this image is $\approx\pm 50°$ measured from the anti-solar point. The *right* image shows the same glory as viewed through a polariser with its transmission axis vertical

Fig. 7.18 van de Hulst's surface wave associated with a $p = 2$ ray with impact parameter $b = 1$ for a sphere with refractive index $n = 1.333$

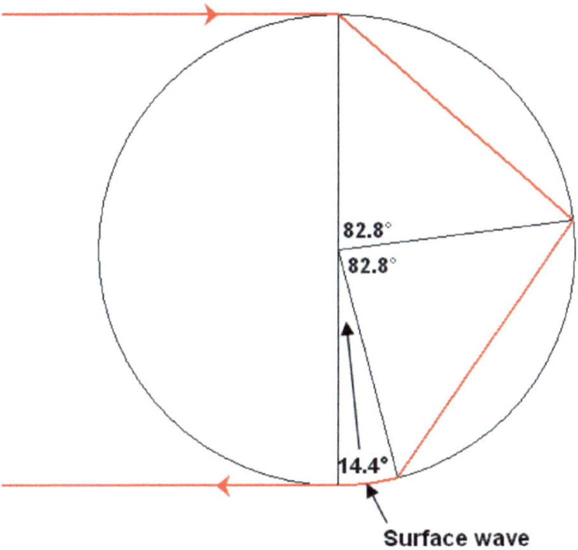

Fahlen and Bryant [14] conducted an experiment in which a water drop of 1.23 mm diameter was suspended in a sound field and illuminated by a He-Ne laser. They reported that, when the drop was viewed from $\theta \approx 180°$, bright patches of light appeared near the centre of the droplet (probably corresponding to light reflected from the front and back of the water drop for $b \approx 0$) and around the circumference of the sphere. Although the latter observation is consistent with van de Hulst's explanation, it does not provide any information about the physical nature of the glory-making rays or about the actual role of surface waves.

The pioneering work of van de Hulst has been endorsed and extended by many other authors [8, 9, 12, 24]. Nussenzveig's many studies of the glory [26, 48–51]

include use of the Debye series to analyse the relative importance of various scattering processes in the formation of the glory. His results indicated that $p = 2$ contributions are dominant for droplets with effective radius $x = 150$, corresponding to $r \approx 13\,\mu\text{m}$ for scattering of white light.

Despite the success of van de Hulst's theory for the formation of the glory, various authors have bemoaned the lack of a simple physical model for the mechanisms causing the glory. For example, Lynch and Livingston [43] remark:

> Although the glory pattern is correctly predicted by Mie theory, a good physical explanation is, in our opinion, lacking. In some way light is back-scattered after traversing the periphery of the droplet. Examined in detail, each drop is found to shine uniformly around its edge with an annulus of light that is coherent (the waves are in phase).

Greenler [21] notes:

> In one sense, the glory is now well understood. A mathematical theory (Mie scattering theory) enables us to calculate the intensity variation in the glory pattern. Unfortunately, it gives us little physical insight into the process that produces the rings ...I wonder if there is no simple model containing the physical essence of the glory.

Similarly, Bohren and Huffmann [6] state:

> Unlike the rainbow, the glory is not easy to explain, other than to say that it is a consequence of all of the thousands of terms in the scattering series, a correct but unsatisfying statement.

In a paper [52] entitled *Does the glory have a simple explanation?* Nussenzveig tried to respond to the above requests, but he concluded that:

> Mie theory describes the glory by the sum of a large number of complicated terms within which the physical mechanisms cannot be discerned. CAM [Complex Angular Momentum] theory brings out the dominant physical effects and provides an accurate representation for each of them. That it does so by analytic continuation seems inevitable. I know of no other way of quantitatively representing tunneling.

The above extracts indicate that the glory cannot be explained (even by eminent scientists), except by "scientific arm waving". Fortunately, the availability of software for Mie and Debye series calculations allow us to investigate the glory in considerable detail. For example, Fig. 7.19 shows that the Mie results are almost identical to the Debye series $p = 2$ contributions (i.e. those that have suffered one internal reflection in the sphere). This indicates that $p = 2$ scattering is the dominant mechanism for the formation of the glory—except in the vicinity of $\theta = 180°$ where $p = 11$, $p = 7$ and $p = 6$ scattering makes a significant contribution. The coloured rings of the glory (e.g. between $\theta = 175°$ and $\theta = 178°$) are obviously due to Debye $p = 2$ rays. Note that the colours of the plotted lines represent the saturated colours. In particular, the $p = 0$ contribution (caused by reflection from the exterior of the sphere and by diffraction) is essentially white, but all of the contributions from higher orders (e.g. $p = 6$, $p = 7$, $p = 11$ and so on) show colours which are essentially identical to those for $p = 2$.

Figure 7.20 shows two simulations of the glory caused by scattering of sunlight from a spherical drop of water with $r = 10\,\mu\text{m}$: the left side shows a simulation

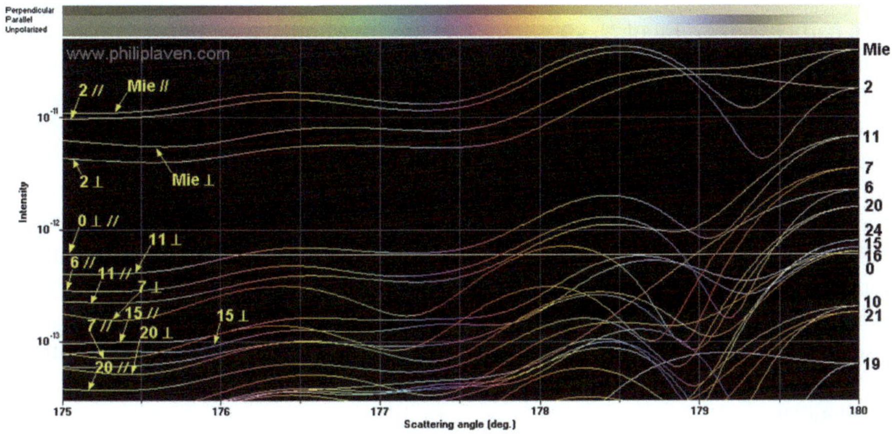

Fig. 7.19 Mie and Debye series calculations showing backscattering of sunlight by a spherical water drop of radius $r = 10\,\mu\text{m}$. The labels on the curves indicate the value of p with the symbols $\perp$ indicating perpendicular polarisation and // denoting parallel polarisation

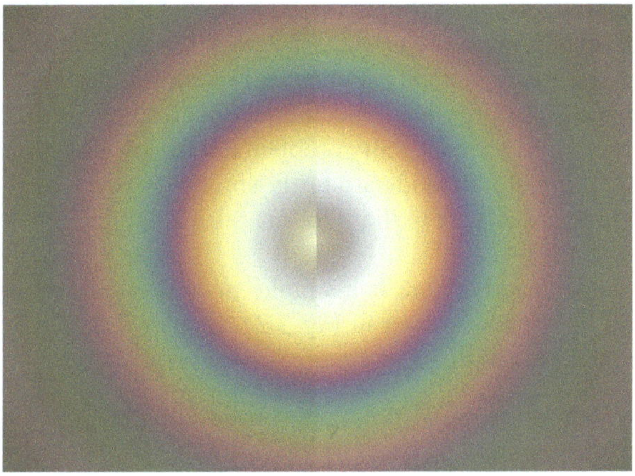

Fig. 7.20 The *left* part of the image shows a simulation of the glory caused by scattering of sunlight from spherical water droplets with $r = 10\,\mu\text{m}$ (based on Mie calculations). The *right* part of the image shows an equivalent simulation based on Debye $p = 2$ contributions

using Mie calculations, whilst the right side considers only $p = 2$ scattering. The key difference is that, as indicated in Fig. 7.19, Mie calculations predict a bright white zone near the centre (i.e. $\theta \rightarrow 180°$). More importantly, both simulations produce essentially identical sequences of coloured rings, thus confirming that the coloured rings of the atmospheric glory are primarily caused by light that has suffered only one reflection within the water drop (i.e. $p = 2$ scattering).

Nussenzveig also suggested that resonances are important in forming the glory; for example, in 2003, he wrote [53]:

> The glory provides direct and visually stunning experimental evidence of the importance of resonances and light tunneling in clouds.

In particular, several of Nussenzveig's papers highlight the importance of $p = 24$ contributions to the glory on the basis that geometrical $p = 24$ rays with impact parameter $b = \pm 1$ result in $\theta = 180°$ when $n = 1.33007$. However, $p = 24$ contributions do not appear to be very important in Fig. 7.19. There are several possible reasons for this disagreement with Nussenzveig's results: first, as shown by Table 7.1, $n = 1.33007$ is outside the range of n for water in the visible spectrum; second, Debye series calculations for $p = 24$ do not show any enhancement in backscattering in the vicinity of $n = 1.33007$. In any event, Nussenzveig's calculations were made for $\theta = 180°$ (i.e. only for the central feature of the glory) and, therefore, do not cover the much more relevant case of the coloured rings of the glory (typically observed when $170° < \theta < 178°$). Given the dominance of $p = 2$ contributions shown in Fig. 7.19, it seems unlikely that resonances are important in the formation of the glory.

Having confirmed that the glory is essentially caused by $p = 2$ scattering, it is necessary to investigate van de Hulst's hypothesis that surface waves play a major rule in formation of the glory. Although Mie and Debye series calculations are rigorous, they do not separate the surface waves from other scattering mechanisms. Regrettably, no rigorous method seems to be available for calculating the intensity of scattering caused by surface waves. Various authors [20, 25, 41, 51] have proposed approximate methods applicable to scattering of light by small spheres, but all warn that their approximations are not valid at $\theta = 180°$ or, indeed, at $\theta \approx 180°$. These limitations are especially problematic for investigations of the glory! Although these approximate methods are fairly similar, Khare's method [25] seems to give results which are closest to the rigorous results obtained by Mie and Debye series calculations [31].

Fortunately, there is a simple method of calculating the maxima resulting from the light paths shown in Fig. 7.21. Scattering at an angle of $\theta = 180° - \delta$ can be considered as the vector sum of two surface waves:

- one following a short path resulting in deviation of the incoming rays by $\theta = 180° - \delta$;
- one following a longer path resulting in deviation of the incoming rays by $\theta = 180° + \delta$.

If the short path involves clockwise deviation, the long path involves counterclockwise deviation (and vice versa). The interference pattern is determined by the phase difference between the short and long paths. Surface waves travel at a speed determined by the refractive index of the medium (not by the refractive index of the sphere) [34]. For water droplets in air, the medium has a refractive index of about $1.003 \approx 1$). Consequently, the difference in path length between the two $p = 2$ contributions at $\theta = 180° - \delta$ is approximately the length of the arc of radius r corresponding to the angle 2δ. Maxima will occur when the path difference $2\pi r(2\delta/360°) = k\lambda$ where k is an integer and λ is wavelength of the light. Rearranging this equation gives $\delta = 90k\lambda/(\pi r)$. Taking a numerical example, when $\lambda = 0.65\,\mu\text{m}$, $\delta = 18.62k/r$ (where r is measured in μm and δ in degrees).

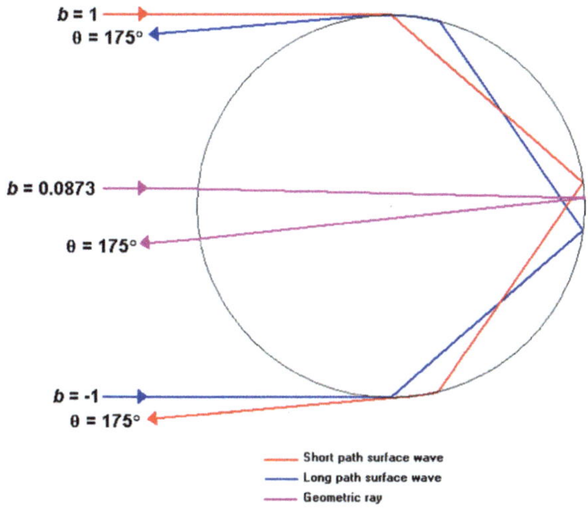

Fig. 7.21 Ray paths contributing to scattering at $\theta = 175°$ by a water drop with refractive index $n = 1.333$. Each of the two edge rays with impact parameters $b = 1$ and $b = -1$ suffer one internal reflection and generate surface waves: in this case, the short path surface wave travels about 10° around the circumference of the sphere before exiting at $\theta = 175°$, whereas the long path surface wave travels about 20° around the circumference of the sphere, thus being deviated by 185° (equivalent to $\theta = 175°$)

If $r = 10\,\mu$m, maxima will occur at $\theta_k = 180° - 1.862k$, corresponding to $\theta_0 = 180°$, $\theta_1 = 178.1°$, $\theta_2 = 176.4°$ and $\theta_3 = 174.1°$. These maxima are the rings of the glory—agreeing well with the results of Mie and Debye series calculations.

Another way to test van de Hulst's hypothesis is to examine the effects of illuminating a droplet of water with a narrow beam of light aligned so that it has an impact parameter $b = \pm 1$, thus generating surface waves. Calculation techniques for scattering of Gaussian beams by homogeneous spherical particles were pioneered by Gouesbet and colleagues [19], resulting in a large body of scientific literature as summarised in [42]. Figure 7.22 shows the results of Gaussian beam calculations based on algorithms described by [39, 40]. The top graph in Fig. 7.22 shows the result of an edge ray with $b = 1$ illuminating a water droplet of radius $r = 10\,\mu$m. The peak intensity for perpendicular polarisation at $\theta \approx 142°$ corresponds to the primary fogbow, whereas scattering at $\theta > 170°$ is due to surface waves which are dominated by parallel polarisation. As surface waves shed light along their path, the intensity declines smoothly with increasing θ. The middle graph in Fig. 7.22 adds results for an incident beam with $b = -1$ which results in propagation in the opposite direction to that caused by the $b = 1$ beam. For the $b = -1$ beam, the primary fogbow occurs at $\theta \approx 218° = 360° - 142°$ and the surface waves result in scattering at $\theta < 190°$ with intensity declining with decreasing θ. We can see that scattering at, say, $\theta = 175°$ is due to two sets of surface waves: the short path corresponding to $b = 1$ and the long path corresponding to $b = -1$. The resulting interference pattern

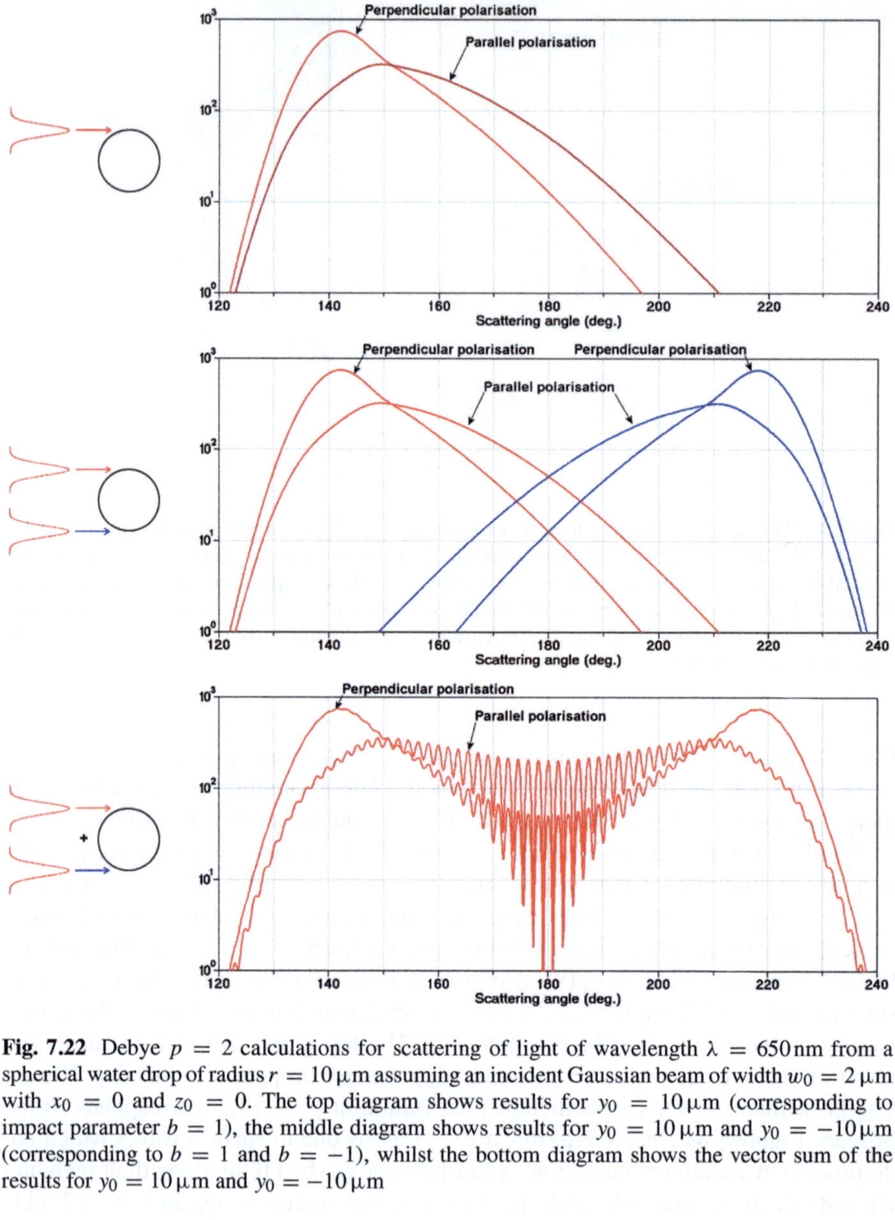

Fig. 7.22 Debye $p = 2$ calculations for scattering of light of wavelength $\lambda = 650\,\text{nm}$ from a spherical water drop of radius $r = 10\,\mu\text{m}$ assuming an incident Gaussian beam of width $w_0 = 2\,\mu\text{m}$ with $x_0 = 0$ and $z_0 = 0$. The top diagram shows results for $y_0 = 10\,\mu\text{m}$ (corresponding to impact parameter $b = 1$), the middle diagram shows results for $y_0 = 10\,\mu\text{m}$ and $y_0 = -10\,\mu\text{m}$ (corresponding to $b = 1$ and $b = -1$), whilst the bottom diagram shows the vector sum of the results for $y_0 = 10\,\mu\text{m}$ and $y_0 = -10\,\mu\text{m}$

due to the vector sum of these two components is shown in the bottom graph of Fig. 7.22.

Various authors have examined the effects of illuminating a spherical particle with very short pulses of light [44, 45]. Bech and Leder [4, 5] extended such work by producing various diagrams showing how the impulse response of a sphere varies

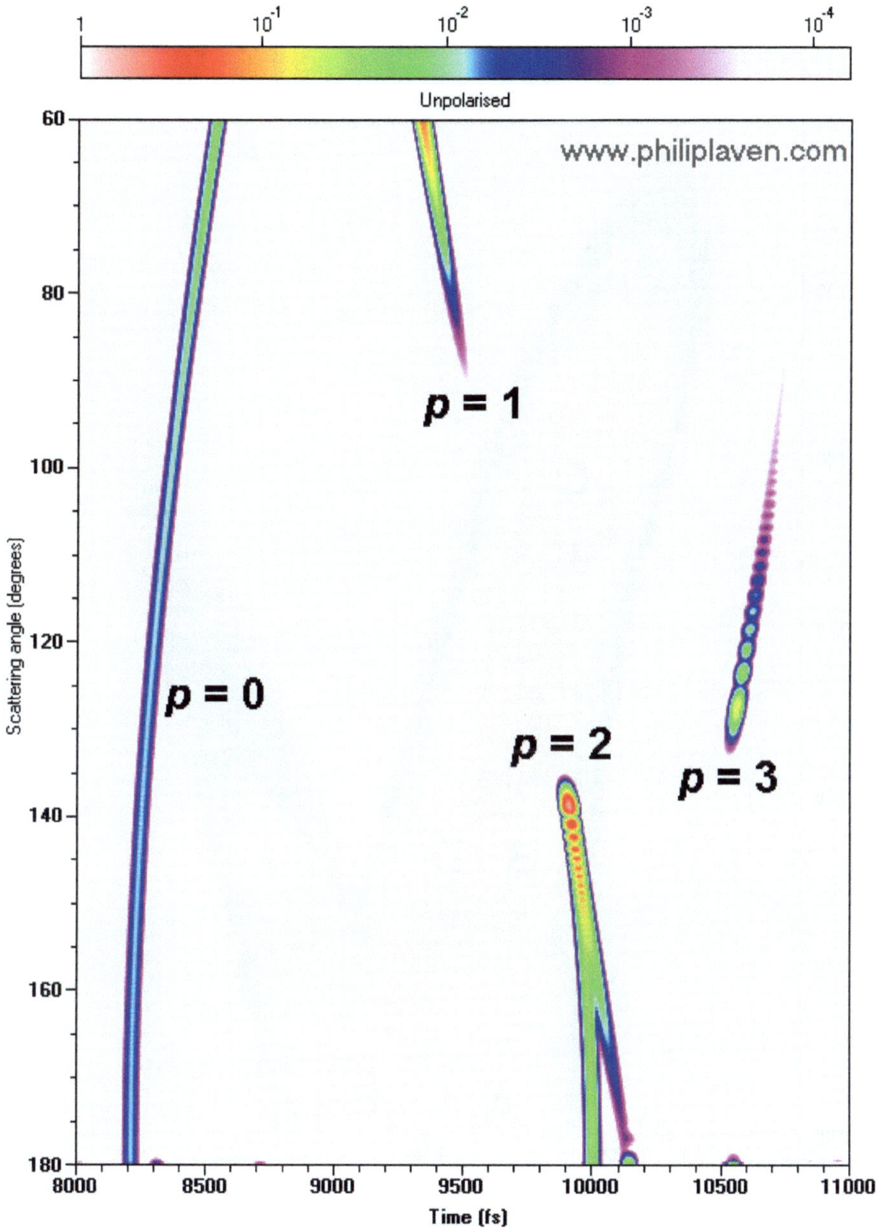

Fig. 7.23 Mie calculations of the impulse response of a spherical water drop of radius $r = 100\,\mu\text{m}$ when illuminated by a 50 femtosecond pulse of light of wavelength $\lambda = 650\,\text{nm}$. This false-colour map shows the intensity of the scattered signal as a function of scattering angle θ and time delay

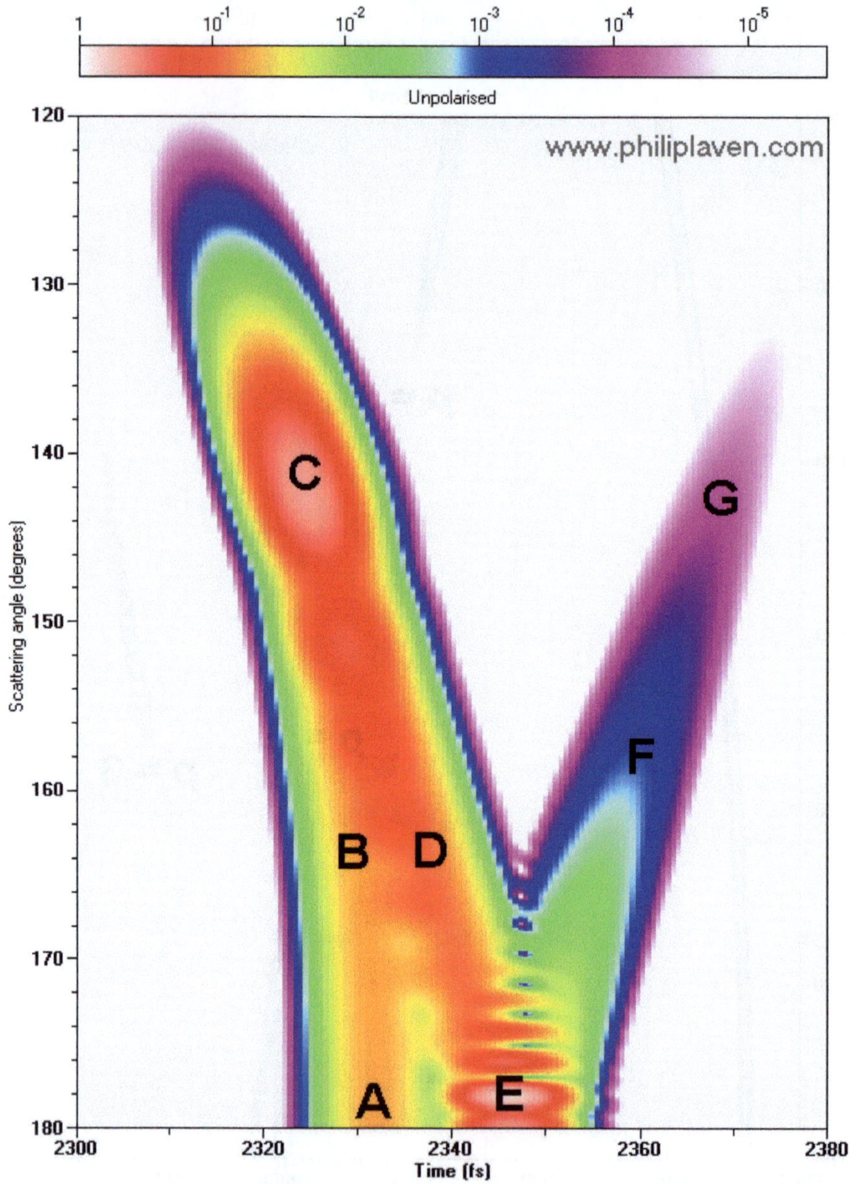

Fig. 7.24 Debye series $p = 2$ calculations of the impulse response of a spherical water drop of radius $r = 10\,\mu\text{m}$ when illuminated by a 10 femtosecond pulse of light of wavelength $\lambda = 650\,\text{nm}$. The zone marked A shows scattering of geometric "central" rays with impact parameter $b < 0.1$. As b increases, the scattered signal moves through B until it reaches the primary rainbow at C caused by rays with $b \approx 0.86$. As $b \rightarrow 1$, the scattered signal moves from C towards D. Between D and E, the scattered signal is due to short-path surface waves. Between E and F, the scattered signal is due to long-path surface waves, which decline in intensity towards G. The maxima and minima near E correspond to the glory, which can be considered as the interference pattern caused by the short path and long path surface waves

as a function of scattering angle θ. Such techniques have been used to generate the false-colour map shown in Fig. 7.23 for scattering by a water drop of radius $r = 100\,\mu$m illuminated by a 100 femtosecond raised-cosine pulse. In this case, it is fairly easy to identify the various contributions to the scattering process. As might be expected from geometrical considerations, the minimum delay corresponds to $p = 0$ scattering (i.e. reflection from the exterior of the sphere) at $\theta = 180°$. Elsewhere, the diagram shows that the maximum $p = 2$ intensity at $\theta \approx 138°$ and time delay $t \approx 10{,}000$ fs corresponds to the primary rainbow, shown with its supernumerary arcs at $\theta > 140°$. Similarly, $p = 3$ scattering at $\theta < 129°$ corresponds to the secondary rainbow and its supernumerary arcs. Such results have already been confirmed by Debye series calculations and, indeed, by geometrical optics. However, the power of this calculation method becomes more obvious in Fig. 7.24 which shows the results of Debye $p = 2$ calculations for a spherical water drop of radius $r = 10\,\mu$m illuminated by a 10 fs pulse. This form of presentation separates the various $p = 2$ scattering mechanisms by the differences in the time delay caused by the propagation paths—in this case, it is possible to identify not just that the rainbow is caused by two geometric rays (as shown in Fig. 7.4), but also the contributions made by short path and long path surface waves.

7.5 Conclusions

The apparently trivial process of scattering of sunlight by spherical drops of water produces beautifully-coloured rainbows, coronas and glories. Mie calculations are necessary to provide accurate simulations of these phenomena. The appearance of rainbows depends crucially on the radius r of the spherical drops of water—ranging from a white fogbow when $r = 10\,\mu$m to a vividly-coloured rainbow when $r = 1{,}000\,\mu$m. At intermediate values of r, supernumerary arcs can appear if the water drops are of uniform size. Care should be taken when performing calculations for scattering by very large drops of water (e.g. $r > 1{,}000\,\mu$m) because such drops are often flattened at the bottom—in these circumstances, Mie calculations are not valid.

The refractive index n of water varies across the the visible spectrum, thus generating the colours of the rainbow. If n was constant across the visible spectrum, rainbows would be essentially white with hints of colour amongst the supernumerary arcs. However, the colours of the corona and glory are almost independent of n. Consequently, these phenomena are, to a first approximation, determined purely by the radius r of the water droplet.

Rainbows and glories are strongly polarised resulting in complex displays when viewed through a polarising filter, but coronas are essentially unpolarised.

Mie calculations do not provide any useful information about the scattering mechanisms responsible for these phenomena. Fortunately, calculations using the Debye series help to identify the scattering mechanisms—for example, by showing that the primary rainbow and the glory are both caused by $p = 2$ scattering (involving one internal reflection in the spherical water drops). To verify van de Hulst's hypothesis that surface waves are responsible for the glory, it is necessary to call on other calculation techniques, such as:

- Mie and Debye series calculations for spherical particles illuminated by Gaussian beams;
- Mie and Debye series calculations of the impulse response of spherical particles illuminated by femtosecond pulses of light.

Although Mie's rigorous solution for the scattering of plane wave light by an homogeneous sphere has been available for more than 100 years, it is surprising to realise that rainbows, coronas and glories are still not entirely understood.

References

1. J.A. Adam, The mathematical physics of rainbows and glories. Phys. Rep. **356**, 229–365 (2002)
2. J.A. Adam, Geometric optics and rainbows: generalization of a result by Huygens. Appl. Opt. **47**, H11–H13 (2008)
3. G.B. Airy, On the intensity of light in the neighbourhood of a caustic. Trans. Camb. Philos. Soc. **6**, 397–403 (1838)
4. H. Bech, A. Leder, Particle sizing by ultrashort laser pulses–numerical simulation. Optik **115**, 205–217 (2004)
5. H. Bech, A. Leder, Particle sizing by time-resolved Mie calculations—A numerical study. Optik **117**, 40–47 (2006)
6. C.F. Bohren, D.R. Huffman, *Absorption and Scattering of Light by Small Particles* (Wiley, New York, 1983)
7. C.B. Boyer, *The Rainbow: From Myth to Mathematics*. (Princeton University, Princeton, 1987), reprint of 1959 Thomas Yoseloff edn
8. H.C. Bryant, A.J. Cox, Mie theory and the glory. J. Opt. Soc. Am. **56**, 1529–1532 (1966)
9. H.C. Bryant, N. Jarmie, The glory. Sci. Am. **231**, 60–71 (1974)
10. L. Cowley, P. Laven, M. Vollmer, Rings around the sun and moon: coronae and diffraction. Phys. Educ. **41**, 51–59 (2005)
11. Iris software http://www.atoptics.co.uk/droplets/iris.htm (Cited 31 October 2009)
12. J.V. Dave, Scattering of visible light by large water spheres. Appl. Opt. **8**, 155–164 (1969)
13. P. Debye, Das elektromagnetische Feld um einen Zylinder und die Theorie des Regenbogens. Physikalische Zeitschrift 9, 775–778 (1908) [N.B. An English translation of this paper entitled The electromagnetic field around a cylinder and the theory of the rainbow is available in Selected Papers on Geometrical Aspects of Scattering, SPIE Milestone Series Volume MS 89 (1993)]
14. T.S. Fahlen, H.C. Bryant, Direct observation of surface waves on droplets. J. Opt. Soc. Am. **56**, 1635–1636 (1966)
15. S.D. Gedzelman, Simulating glories and cloudbows in color. Appl. Opt. **42**, 429–435 (2003)
16. S.D. Gedzelman, Simulating rainbows in their atmospheric environment. Appl. Opt. **47**, H176–H181 (2008)
17. S.D. Gedzelman, Simulating halos and coronas in their atmospheric environment. Appl. Opt. **47**, H157–H166 (2008)
18. S.D. Gedzelman, J.A. Lock, Simulating Coronas in Color. Appl. Opt. **42**, 497–504 (2003)
19. G. Gouesbet, B. Maheu, G. Grehan, Light scattering from a sphere arbitrarily located in a Gaussian beam, using a Bromwich formalism. J. Opt. Soc. Am. A **5**, 1427–1443 (1988)

20. W.T. Grandy, *Scattering of Waves from Large Spheres* (Cambridge University, Cambridge, 2001)
21. R. Greenler, *Rainbows* (Halos and Glories. Cambridge University, Cambridge, 1980)
22. R.B. Hoover, F.S. Harris, Die Beugungserscheinungen: a Tribute to F. M. Schwerd's Monumental Work on Fraunhofer Diffraction. Appl. Opt. **8**, 2161–2164 (1969)
23. E.A. Hovenac, J.A. Lock, Assessing the contributions of surface waves and complex rays to far-field Mie scattering by use of the Debye series. J. Opt. Soc. Am. A **9**, 781–795 (1992)
24. H. Inada, New calculation of surface wave contributions associated with Mie backscattering. Appl. Opt. **12**, 1516–1523 (1973)
25. V. Khare, *Short-wavelength scattering of electromagnetic waves by a homogeneous dielectric sphere*. Ph.D. thesis (University of Rochester, Rochester, 1976). This reference may not be readily available, but the calculation method is summarized in [23]
26. V. Khare, H.M. Nussenzveig, Theory of the glory. Phys. Rev. Lett. **38**, 1279–1282 (1977)
27. G.P. Können, J.H. de Boer, Polarized rainbow. Appl. Opt. **18**, 1961–1965 (1979)
28. G.P. Können, *Polarized light in nature* (Cambridge University Press, Cambridge, 1985)
29. P. Laven, Simulation of rainbows, coronas, and glories by use of Mie theory. Appl. Opt. **42**, 436–444 (2003)
30. P. Laven, Simulation of rainbows, coronas and glories using Mie theory and the Debye series. J. Quant. Spectrosc. Radiat. Transf. **89**, 257–269 (2004)
31. P. Laven, How are glories formed? Appl. Opt. **44**, 5675–5683 (2005)
32. P. Laven, Atmospheric glories: simulations and observations. Appl. Opt. **44**, 5667–5674 (2005)
33. P. Laven, Noncircular glories and their relationship to cloud droplet size. Appl. Opt. **47**, H25–H30 (2008)
34. P. Laven, Effects of refractive index on glories. Appl. Opt. **47**, H133–H142 (2008)
35. MiePlot software, http://www.philiplaven.com/MiePlot.htm (Cited 31 October 2009)
36. R.L. Lee, Mie theory, Airy theory, and the natural rainbow. Appl. Opt. **37**, 1506–1519 (1998)
37. R.L. Lee, A.B. Fraser, *The Rainbow Bridge: Rainbows in Art, Myth and Science* (Pennsylvania State University Press, Pennsylvania, 2001)
38. J.A. Lock, L. Yang, Mie theory model of the corona. Appl. Opt. **30**, 3408–3414 (1991)
39. J.A. Lock, Contribution of high-order rainbows to the scattering of a Gaussian laser beam by a spherical particle. J. Opt. Soc. Am. A **10**, 693–706 (1993)
40. J.A. Lock, Improved Gaussian beam-scattering algorithm. Appl. Opt. **34**, 559–570 (1995)
41. J.A. Lock, Role of the tunneling ray in near-critical-angle scattering by a dielectric sphere. J. Opt. Soc. Am. A **20**, 499–507 (2003)
42. J.A. Lock, G. Gouesbet, Generalized Lorenz—Mie theory and applications. J. Quant. Spectrosc. Radiat. Transf. **110**, 800–807 (2009)
43. D.K. Lynch, W. Livingston, *Color and Light in Nature* (Cambridge University, Cambridge, 2001)
44. L. Méès, G. Gouesbet, G. Gréhan, Time-resolved scattering diagrams for a sphere illuminated by plane wave and focused short pulses. Opt. Commun. **194**, 59–65 (2001)
45. L. Méès, G. Gouesbet, G. Gréhan, Scattering of Laser Pulses (Plane Wave and Focused Gaussian Beam) by Spheres. Appl. Opt. **40**, 2546–2550 (2001)
46. G. Mie, Beitrage zur Optik trüber Medien, speziell kolloidaler Metallosungen. Ann. Phys. Leipzig **25**, 377–445 (1908)
47. M. Minnaert, *The Nature of Light and Colour in the Open Air* (Dover Publications, New York, 1954)
48. H.M. Nussenzveig, High-frequency scattering by a transparent sphere. I. Direct reflection and transmission. J. Math. Phys. **10**, 82–124 (1969)
49. H.M. Nussenzveig, High-frequency scattering by a transparent sphere. II. Theory of the rainbow and the glory. J. Math. Phys. **10**, 125–176 (1969)
50. H.M. Nussenzveig, Complex angular momentum theory of the rainbow and the glory. J. Opt. Soc. Am. **69**, 1068–1079 (1979)
51. H.M. Nussenzveig, *Diffraction Effects in Semiclassical Scattering* (Cambridge University, Cambridge, 1992)

52. H.M. Nussenzveig, Does the glory have a simple explanation? Opt. Lett. **27**, 1379–1381 (2002)
53. H.M. Nussenzveig, Light tunneling in clouds. Appl. Opt. **42**, 1588–1593 (2003)
54. J.A. Shaw, P.J. Neiman, Coronas and iridescence in mountain wave clouds. Appl. Opt. **42**, 476–485 (2003)
55. R.A.R. Tricker, *Introduction to Meteorological Optics* (American-Elsevier, New York, 1970)
56. H.C. van de Hulst, A theory of the anti-coronae. J. Opt. Soc. Am. **37**, 16–22 (1947)
57. H.C. van de Hulst, *Light Scattering by Small Particles* (Dover, New York, 1981), reprint of 1957 Wiley edition
58. M. Vollmer, Effects of absorbing particles on coronas and glories. Appl. Opt. **44**, 5658–5666 (2005)
59. M. Vollmer, *Lichtspiele in der Luft* (Elsevier, München, 2006)
60. R.T. Wang, H.C. van de Hulst, Rainbows: Mie computations and the Airy approximation. Appl. Opt. **30**, 106–117 (1991)
61. T. Young, Experiments and calculations relative to physical optics. A Bakerian Lecture read on November 24, 1803. Phil. Trans. Roy. Soc. 1–16 (1804)

Chapter 8
The Extension of Mie Theory to Multiple Spheres

Daniel Mackowski

Abstract A generalized mathematical formulation is presented for the scattering and absorption of electromagnetic time harmonic waves by multiple spherical particles. A central element of this formulation is the addition theorem for vector wave functions, which allows a scattered field from one sphere to be represented as an exciting field about another sphere. A simplified derivation of the addition theorem, and important characteristics of where it can and can not be used, are developed. The Mie solution, coupled with the addition theorem, results in a system of linear interaction equations for the multipole coefficients that describe the scattered field from each sphere in the system. In this regard, the multiple sphere formulation results in an implicit, rather than explicit, solution for the scattered field; numerical methods (i.e., linear equation solvers) must be applied to obtain numerical results. The calculation of the T matrix of the multiple sphere system, from which orientation averaged scattering and absorption properties can be obtained, is described. The presentation ends with a discussion on the application of the multiple sphere formulation to describing the propagation of electromagnetic waves in discretely inhomogeneous media.

8.1 Introduction

As the very presence of this text makes clear, analytical solutions to light scattering by objects are both rare and historically significant. Indeed, it can be argued that Mie's solution for the sphere is the only analytical light scattering solution for a 3-D object that is routinely applied in scientific and engineering disciplines. Of course, analytical solutions do exist for other shapes—most notably the ellipsoids—yet a review of the literature will reveal that most of the theoretical examinations

---

D. Mackowski (✉)
Mechanical Engineering Department, Auburn University,
Auburn, AL 36849, USA
e-mail: mackodw@auburn.edu

W. Hergert and T. Wriedt (eds.), *The Mie Theory*, Springer Series in Optical Sciences 169, DOI: 10.1007/978-3-642-28738-1_8,

of scattering by such objects are conducted using the numerical extended boundary condition method (EBCM), a.k.a., T matrix or discrete dipole approximation (DDA) procedures. That such is the case has much to do with the relative inaccessibility of the analytical solutions for ellipsoids as well as the relative ease, accuracy, and computational efficiency of the numerical procedures.

Mie, in developing his solution, was concerned primarily with the scattering behavior of a single sphere when exposed to a plane–wave source of illumination. His solution, however, is recognized today as representing a general relationship between the incident and scattered field for a sphere. That is, the solution will provide a description of the scattered field for an *arbitrary* incident field; the only requirement being the ability to represent the incident field in a vector wave harmonic (VWH) expansion. One specific extension of Mie theory in this regard is the description of scattering of light by spheres when exposed to focussed, Gaussian–profile beams; the effort behind this theory has not been spent to determine the response of the sphere to the incident field—as this is already known from Mie's solution–rather, it has been directed to approximating the Gaussian beam in a VWH expansion [1, 2].

This paper addresses a similar extension—or generalization—of Mie theory; that being the scattering response of a group of two or more spheres, in close proximity to each other, that are collectively exposed to an incident field. By 'close proximity', I am referring to situations in which the scattered field from one sphere becomes a significant component to the exciting field of a neighboring sphere. That is, the sphere system exhibits near-field coupling. The presence of such effects would be expected to render inaccurate an independent-scattering model of the sphere system, i.e., a model which assumes that the spheres interact solely with the incident field.

The multiple-sphere scattering problem can be properly viewed as an extension of Mie theory, in that the crux of the problem is the representation of the exciting field at each of the spherical surfaces comprising the system. Once such information is known, the scattering response follows from a simple application of Mie theory. However, the scattered fields from neighboring spheres will contribute to the exciting fields, and in this sense the scattered fields from all the spheres are coupled. As will be described herein, this coupling results in an *implicit* solution for the scattering behavior of the multiple-sphere system as opposed to the *explicit* formulas for the single-sphere case; practical calculation of the multiple-sphere solution will invariably involve solutions to linear equations which describe the intersphere coupling. Nevertheless, the multiple-sphere solution should be considered an analytical solution to the boundary-value problem in the same sense as Mie theory is; the solution identically satisfies the governing differential equations and far-field radiation conditions, and matches the continuity conditions at the sphere surfaces to an arbitrary precision based entirely on the truncation of the VWH expansions used to represent the fields.

The most important element in constructing the multiple sphere solution—apart from the underlying Mie theory—is the ability to represent an outgoing VWH centered about one origin as an expansion of regular VWH centered about another origin. This transformation, which is generally known as an addition theorem, was first derived by Stein [3] and, shortly thereafter, improved by Cruzan [4]; all subsequent contributions to the multiple sphere scattering theory can trace their lineage, either directly or indirectly, back to these two seminal works. Liang and Lo, and Brunning and Lo [5, 6], were the first to combine the addition theorems with Mie theory to develop an exact solution for a pair of spheres. Significant advances in, and applications of, the formulation were subsequently developed by Borghese et al. [7, 8], Fuller et al. [9, 10], and Mackowski et al. [11–13]. Comprehensive reviews of the historical development of the method can be found in [14, 15].

At the time of this report the multiple sphere solution can be considered well established and mature; little remains to be done in the way of mathematical work to uncover hidden features of the formulation, and most of the current work is aimed at applying the solution method to predict the scattering and absorption properties of aggregated and cluster particles. A measure of this progress can be seen in the recent compendia of references related to the T matrix method [16, 17]; listed are over 200 citations under the category of multiple sphere scattering, and the vast majority of these works deal with the application of the solution method—in a diversity of fields ranging from nanotechnology to astrophysics—as opposed to mathematical/theoretical developments. Indeed, the accessibility and efficiency of the multiple sphere formulation and codes—combined with the increasing capabilities of modern computational machinery—are now enabling the direct simulation (or, alternatively, numerical experiments) of radiation propagation in particulate media via calculations involving on the order of 10^3 spheres [18, 19].

In keeping with the historical occasion and context of this volume, my report will revisit the mathematical side of the multiple sphere scattering solution. One objective of the report is to show that the entire solution—including the details of the VWH addition theorem—can be derived using the basic properties of the VWH functions. The basic procedures for applying the solution to obtain the observable properties of a cluster of spheres, in either a fixed orientation with respect to the incident field or in a random orientation, will be reviewed. The report will finish with a brief discussion of future directions and applications of the multiple sphere formulation.

8.2 The Multiple Sphere Solution

8.2.1 The Superposition Strategy

A general configuration for the multiple sphere problem is illustrated in Fig. 8.1. The ensemble consists of N_S nonoverlapping spheres, each characterized by a radius a_i, refractive index m_i, and position $\mathbf{r}_i$ relative to a common origin, for $i = 1, 2, \ldots N_S$. Incident upon the ensemble is a monochromatic, time harmonic electromagnetic field of wavelength λ, typically in the form of a plane, linearly polarized wave.

The objective of the problem, in the most general sense, is to obtain a solution to Maxwell's time harmonic equations which satisfies the continuity conditions on all

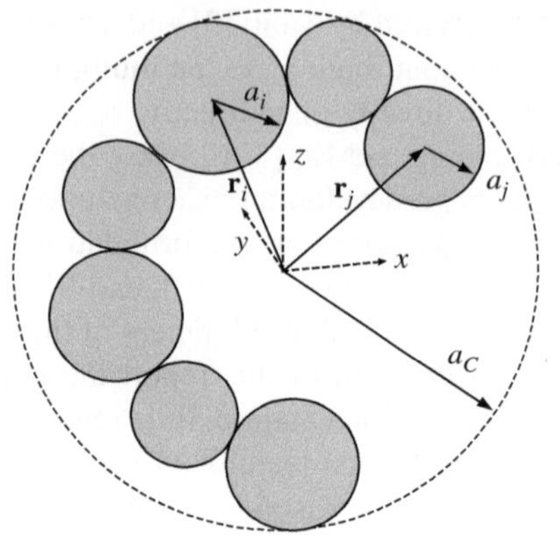

Fig. 8.1 Ensemble configuration

surfaces. This problem would likely be intractable if the ensemble was mathematically represented as a single particle with a single—albeit discontinuous—surface. On the other hand, the problem becomes relatively simple by adopting a superposition strategy. In this approach, the field external to the spheres is represented by the superposition of the incident field and the fields scattered from each sphere in the ensemble;

$$\mathbf{E}_{ext} = \mathbf{E}_{inc} + \mathbf{E}_{sca} = \mathbf{E}_{inc} + \sum_{i=1}^{N_S} \mathbf{E}_{sca,i} \tag{8.1}$$

And as was done for the single sphere problem, the incident and scattered fields, at the ith sphere in the cluster, can be represented by regular and outgoing vector spherical harmonic (VWH) expansions, centered about the origin of the sphere;

$$\mathbf{E}_{inc} = \sum_{n=1}^{L_i} \sum_{m=-n}^{n} \sum_{p=1}^{2} p^i_{mnp} \mathbf{N}^{(1)}_{mnp}(\mathbf{r} - \mathbf{r}_i) \tag{8.2}$$

$$\mathbf{E}_{sca,i} = \sum_{n=1}^{L_i} \sum_{m=-n}^{n} \sum_{p=1}^{2} a^i_{mnp} \mathbf{N}^{(3)}_{mnp}(\mathbf{r} - \mathbf{r}_i) \tag{8.3}$$

In the above, $\mathbf{N}_{mnp}$ denotes the VWH of either type 1 (regular) or 3 (outgoing), of order n, degree m, and mode $p = 1$ (TM) or 2 (TE). The VWH functions used in this report are defined by

$$\mathbf{N}^{(\nu)}_{mn2}(\mathbf{r}) = \left(\frac{1}{n(n+1)}\right)^{1/2} \nabla \times \left(\mathbf{r}\psi^{(\nu)}_{mn}(\mathbf{r})\right) \tag{8.4}$$

$$\mathbf{N}^{(\nu)}_{mn1}(\mathbf{r}) = \frac{1}{\mathsf{k}} \nabla \times \mathbf{N}^{(\nu)}_{mn2}(\mathbf{r}) \tag{8.5}$$

where ψ denotes the scalar wave harmonic;

$$\psi_{mn}^{(\nu)}(\mathbf{r}) = \begin{cases} j_n(\mathrm{k}r)\, Y_{mn}(\cos\theta, \phi) & \nu = 1 \\ h_n(\mathrm{k}r)\, Y_{mn}(\cos\theta, \phi) & \nu = 3 \end{cases} \tag{8.6}$$

with j_n and $h_n = j_n + i\, y_n$ representing the spherical Bessel and Hankel functions and Y_{mn} denoting the spherical harmonic,

$$Y_{mn}(\cos\theta, \phi) = \left(\frac{2n+1}{4\pi} \frac{(n-m)!}{(n+m)!} \right)^{1/2} P_n^m(\cos\theta)\, e^{\mathrm{i}\, m\, \phi} \tag{8.7}$$

The order truncation limit L_i in Eq. (8.3) is chosen to provide an acceptable precision in the calculated scattering properties of the cluster; L_i can typically be set using the Lorenz/Mie criterion, although in certain cases the neighboring spheres can have a significant effect on the convergence of Eq. (8.3) [12]. The incident field expansion coefficients p_{mnp}^i, for sphere i, will depend on the characteristics of the incident field and the relative position of sphere i, whereas the scattering coefficients a_{mnp}^i for each sphere are initially unknown and are sought from the analysis.

8.2.2 Translated Fields and the Addition Theorem

Lorenz/Mie theory provides a means of relating the scattering coefficients for sphere i to those representing the *exciting* field at i, yet this exciting field will consist not only of the incident field, but also the scattered fields from the other spheres. In this respect, what is needed to close the problem is a means of casting the outgoing VWH about some origin j—which describes the scattered field from j—into an expansion of VWH about origin i. This feat is accomplished with an *addition theorem.*

In a general sense, an addition theorem transforms an eigenfunction for a given differential equation, referenced to a given coordinate origin, into an expansion of eigenfunctions *satisfying the same DE* about a different coordinate origin. The existence of such a transformation can be anticipated simply from the fact that the vector harmonics form a complete basis for representation of fields which satisfy (with the exception of singular points) the vector wave equation. Consequently, a harmonic centered about one origin describes a field which must be representable—within a given radius of convergence—by an expansion of harmonics centered about a different origin.

With regard to an outgoing VWH centered about an origin j, the form and convergence of the addition theorem is dictated by whether the singularity at origin j lies inside or outside the radial bounds of the translated VWH expansion at some new origin i. For the latter case, which corresponds to the region defined by $|\mathbf{r}-\mathbf{r}_i| < |\mathbf{r}_{i-j}|$, where $\mathbf{r}$ and $\mathbf{r}_{i-j}$ denote the evaluation point of the harmonics and the position vector between origins i and j (case 1 in Fig. 8.2), the field generated by the outgoing harmonic will be represented entirely within the region by an expansion of regular VWH centered about i. Accordingly, the addition theorem for this case will

Fig. 8.2 Regions of reexpansion for an outgoing harmonic

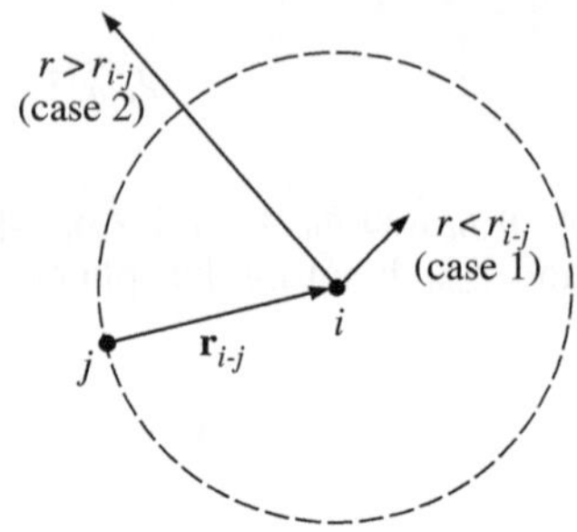

appear as

$$\mathbf{N}^{(3)}_{mnp}(\mathbf{r}-\mathbf{r}_j) = \sum_{l=1}^{\infty}\sum_{k=-l}^{l}\sum_{q=1}^{2} H_{klq\,mnp}(\mathbf{r}_{i-j})\,\mathbf{N}^{(1)}_{klq}(\mathbf{r}-\mathbf{r}_i), \quad \left|\mathbf{r}_{i-j}\right| > |\mathbf{r}-\mathbf{r}_i| \quad (8.8)$$

in which the as-of-yet-undetermined translation matrix H would be a function solely of the distance and direction between origins i and j. The derivation and properties of the translation matrix will be examined in a subsequent section; addressed now will be the incorporation of Eq. (8.8) into the formal solution for the multiple sphere scattering problem.

Since the spheres in the ensemble are assumed not to overlap, Eq. (8.8) will be valid for representing the scattered field from sphere j in terms of a regular VWH expansion about i that is convergent on all points on the surface of i. Combining the scattered fields from all spheres in the ensemble and the incident field, the exciting field at i can be described by a single, regular VWH expansion given by

$$\mathbf{E}_{exi,i} = \sum_{n=1}\sum_{m=-n}^{n}\sum_{p=1}^{2} g^i_{mnp}\,\mathbf{N}^{(1)}_{mnp}(\mathbf{r}-\mathbf{r}_i)$$

with

$$g^i_{mnp} = p^i_{mnp} + \sum_{\substack{j=1\\ j\neq i}}^{N_S}\sum_{l=1}\sum_{k=-l}^{l}\sum_{q=1}^{2} H^{i-j}_{mnp\,klq}\,a^j_{klq}$$

The scattering coefficients for sphere i will then be given by the Mie formula,

$$a^i_{mnp} = \overline{a}^i_{np}\,g^i_{mnp} \quad (8.9)$$

in which $\overline{a}^i_{np}$ denote the Mie coefficients of the sphere and depend on the sphere size parameter $x_i = \mathsf{k}\,a_i$ and the refractive index m_i. Equation (8.9) is recognized as a T

matrix relationship; made specific for the special case of the sphere in which the T matrix is diagonal and azimuthal–degree independent.[1]

8.2.3 The Interaction Equations

By combining the previous two equations, one arrives at the *interaction* equations for the sphere ensemble;

$$a^i_{mnp} - \overline{a}^i_{np} \sum_{\substack{j=1 \\ j\neq i}}^{N_S} \sum_{l=1}^{l} \sum_{k=-l}^{l} \sum_{q=1}^{2} H^{i-j}_{mnp\,klq}\, a^j_{klq} = \overline{a}^i_{np}\, p^i_{mnp} \tag{8.10}$$

Equation (8.10), in conjunction with Eqs. (8.1–8.3), represents the formal solution for the scattered field produced by the sphere ensemble. In the case of equal-sized spheres with equal truncation limits L_S, Eq. (8.10) forms a system of $2N_S\,L_S$ $(L_S + 2)$ linear equations for the set of scattering coefficients. The matrix H^{i-j} will be fully populated for an arbitrary translation between j and i, and correspondingly all orders/degrees/modes of the scattered field from a sphere j will (in general) influence a particular order/degree/mode of the field at i. This is in stark contrast to the isolated sphere case, in which each scattering order/degree/mode is excited only by the same harmonic component for the incident field. And it is in this respect that the multiple sphere solution departs—in a practical sense—from the single sphere Mie theory: the latter provides an *explicit* formula for the scattered field, whereas the former gives only an *implicit* relationship. That is, numerical methods (in the form of linear equation solvers) are needed to produce a useable solution.

Important properties of the interaction equations, and the calculation of relevant quantities (scattering matrix, cross sections) from the solution, will be presented in a subsequent section. Before getting to this material, it is necessary to develop formulas and properties for the translation matrix elements.

8.3 The Addition Theorem

8.3.1 Forms of the Addition Theorem

The same arguments that were used to pose Eq. (8.8) can be applied to the transformation of a regular VWH, centered about origin j, into an expansion of harmonics about some new origin i. Since the regular harmonic contains no singularities, its expansion about a new origin will not be constrained to a specified convergence radius. Accordingly, the addition theorem for regular harmonics will appear as

---

[1] The $\overline{a}_{np}$ coefficients defined by Eq. (8.9) are the negative of those formulated in previous works, e.g., Bohren and Huffman.

$$\mathbf{N}^{(1)}_{mnp}(\mathbf{r}-\mathbf{r}_j) = \sum_{l=1}^{\infty}\sum_{k=-l}^{l}\sum_{q=1}^{2} J_{klq\,mnp}(\mathbf{r}_{i-j})\,\mathbf{N}^{(1)}_{klq}(\mathbf{r}-\mathbf{r}_i) \tag{8.11}$$

where J denotes the regular harmonic translation matrix.

Some basic properties of the translation matrices H and J can be obtained directly from Eqs. (8.8) and (8.11). Since both sides of Eqs. (8.8) and (8.11) must satisfy the vector wave equation, and since application of the differential operators to the right hand sides can be taken with either $\mathbf{r}_{i-j}$ or $\mathbf{r}-\mathbf{r}_i$ fixed, the elements of the translation matrices H and J must satisfy the scalar wave equation. It follows that these elements will be represented as expansions of the scalar wave harmonic functions ψ defined in Eq. (8.6). In addition, by taking the limit of $|\mathbf{r}-\mathbf{r}_i| \to 0$ in Eqs. (8.8) and (8.11)—for which the only surviving term on the right-hand side will involve $l = 1$ and $q = 1$—the outgoing and regular VWH functions can be related to the translation matrix elements by

$$\mathbf{N}^{(3)}_{mnp}(\mathbf{r}) = \sum_{k=-1}^{1} \mathbf{P}_k\, H_{k11\,mnp}(\mathbf{r}) \tag{8.12}$$

$$\mathbf{N}^{(1)}_{mnp}(\mathbf{r}) = \sum_{k=-1}^{1} \mathbf{P}_k\, J_{k11\,mnp}(\mathbf{r}) \tag{8.13}$$

with

$$\mathbf{P}_k = \frac{1}{2\sqrt{3\pi}}\left[\delta_{|k|-1}\left((-1)^{(k+1)/2}\,\hat{\mathbf{x}} - \mathrm{i}\,\hat{\mathbf{y}}\right) + \sqrt{2}\,\delta_k\,\hat{\mathbf{z}}\right] \tag{8.14}$$

in which $\delta_k = (0, 1)$ for $k(\neq, =)0$ is the Kronecker delta function. By using the above formulas along with the definitions of the vector and scalar harmonics in Eqs. (8.4) and (8.6), it can be deduced that $\psi^{(3)}(\mathbf{r}_{ij})$ and $\psi^{(1)}(\mathbf{r}_{ij})$ are the basis functions for H and J, respectively. Since $\mathbf{N}^{(3)} = \mathbf{N}^{(1)} + \mathrm{i}\mathbf{N}^{(2)}$, it follows that $H + J + \mathrm{i}\,Y$, where $\mathbf{N}^{(2)}$ and Y are the VWH and the translation matrix based on the spherical Neumann function y_n. Consequently, the only difference in the formulas for the H and J must be the basis functions $\psi^{(3)}$ and $\psi^{(1)}$.

The remaining form of the addition theorem that needs to be identified concerns the translation of an outgoing harmonic from j to i for the condition $|\mathbf{r}-\mathbf{r}_i| > |\mathbf{r}_{ij}|$, i.e., when the spherical surface, centered about i, upon which the harmonics are evaluated encloses the singularity at origin j (case 2 in Fig. 8.2). In order to maintain the correct far-field behavior ($\mathsf{k}\,|\mathbf{r}-\mathbf{r}_i| \to \infty$), the translated harmonic must appear as an expansion of outgoing harmonics about the new origin. The translation must also yield an identity relation when $|\mathbf{r}_{ij}| \to 0$, which implies that the translation matrix will be generated from the regular scalar harmonic basis functions $\psi^{(1)}$. And since the regular components of the outgoing harmonics must translate according to Eq. (8.11), the translation matrix must be identical to that appearing in Eq. (8.11). Therefore, the addition theorem for this case will appear as

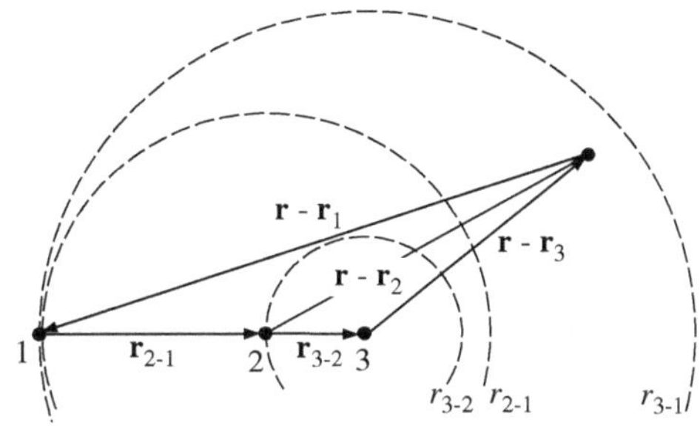

Fig. 8.3 Recursive translations: the addition theorem cannot be cheated!

$$\mathbf{N}^{(3)}_{mnp}(\mathbf{r}-\mathbf{r}_j) = \sum_{l=1}^{\infty}\sum_{k=-l}^{l}\sum_{q=1}^{2} J_{klq\,mnp}(\mathbf{r}_{i-j})\,\mathbf{N}^{(3)}_{klq}(\mathbf{r}-\mathbf{r}_i), \quad |\mathbf{r}_{i-j}| < |\mathbf{r}-\mathbf{r}_i| \quad (8.15)$$

8.3.2 Recursive Translations

Equations (8.8), (8.11), and (8.15) can be used to infer the following translation and factorization properties of the translation matrices,

$$J_{mnp\,klq}(\mathbf{r}_{i-j}) = \sum_{n'=1}^{\infty}\sum_{m'=-n'}^{n'}\sum_{p'=1}^{2} J_{mnp\,m'n'p'}(\mathbf{r}_{i-i'})\,J_{m'n'p'\,klq}(\mathbf{r}_{i'-j}) \quad (8.16)$$

$$H_{mnp\,klq}(\mathbf{r}_{i-j}) = \sum_{n'=1}^{\infty}\sum_{m'=-n'}^{n'}\sum_{p'=1}^{2} J_{mnp\,m'n'p'}(\mathbf{r}_{i-i'})\,H_{m'n'p'\,klq}(\mathbf{r}_{i'-j}),$$
$$|\mathbf{r}_{i-i'}| < |\mathbf{r}_{i'-j}| \quad (8.17)$$

$$H_{mnp\,klq}(\mathbf{r}_{i-j}) = \sum_{n'=1}^{\infty}\sum_{m'=-n'}^{n'}\sum_{p'=1}^{2} H_{mnp\,m'n'p'}(\mathbf{r}_{i-i'})\,J_{m'n'p'\,klq}(\mathbf{r}_{i'-j}),$$
$$|\mathbf{r}_{i-i'}| > |\mathbf{r}_{i'-j}| \quad (8.18)$$

Care must be exercised when applying Eqs. (8.8) and/or (8.15) recursively via the above three formulas. In particular, it is not possible to extend, or to bypass, the convergence radius criteria of the translations via a successive set of translations. One such fallacious approach is illustrated using Fig. 8.3. In this example, an outgoing harmonic centered about origin 1 and evaluated at point $\mathbf{r}$ is first translated to origin 2. Because $|\mathbf{r}-\mathbf{r}_2| > |\mathbf{r}_{2-1}|$, the expansion will be in the form of Eq. (8.15), i.e., with outgoing VWH centered about origin 2. One might then attempt to apply a second translation, in which the outgoing VWH, centered about 2, are expanded about origin 3. And since $|\mathbf{r}-\mathbf{r}_3| > |\mathbf{r}_{3-2}|$, it would appear that Eq. (8.15) would again apply, leading to outgoing VWH centered about 3. The net result would be

Fig. 8.4 Application of the addition theorem to a far-field point

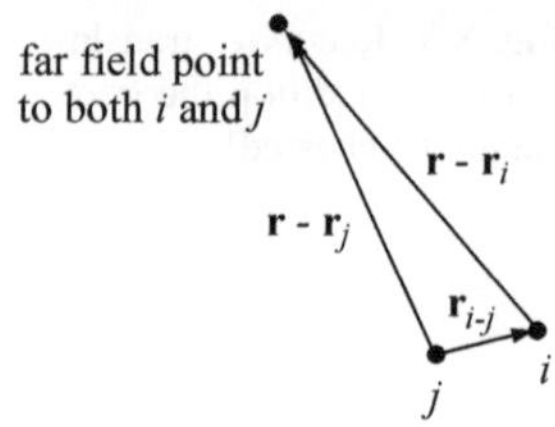

$$\begin{aligned}\mathbf{N}_{\mu}^{(3)}(\mathbf{r}-\mathbf{r}_1) &= \sum_{\nu} J_{\nu\,\mu}(\mathbf{r}_{2-1})\,\mathbf{N}_{\nu}^{(3)}(\mathbf{r}-\mathbf{r}_2)\\ &= \sum_{\nu} J_{\nu\,\mu}(\mathbf{r}_{2-1}) \sum_{\nu'} J_{\nu'\,\nu}(\mathbf{r}_{3-2})\,\mathbf{N}_{\nu'}^{(3)}(\mathbf{r}-\mathbf{r}_3)\\ &= \sum_{\nu'} J_{\nu'\,\mu}(\mathbf{r}_{3-1})\,\mathbf{N}_{\nu'}^{(3)}(\mathbf{r}-\mathbf{r}_3) \longrightarrow \pm\infty \end{aligned} \tag{8.19}$$

In the above Greek subscripts are shorthand for order/degree/mode, i.e., $\nu = (klq)$. Although the first equation in the above is valid, the second and third are not because the implied result would contradict the convergence criterion of Eq. (8.15): the translations shown in Fig. 8.3 have $|\mathbf{r}-\mathbf{r}_3| < |\mathbf{r}_{3-1}|$, and the translation of $\mathbf{N}_{\mu}^{(3)}(\mathbf{r}-\mathbf{r}_1)$ *can only* appear as an expansion of regular harmonics centered about 3.

8.3.3 Derivation of the Addition Theorem

It was demonstrated in [14] that the translation coefficients can be derived from the expansion of a plane wave in regular VWH. Alternatively, a more compact derivation can be obtained by consideration of the far-field behavior of outgoing VWH. This derivation begins with the translation of an outgoing VWH from origin j to i for $|\mathbf{r}-\mathbf{r}_i| > |\mathbf{r}_{i-j}|$ given in Eq. (8.15). Referring to Fig. 8.4, and recognizing that the translation coefficients are independent of the location of the evaluation point $\mathbf{r}$, the distances $\mathsf{k}|\mathbf{r}-\mathbf{r}_i|$ and $\mathsf{k}|\mathbf{r}-\mathbf{r}_j|$ can be made arbitrarily large so that the VWH attain their far-field asymptotic forms, which are represented by

$$\mathbf{N}_{mnp}^{(3)}(\mathbf{r}-\mathbf{r}_j)\Big|_{\mathsf{k}|\mathbf{r}-\mathbf{r}_j|\to\infty} \to \frac{1}{\mathrm{i}\mathsf{k}\,|\mathbf{r}-\mathbf{r}_j|}\, e^{\mathrm{i}\mathsf{k}\,|\mathbf{r}-\mathbf{r}_j|}\,\boldsymbol{\Pi}_{mnp}(\cos\theta_j,\,\phi_j) \tag{8.20}$$

and likewise for $\mathbf{N}_{klq}^{(3)}(\mathbf{r}-\mathbf{r}_i)$. In the above $\boldsymbol{\Pi}$ represent the vector spherical harmonic (VSH) functions, which are formally defined by

$$\boldsymbol{\Pi}_{mn1}(\cos\theta,\,\phi) = (-\mathrm{i})^{n+1}\, r\left(\frac{1}{n(n+1)}\right)^{1/2} \nabla\, Y_{mn}(\cos\theta,\,\phi) \tag{8.21}$$

$$\boldsymbol{\Pi}_{mn2}(\cos\theta,\,\phi) = \mathrm{i}\,\hat{\mathbf{r}}\times\boldsymbol{\Pi}_{mn1}(\cos\theta,\,\phi) \tag{8.22}$$

and are functionally given by

$$\boldsymbol{\Pi}_{mnp}(\cos\theta,\phi) = (-\mathrm{i})^{n+1}\left(\tau_{mnp}(\cos\theta)\,\hat{\mathbf{e}}_\theta + \mathrm{i}\,\tau_{mn3-p}(\cos\theta)\,\hat{\mathbf{e}}_\phi\right)\,e^{\mathrm{i}m\phi} \tag{8.23}$$

$$\tau_{mn1}(x) = -(1-x^2)^{1/2}\left(\frac{2n+1}{4\pi n(n+1)}\,\frac{(n-m)!}{(n+m)!}\right)^{1/2}\,\frac{d\,P_n^m(x)}{dx} \tag{8.24}$$

$$\tau_{mn2}(x) = m\,(1-x^2)^{-1/2}\left(\frac{2n+1}{4\pi n(n+1)}\,\frac{(n-m)!}{(n+m)!}\right)^{1/2}\,P_n^m(x) \tag{8.25}$$

The origin separation distance $|\mathbf{r}_{i-j}|$ can also be made small with respect to both $|\mathbf{r}-\mathbf{r}_i|$ and $|\mathbf{r}-\mathbf{r}_j|$ so that $\theta_i \approx \theta_j = \theta$, $\phi_i \approx \phi_j = \phi$, and $|\mathbf{r}-\mathbf{r}_i| \approx |\mathbf{r}-\mathbf{r}_j|$. By incorporating these limits into Eq. (8.15), and accounting for the phase shift due to the displacement of the origin, one obtains

$$e^{\mathrm{i}\,\mathbf{k}\cdot\mathbf{r}_{i-j}}\,\boldsymbol{\Pi}_{mnp}(\cos\theta,\phi) = \sum_{l=1}^{\infty}\sum_{k=-l}^{l}\sum_{q=1}^{2} J_{klq\,mnp}(\mathbf{r}_{i-j})\,\boldsymbol{\Pi}_{klq}(\cos\theta,\phi) \tag{8.26}$$

in which the argument of the exponential is

$$\mathbf{k}\cdot\mathbf{r}_{i-j} = \mathsf{k}\left[\left(x_{i-j}\cos\phi + y_{i-j}\sin\phi\right)\sin\theta + z_{i-j}\cos\theta\right] \tag{8.27}$$

$$= \mathsf{k}\,r_{i-j}\left[\sin\theta\,\sin\theta_{i-j}\,\cos(\phi-\phi_{i-j}) + \cos\theta\,\cos\theta_{i-j}\right] \tag{8.28}$$

Inversion of Eq. (8.26) can be accomplished by use of the VSH orthogonality relations;

$$\int_0^{2\pi}\int_{-1}^{1} \boldsymbol{\Pi}_{mnp}(x,\phi)\cdot\boldsymbol{\Pi}^*_{klq}(x,\phi)\,dx\,d\phi = \delta_{n-l}\delta_{m-k}\,\delta_{p-q} \tag{8.29}$$

and the formula for the regular VWH translation matrix becomes

$$J_{klq\,mnp}(\mathbf{r}_{ij}) = \int_0^{2\pi}\int_{-1}^{1} e^{\mathrm{i}\mathbf{k}\cdot\mathbf{r}_{i-j}}\,\boldsymbol{\Pi}_{mnp}(\cos\theta,\phi)\cdot\boldsymbol{\Pi}^*_{klq}(\cos\theta,\phi)\,d\cos\theta\,d\phi \tag{8.30}$$

Evaluation of the integrals in Eq. (8.30) can be performed by utilizing the linearization expansion for VSH functions,

$$\boldsymbol{\Pi}_{mnp}\cdot\boldsymbol{\Pi}^*_{klq} = \mathrm{i}^{l-n}\left[\frac{(2n+1)(2l+1)}{4\pi}\right]^{1/2} \times \sum_{w=w_1,2}^{w_2} \frac{1}{\sqrt{2w+1}}\,C^w_{-kl\,mn}\,C^w_{-1l\,1n}\,Y_{m-k\,w}(\cos\theta,\phi) \tag{8.31}$$

in which $C^{w}_{kl\,mn}$ is shorthand for the Clebsch-Gordon coefficient $C(k, l; m, n; k + m, w)$. The limits on the summation are $w_1 = |n - l| + |p - q|$, $w_2 = n + l - |p - q|$, and the notation $w = w_1, 2$ indicates that every other term is added, i.e., $w = w_1, w_1 + 2, \ldots, w_2$. The exponential can also be expanded via

$$e^{\mathrm{i}\mathbf{k}\cdot\mathbf{r}_{i-j}} = 4\pi \sum_{w=0}^{\infty} \sum_{v=-w}^{w} \mathrm{i}^w\, j_w(\mathrm{k}r_{i-j})\, Y_{vw}(\cos\theta_{i-j}, \phi_{i-j})\, Y_{-vw}(\cos\theta, \phi) \tag{8.32}$$

The orthogonality properties of the spherical harmonics can now be employed to integrate Eq. (8.30), and the formula for the regular VWH translation matrix becomes

$$\begin{aligned} J_{klq\,mnp}(\mathbf{r}_{i-j}) &= -(-1)^k\, \mathrm{i}^{l-n}\, [4\pi\, (2n+1)(2l+1)]^{1/2} \\ &\quad \times \sum_{w=w_1,2}^{w_2} \frac{\mathrm{i}^w}{\sqrt{2w+1}}\, C^{w}_{-kl\,mn}\, C^{w}_{-1l\,1n}\, \psi^{(1)}_{m-k\,w}(\mathbf{r}_{i-j}) \end{aligned} \tag{8.33}$$

It is implicitly understood that $C^{w}_{kl\,mn}$ is zero for $w < w_s = \mathrm{Max}(|n-l|, |m+k|)$. In this sense, the evaluation of the sum in Eq. (8.33) should begin with $w_2 = n+l-|p-q|$ and step downwards by two until w_s is reached. The formula for the outgoing harmonic translation matrix is obtained simply by replacing $\psi^{(1)}$ with $\psi^{(3)}$ in the above;

$$\begin{aligned} H_{klq\,mnp}(\mathbf{r}_{i-j}) &= -(-1)^k\, \mathrm{i}^{l-n}\, [4\pi\, (2n+1)(2l+1)]^{1/2} \\ &\quad \times \sum_{w=w_1,2}^{w_2} \frac{i^w}{\sqrt{2w+1}}\, C^{w}_{-kl\,mn}\, C^{w}_{-1l\,1n}\, \psi^{(3)}_{m-k\,w}(\mathbf{r}_{i-j}) \end{aligned} \tag{8.34}$$

Calculation of the translation matrix elements is discussed in the Appendix.

8.3.4 Application to the Plane Wave Expansion

Equation (8.26) shows that the VSH functions and the exponential phase factor represent the eigenvectors and eigenvalues to the regular VWH translation matrix. This feature, along with the translation properties of the J matrix, can be used to identify the coefficients for the VWH expansion of a plane, linearly polarized wave. By taking the magnitude of both sides of Eq. (8.26) and applying Eq. (8.29), it can be shown that the J matrix is unitary,

$$\sum_{l=1} \sum_{k=-l}^{l} \sum_{q=1}^{2} J_{klq\,mnp}(\mathbf{r}_{j-i})\, J^{*}_{klq\,m'n'p'}(\mathbf{r}_{j-i}) = \delta_{m-m'}\delta_{n-n'}\delta_{p-p'} \tag{8.35}$$

and by using the recursive translation properties of J in combination with the above formula, it follows that

$$J_{klq\,mnp}(\mathbf{r}_{j-i}) = J^{*}_{mnp\,klq}(\mathbf{r}_{i-j}) \tag{8.36}$$

Consequently, Eq. (8.26) has the alternative form

$$e^{\mathrm{i}\,\mathbf{k}\cdot\mathbf{r}_{i-j}}\,\boldsymbol{\Pi}^{*}_{mnp}(\cos\theta,\phi) = \sum_{l=1}^{\infty}\sum_{k=-l}^{l}\sum_{q=1}^{2} J_{mnp\,klq}(\mathbf{r}_{i-j})\,\boldsymbol{\Pi}^{*}_{klq}(\cos\theta,\phi) \tag{8.37}$$

This formula is now multiplied by $\mathbf{N}^{(1)}_{mnp}(\mathbf{r}-\mathbf{r}_i)$, where $\mathbf{r}$ denotes an arbitrary point, and summed over (m,n,p). Application of the translation properties in Eq. (8.11) results in

$$\begin{aligned} e^{\mathrm{i}\,\mathbf{k}\cdot\mathbf{r}_{i-j}} &\sum_{n=1}^{\infty}\sum_{m=-n}^{n}\sum_{p=1}^{2} \boldsymbol{\Pi}^{*}_{mnp}(\cos\beta,\alpha)\,\mathbf{N}^{(1)}_{mnp}(\mathbf{r}-\mathbf{r}_i) \\ &= \sum_{l=1}^{\infty}\sum_{k=-l}^{l}\sum_{q=1}^{2} \boldsymbol{\Pi}^{*}_{klq}(\cos\beta,\alpha)\,\mathbf{N}^{(1)}_{klq}(\mathbf{r}-\mathbf{r}_j) \end{aligned} \tag{8.38}$$

The angles β and α, which represent the polar and azimuthal direction of the vector $\mathbf{k}$, have been introduced to avoid confusion with the coordinates of the vector $\mathbf{r}$. Equation (8.38) indicates that the VWH expansions centered about origins i and j must each be proportional to $\exp(\mathbf{k}\cdot(\mathbf{r}-\mathbf{r}_i))$ and $\exp(\mathbf{k}\cdot(\mathbf{r}-\mathbf{r}_j))$. In this context, Eq. (8.38) represents a generalized plane wave VWH expansion for a wave that is propagating in the $\theta=\beta$, $\phi=\alpha$ direction as illustrated in Fig. 8.5. The vector $\boldsymbol{\Pi}$ in this expansion defines the direction of the electric field vector in the transverse plane; the two cases of $\hat{\beta}\cdot\boldsymbol{\Pi}$ and $\hat{\alpha}\cdot\boldsymbol{\Pi}$ would represent plane wave polarization parallel and perpendicular to the $\hat{\mathbf{z}}-\mathbf{k}$ plane. A unit amplitude can established by multiplying the coefficients in the expansion by a constant, which can be obtained evaluating the right-hand side in the limit of $|\mathbf{r}-\mathbf{r}_j|\to 0$.

In the general case, the electric field vector can be defined by the angle γ relative to the $\hat{\mathbf{z}}-\mathbf{k}$ plane. The expansion coefficients for the field, referenced to the target origin, will then be given by

$$p_{mnp} = \cos\gamma\,p_{mnp,\,1} + \sin\gamma\,p_{mnp,\,2} \tag{8.39}$$

with

$$p_{mnp,\,1} = -4\pi\,\mathrm{i}^{n+1}\,\tau_{mnp}(\cos\beta)\,e^{-\mathrm{i}\,m\,\alpha} \tag{8.40}$$

$$p_{mnp,\,2} = -4\pi\,\mathrm{i}^{n}\,\tau_{mn3-p}(\cos\beta)\,e^{-\mathrm{i}\,m\,\alpha} \tag{8.41}$$

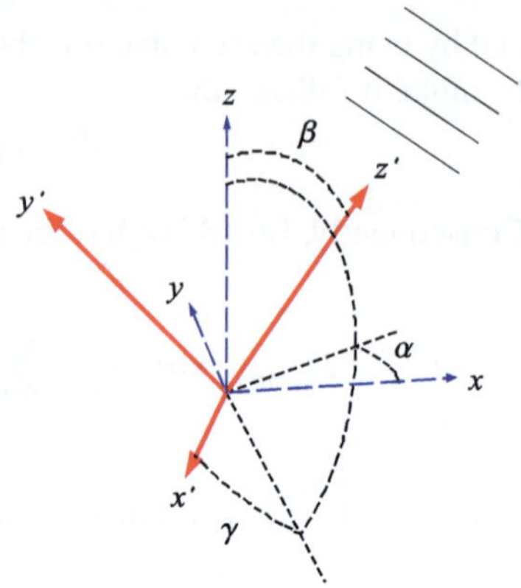

Fig. 8.5 Target and incident field frames

As illustrated in Fig. 8.5, the angles α, β, and γ, define an incident field frame, obtained by Euler rotation of the target frame, so that z' denotes the propagation direction.The $p_{mnp,1}$ and $p_{mnp,2}$ expansion coefficients describe an incident plane wave that is polarized either in the x' or y' directions.

8.3.5 Far-Field Limit of the Translation Matrix

When a pair of spheres are widely separated, so that the scattered field from one attains the far-field form about the other, the translation operation in Eq. (8.8) can be reduced to a much simpler form. The approach is to replace the Hankel function appearing in the scalar wave harmonic in Eq. (8.34) with the asymptotic limit

$$h_n(\mathrm{k}r) \sim \frac{(-\mathrm{i})^n}{\mathrm{ik}r} e^{\mathrm{ik}r}, \quad \mathrm{k}\, r \gg n^2 \tag{8.42}$$

By employing the linearization relation in Eq. (8.31), the outgoing wave translation matrix will reduce to

$$H_{klq\,mnp}(\mathbf{r}_{i-j}) \sim \frac{4\pi}{\mathrm{ik}r_{i-j}} e^{\mathrm{ik}r_{i-j}}\, \boldsymbol{\Pi}_{mnp}(\cos\theta_{i-j}, \phi_{i-j}) \cdot \boldsymbol{\Pi}^*_{klq}(\cos\theta_{i-j}, \phi_{i-j}) \tag{8.43}$$

8.4 Properties of the Multiple Sphere Solution

8.4.1 Numerical Issues

All the necessary details underlying Eq. (8.10) are now in place, and we can now turn to the problem of solving Eq. (8.10) for the scattering coefficients and, from the solution, determining the observable scattering and absorption properties for the given cluster of spheres. Linear system of equations such as Eq. (8.10) can be solved using either iteration or direct (i.e., inversion) methods; the optimum strategy will depend on the properties of the system (such as the condition number which will affect the iterative convergence rate), the requirements of the solution (e.g., solutions for one versus multiple right-hand sides), and the experience of the user (perhaps the most important criterion of all). Iteration methods will typically be more computationally efficient than direct inversion when a solution is needed for a small number of specified incident directions. Our experience, and that of others, is that the biconjugate gradient method provides the most reliable, and fastest, solution as compared to over/under relaxation and order-of-scattering methods [20]. The number of iterations required for a solution depends on a host of parameters; i.e., the number, size parameters, and refractive indices of the spheres, and the proximity of the spheres to each other. In general, as the spheres become more widely separated the solution will converge faster. An important factor affecting convergence is whether any of the spheres is at or near a resonance mode; such conditions can lead to extremely small convergence rates and may be more effectively solved using direct methods [21].

In situations where the scattering properties are needed for a relatively large number of incident directions, or where the random-orientation properties of the cluster are sought, it becomes most efficient to calculate a T matrix for the cluster. The T matrix will provide the linear relationship between the VWH coefficients representing the incident and the scattered field, and the matrix can be obtained via direct inversion of Eq. (8.10). It turns out, however, that the multiple-spheres problem allows for a computationally efficient, iteration-based method for calculating the cluster T matrix.

8.4.2 The T Matrix Relationships

There are three levels of T matrix relationships for a cluster of spheres. The first, referred to as the sphere–sphere T matrix, represents the formal solution to Eq. (8.10) and appears as

$$a_\mu^i = \sum_{j=1}^{N_S} \sum_\nu T_{\mu\nu}^{i-j}\, p_\nu^j \tag{8.44}$$

in which the sphere–sphere T^{i-j} matrix is the solution to

$$\frac{1}{\bar{a}_\mu} T_{\mu\nu}^{i-j} - \sum_{\substack{j'=0 \\ j'\neq i}}^{N_S} \sum_{\nu'} H_{\mu\nu'}^{i-j'}\, T_{\nu'\nu}^{j'-j} = \delta_{i-j}\, \delta_{\mu-\nu} \tag{8.45}$$

In the above and what follows, Greek subscripts are a shorthand for order/degree /mode, i.e., $\mu = (m, n, p)$. The T^{j-i} matrix describes the scattered field produced at sphere i due to an incident field at j; all the effects of multiple scattering are embedded within the matrix.

The next T matrix relationship describes the scattered field at sphere i due to the entire effect of the incident field on the ensemble.This description begins with an expansion of the incident field about the coordinate origin of the target;

$$\mathbf{E}_{inc} = \sum_{n=1}^{L} \sum_{m=-n}^{n} \sum_{p=1}^{2} p_{mnp} \, \mathbf{N}^{(1)}_{mnp}(\mathbf{r}) \tag{8.46}$$

The order truncation limit L in this expansion—which is chosen so that the expansion will yield an acceptable description of the incident field *on each sphere in the ensemble*—will typically depend on the size parameter ka_C based on the circumscribing radius a_C (illustrated in Fig. 8.1).As was the case with the individual sphere limits L_i, the target L can typically be estimated using the Mie criterion applied to $k a_C$. The sphere centered incident field expansion coefficients can then be obtained from the target coefficients via the translation operation,

$$p^i_{mnp} = \sum_{l=1}^{L} \sum_{k=-l}^{l} \sum_{q=1}^{2} J^{i-0}_{mnp\,klq} \, p_{klq} \tag{8.47}$$

where origin 0 denotes the target origin. Replacing this into Eq. (8.44) results in

$$\begin{aligned} a^i_\mu &= \sum_{i=1}^{N_S} \sum_{\nu} \sum_{\nu'} T^{i-j}_{\mu\nu'} \, J^{j-0}_{\nu'\nu} \, p_\nu \\ &= \sum_{\nu} T^i_{\mu\nu} \, p_\nu \end{aligned} \tag{8.48}$$

In general, the sphere-cluster matrix $T^i_{\mu\nu}$ will have more columns than rows; the largest row and column order will be L_i and L, respectively.

A set of interaction equations for T^i—which obviate the need to directly perform the contraction in Eq. (8.48)—can be obtained by substituting Eqs. (8.47) and (8.48) into Eq. (8.10), leading to

$$\frac{1}{\overline{a}^i_\mu} T^i_{\mu\,\nu} - \sum_{\substack{j=1 \\ j\neq i}}^{N_S} \sum_{\nu'} H^{i-j}_{\mu\,\nu'} \, T^j_{\nu'\nu} = J^{i-0}_{\mu\nu} \tag{8.49}$$

This system can be solved by iteration for successive values of $\nu = (klq)$ to build up the T^i matrices for the spheres. Such a method is often more computationally efficient—and less requiring of memory resources—than direct inversion of Eq. (8.45) followed by Eq. (8.48) [13].

The third T matrix relationship performs a contraction on T^i, so that the scattered field from the cluster—which, per Eq. (8.1), is the sum of the individual-sphere direct scattered fields—is represented by a single outgoing VWH expansion centered about the target origin. The coefficients for the target-centered expansion are obtained by applying Eq. (8.15) to Eq. (8.3) for each sphere, and results in

$$\mathbf{E}_{sca} = \sum_{n=1}^{L} \sum_{m=-n}^{n} \sum_{p=1}^{2} a_{mnp} \, \mathbf{N}_{mnp}^{(3)}(\mathbf{r}) \tag{8.50}$$

$$a_\mu = \sum_{i=1}^{N_S} \sum_{\nu} J_{\mu\nu}^{0-i} \, a_\nu^i \tag{8.51}$$

And by using Eq. (8.48) in Eq. (8.51), one obtains,

$$\begin{aligned} a_\mu &= \sum_{i=1}^{N_S} \sum_{\nu} \sum_{\nu'} J_{\mu\nu'}^{0-i} \, T_{\nu'\nu}^{i} \, p_\nu \\ &= \sum_{\nu} T_{\mu\nu} \, p_\nu \end{aligned} \tag{8.52}$$

The cluster-centered T matrix treats the ensemble of spheres as a single—albeit nonspherical—particle, and in this respect it is directly analogous to the Mie relation of Eq. (8.9). It is a very useful quantity, for the cross sections of the ensemble and the scattering matrix—in either fixed or random orientation—can be obtained analytically from its properties. However, the T matrix cannot predict the fields within the cluster or the cross sections of the individual spheres; realize that, as per the restriction on the addition theorem in Eq. (8.15), and (8.50) will be valid only for radii that exceed the largest sphere-target origin distance. The detailed individual-sphere and near-field information—which is inaccessible from Eq. (8.15)—can, however, be obtained from the original superposition model.

8.4.3 Coordinate Rotation and the Amplitude and Scattering Matrix

Calculation of the full polarimetic scattering properties of the ensemble for a fixed incident direction requires the solution of Eq. (8.10) for two mutually perpendicular linear polarization states of the incident wave. Referring again to Fig. 8.5, the polarization state of the incident beam is specified by the angle γ. Since the incident field coefficients defined in Eq. (8.39) are a linear combination of $\cos\gamma$ and $\sin\gamma$ components, a solution to Eq. (8.10) for an arbitrary γ can be obtained via

$$a^i_{mnp,\parallel} = a^i_{mnp,1} \cos\gamma + a^i_{mnp,2} \sin\gamma \tag{8.53}$$

where $a^i_{mnp,1}$ and $a^i_{mnp,2}$ denote the two solutions to Eq. (8.10) for incident field coefficients of $p_{mnp,1}$ and $p_{mnp,2}$ given by Eqs. (8.40) and (8.41), i.e., $\gamma = 0$ and $\pi/2$. The parallel ($\|$) notation will be described shortly. The scattering plane used to define the amplitude matrix will now be adopted as the $x' - z'$ plane, and following the usual convention, the parallel and perpendicular incident polarization states will correspond to the electric field pointing in the x' and the $-y'$ directions, respectively. The solution corresponding to the parallel case will therefore be that given in Eq. (8.53), and the perpendicular solution will be that obtained by replacing γ with $\gamma - \pi/2$;

$$a^i_{mnp,\perp} = a^i_{mnp,1} \sin\gamma - a^i_{mnp,2} \cos\gamma \tag{8.54}$$

The amplitude and scattering matrix of the ensemble can be calculated using either the superposition representation in Eq. (8.1), which would retain the scattering coefficients for the individual spheres, or the single-origin expansion in Eq. (8.50). In the far-field limit, the former method is made complicated by the need to explicitly account for the relative phase shift among the different scattering origins as a function of the scattering angle, yet this method can also be more computationally efficient in situations in which the spheres are relatively far apart (i.e., large $\mathsf{k}\, a_C$), for which a large value of L would be required for convergence of Eq. (8.50) [22]. However, the latter method retains a spherical harmonic representation of the scattered field, and as such allows for a more direct identification of various moments of the scattered intensity (such as the scattering efficiency and the asymmetry parameter). Our experience is that advantages of the single-expansion method outweigh the disadvantages; it should be noted as well that convergence or loss-of-precision problems associated with the numerical implementation of Eq. (8.51) have not been encountered in our codes.

Equations (8.53) and (8.54) will hold equally for the total scattered field coefficients $a_{mnp,1}$ and $a_{mnp,2}$ that are obtained from Eq. (8.51). The expansion in Eq. (8.50) that is obtained from these coefficients would describe the scattered field in a target-based reference frame; to conform to the conventional, incident-based frame it is necessary to perform a rotation transformation on Eq. (8.50). The incident frame is obtained by a Euler rotation of the target frame through $(\alpha, \beta\, \gamma)$, and the corresponding rotation transformation on the harmonic functions is given by

$$\mathbf{N}^{(3)}_{mnp}(\mathsf{k}r, \theta, \phi) = e^{\mathrm{i}\, m\,\alpha} \sum_{m'=-n}^{n} \mathscr{D}^{(n)}_{m'm}(\cos\beta)\, e^{\mathrm{i}\, m'\,\gamma}\, \mathbf{N}^{(3)}_{m'np}(\mathsf{k}r, \theta', \phi') \tag{8.55}$$

where θ', ϕ' denote the polar and azimuthal angles in the incident frame; $\theta' = 0$ (corresponding to $\theta = \beta$) now represents the forward scattering direction. The generalized spherical functions $\mathscr{D}^{(n)}_{mk}$ are related to the Jacobi polynomials; definitions and properties of the functions are presented in the Appendix. Application of Eq. (8.55) to (8.50) will transform the expansion so that the VWH become based on the incident frame (i.e., $\mathbf{r}$ replaced with $\mathbf{r}'$); the total scattering coefficients appearing in the rotated expansion will be given by

$$a'_{mnp,s} = e^{\mathrm{i}\,m\,\gamma} \sum_{m'=-n}^{n} \mathscr{D}^{(n)}_{mm'}(\cos\beta)\, e^{\mathrm{i}\,m'\,\alpha}\, a_{m'np,s} \tag{8.56}$$

where $s = 1, 2$ denotes the incident polarization state.

Following the convention of Bohren and Huffman [23], the amplitude matrix is defined so that

$$\begin{pmatrix} E_{sca,\parallel} \\ E_{sca,\perp} \end{pmatrix} = -\frac{1}{\mathrm{i}\,k\,r} e^{\mathrm{i}\,k(r-z')} \begin{pmatrix} S_2 & S_3 \\ S_4 & S_1 \end{pmatrix} \begin{pmatrix} E_{inc,\parallel} \\ E_{inc,\perp} \end{pmatrix} \tag{8.57}$$

where $E_{inc,\parallel}$ and $E_{inc,\perp}$ denote the magnitudes of the incident electric field amplitude parallel and perpendicular to the scattering plane. The explicit formulas for the amplitude matrix elements are obtained by applying to the rotated Eq. (8.50) the far-field asymptotic form of the outgoing VWH in Eq. (8.20), and noting that the scattering plane corresponds to $\phi' = 0$.This gives

$$S_1 = \sum_{n=1}^{L} \sum_{m=-n}^{n} \sum_{p=1}^{2} (-\mathrm{i})^n a'_{mnp,\perp} \tau_{mn3-p}(\cos\theta') \tag{8.58}$$

$$S_2 = \sum_{n=1}^{L} \sum_{m=-n}^{n} \sum_{p=1}^{2} (-\mathrm{i})^{n+1} a'_{mnp,\parallel} \tau_{mnp}(\cos\theta') \tag{8.59}$$

$$S_3 = \sum_{n=1}^{L} \sum_{m=-n}^{n} \sum_{p=1}^{2} (-\mathrm{i})^{n+1} a'_{mnp,\perp} \tau_{mnp}(\cos\theta') \tag{8.60}$$

$$S_4 = \sum_{n=1}^{L} \sum_{m=-n}^{n} \sum_{p=1}^{2} (-\mathrm{i})^n a'_{mnp,\parallel} \tau_{mn3-p}(\cos\theta') \tag{8.61}$$

Elements of the scattering matrix can be obtained directly from those of the amplitude matrix following the formulas presented in Bohren and Huffman [23].

8.4.4 Cross Sections and Energy Conservation

The absorption cross section of sphere i is defined so that $C_{abs,i} I_0$ is the rate at which the sphere absorbs energy from the incident wave of irradiance I_0. This quantity can be obtained by integration of the Poynting vector over the inside surface of the particle. By employing the formulas of Lorenz/Mie theory to relate the internal and external fields, the expression for the absorption cross section is

$$C_{abs,\,i} = \frac{2\pi}{\mathsf{k}^2} \sum_{n=1}^{L_i} \sum_{m=-n}^{n} \sum_{p=1}^{2} \overline{b}_{np}^{i} \left| a_{mnp}^{i} \right|^2 \tag{8.62}$$

in which $\overline{b}_{np}^{i}$ is a positive (or zero) real valued property solely of sphere i and is defined by

$$\overline{b}_{np}^{i} = -\text{Re}\left(\frac{1}{\overline{a}_{np}^{i}} + 1 \right) \tag{8.63}$$

The scattering coefficients in the above formula would correspond to either the parallel or perpendicular polarization values; the absorption cross section for unpolarized incident radiation would be the average of the two. By simple conservation-of-energy considerations, the absorption cross section of the entire ensemble is the sum of the individual sphere cross sections;

$$C_{abs} = \sum_{i=1}^{N_S} C_{abs,i} \tag{8.64}$$

In a similar manner, an extinction cross section of an individual sphere can be defined so that $I_0\, C_{ext,i}$ is the rate at which the sphere removes energy from the incident wave. The optical theorem applied to the field scattered from the sphere gives

$$C_{ext,\,i} = -\frac{2\pi}{\mathsf{k}^2} \text{Re} \sum_{n=1}^{L_i} \sum_{m=-n}^{n} \sum_{p=1}^{2} a_{mnp}^{i} p_{mnp}^{i*} \tag{8.65}$$

As before, the scattering and incident field coefficients would correspond to the particular polarization state. The total ensemble extinction cross section would also be obtained from the sum of the parts;

$$C_{ext} = \sum_{i=1}^{N_S} C_{ext,i} \tag{8.66}$$

Unlike the absorption cross section, the extinction cross section for the individual sphere would be difficult—if not impossible—to experimentally measure. The definition of this quantity relies on the superposition model of the scattered field, and although this model serves perfectly well as a means to solve Maxwell's equations for the ensemble, it is not obvious how the 'partial' fields scattered from the individual spheres could be discriminated in an experiment.

The individual sphere extinction cross section is also not bounded by the usual behavior for an isolated particle; although the total cross sections must satisfy $C_{ext} \geq C_{abs}$, it is entirely possible for $C_{ext,i} < C_{abs,i}$ and even for $C_{ext,i}$ to be negative. The precise balance of radiant energy, at the individual sphere level, can be identified by multiplying Eq. (8.10) through by $2\pi/\mathsf{k}^2\, a^{i*}_{mnp}/\overline{a}^i_{np}$, summing over m, n, p, and rearranging to isolate Eqs. (8.62) and (8.65). This results in

$$\begin{aligned} C_{ext,i} - C_{abs,i} &= \frac{2\pi}{\mathsf{k}^2} \sum_{\mu} \left| a^i_{\mu} \right|^2 + \frac{2\pi}{\mathsf{k}^2} \operatorname{Re} \sum_{\mu} a^{i*}_{\mu} \sum_{\substack{j=1 \\ j \neq i}}^{N_S} \sum_{\nu} H^{i-j}_{\mu\,\nu}\, a^j_{\nu} \\ &= C_{i-sca,i} + C_{d-sca,i} \end{aligned} \tag{8.67}$$

The quantities appearing on the right-hand side are referred to as the independent and the dependent scattering cross sections of the sphere.The former will always be positive, yet the latter—which accounts for energy transfer via the interference between the fields scattered from i and the neighboring spheres—can take on positive or negative values. And since $C_{abs,i} + C_{i-sca,i} > 0$, the above balance implies that $C_{ext,i} > C_{d-sca,i}$.

By using $H^{i-j} = J^{i-j} + \mathrm{i}Y^{i-j}$ in Eq. (8.67), and applying Eq. (8.104) to both J and Y, it follows that the total dependent scattering cross section is given by

$$\begin{aligned} C_{d-sca} &= \sum_{i=1}^{N_S} C_{d-sca,i} \\ &= \frac{2\pi}{\mathsf{k}^2} \sum_{i=1}^{N_S} \sum_{j=i+1}^{N_S} \sum_{\mu,\nu} \operatorname{Re} \left[a^{i*}_{\mu}\, H^{i-j}_{\mu\,\nu}\, a^j_{\nu} + a^{j*}_{\mu}\, H^{j-i}_{\mu\,\nu}\, a^i_{\nu} \right] \\ &= \frac{2\pi}{\mathsf{k}^2} \sum_{i=1}^{N_S} \sum_{j=i+1}^{N_S} \sum_{\mu,\nu} \operatorname{Re} \Big[a^{i*}_{\mu} \left(J^{i-j}_{\mu\,\nu} + \mathrm{i}\, Y^{i-j}_{\mu\,\nu} \right) a^j_{\nu} \\ &\quad + a^i_{\mu} \left[\left(J^{i-j}_{\mu\,\nu} - \mathrm{i}\, Y^{i-j}_{\mu\,\nu} \right) a^j_{\nu} \right]^* \Big] \\ &= \frac{2\pi}{\mathsf{k}^2} \sum_{i=1}^{N_S} \sum_{\substack{j=1 \\ j \neq i}}^{N_S} \sum_{\mu,\nu} a^{i*}_{\mu}\, J^{i-j}_{\mu\,\nu}\, a^j_{\nu} \end{aligned} \tag{8.68}$$

Note that it is no longer necessary to indicate the real part in the last line, because the quantity will be identically real. The regular–wave translation matrix can now be reexpanded about the target origin, via

$$\begin{aligned} J^{i-j}_{mnp\,klq} &= \sum_{n'=1}^{L} \sum_{m'=-n'}^{n'} \sum_{p'=1}^{2} J^{i-0}_{mnp\,m'n'p'}\, J^{0-j}_{m'n'p'\,klq} \\ &= \sum_{n'=1}^{L} \sum_{m'=-n'}^{n'} \sum_{p'=1}^{2} \left(J^{0-i}_{m'n'p'\,mnp} \right)^* J^{0-j}_{m'n'p'\,klq} \end{aligned} \tag{8.69}$$

Substituting the above into the previous equation and using Eq. (8.51) gives the expected result for energy conservation,

$$C_{ext} - C_{abs} = \sum_{i=1}^{N_S} \left(C_{i-sca,i} + C_{d-sca,i}\right) = C_{sca} \tag{8.70}$$

with the total scattering cross section given by

$$C_{sca} = \frac{2\pi}{\mathrm{k}^2} \sum_{\mu} \left|a_\mu\right|^2 \tag{8.71}$$

8.4.5 Random Orientation

The random orientation cross sections can be obtained by using the matrix relationships for the scattered and incident field (Eqs. 8.44, 8.47, 8.48) in Eqs. (8.62) and (8.65) and integrating the incident field over all propagation and polarization directions. For a plane wave, this integration results in

$$\left\langle p^i_\nu \left(p^i_{\nu'}\right)^* \right\rangle = \delta_{\nu-\nu'} \tag{8.72}$$

where < ... > denotes orientation averaging. Consequently,

$$\begin{aligned}\left\langle C_{abs,i}\right\rangle &= \frac{2\pi}{\mathrm{k}^2} \sum_\mu \sum_\nu \overline{b}^i_\mu \left|T^i_{\mu\,\nu}\right|^2 \\ &= \frac{2\pi}{\mathrm{k}^2} \sum_\mu \sum_\nu \overline{b}^i_\mu \sum_{j=1}^{N_S} \sum_{j'=1}^{N_S} \sum_{\nu'} T^{i-j}_{\mu\,\nu'} J^{j-j'}_{\nu'\,\nu} \left(T^{i-j'}_{\mu\,\nu}\right)^* \end{aligned} \tag{8.73}$$

$$\begin{aligned}\left\langle C_{ext,i}\right\rangle &= -\frac{2\pi}{\mathrm{k}^2} \mathrm{Re} \sum_\mu \sum_\nu J^{0-i}_{\mu\,\nu} T^i_{\nu\,\mu} \\ &= -\frac{2\pi}{\mathrm{k}^2} \mathrm{Re} \sum_\mu \sum_{j=1}^{N_S} \sum_\nu T^{i-j}_{\mu\,\nu} J^{j-i}_{\nu\,\mu} \end{aligned} \tag{8.74}$$

As before, the total orientation–averaged absorption and extinction cross sections for the cluster will be the sum of the individual sphere values, and the total scattering cross section will be the difference between the total extinction and absorption cross sections. Alternatively, the conservation-of-energy properties of Eq. (8.49) can be used to identify the traditional formula for the scattering cross section, which is based on the total scattered power from the ensemble. By multiplying Eq. (8.49) through by $T^i_{\mu\,\nu}$, contracting over all indices, and applying the relations used in Eqs. (8.68) and (8.69), it can be shown that

$$
\begin{aligned}
\frac{2\pi}{\mathsf{k}^2}\sum_{\mu}\sum_{\nu}\left|T_{\mu\,\nu}\right|^2 &= \frac{2\pi}{\mathsf{k}^2}\sum_{i=1}^{N_S}\left[-\mathrm{Re}\sum_{\mu}\sum_{\nu}J^{0-i}_{\mu\,\nu}\,T^{i}_{\nu\,\mu}-\sum_{\mu}\sum_{\nu}\overline{b}^{i}_{\mu}\left|T^{i}_{\mu\,\nu}\right|^2\right] \\
&= \frac{2\pi}{\mathsf{k}^2}\left[-\mathrm{Re}\sum_{\mu}T_{\mu\,\mu}-\sum_{\mu}\sum_{\nu}\sum_{i=1}^{N_S}\overline{b}^{i}_{\mu}\left|T^{i}_{\mu\,\nu}\right|^2\right] \\
&= \langle C_{ext}\rangle - \langle C_{abs}\rangle = \langle C_{sca}\rangle
\end{aligned}
\qquad (8.75)
$$

Explicit formulas for the random–orientation scattering matrix elements, which involve operations on the cluster T matrix and yield expansions for the matrix elements in terms of generalized spherical functions, have been derived and are presented in [13].

8.4.6 *Emission Cross Sections and Inter–Sphere Energy Transfer*

A careful distinction must be made when interpreting the sphere absorption cross section in the context of Kirchoff's law. Consider the situation in which each sphere in the ensemble is at a uniform temperature of T_i and the ensemble is surrounded by a black environment at T_e. As per the definition of the absorption cross section, and considering the isotropic environmental radiation, the rate at which sphere i absorbs radiant energy from the environment about a wavelength interval $d\lambda$ will be $q_{e-i} = 4\pi \left\langle C_{abs,i}\right\rangle I_{b,\lambda}(T_e)\, d\lambda$, in which $I_{b,\lambda}$ denotes the blackbody intensity function. Consequently, Kirchoff's law states that $q_{i-e} = 4\pi \left\langle C_{abs,i}\right\rangle I_{b,\lambda}(T_i)\, d\lambda$ will be the rate at which emission from sphere i is transferred to the environment. That is, $\left\langle C_{abs,i}\right\rangle$ does *not* directly measure the rate of energy emission from sphere i; only how much of this rate escapes the confines of the ensemble.On the other hand, the detailed transfer of radiant energy among the particles should be embedded within the formula for $\left\langle C_{abs,i}\right\rangle$. This structure is not obvious in the definition given in Eq. (8.73), and an alternative definition is needed to elucidate the interactive form of the absorption cross section.

The first step in this process is to combine Eqs. (8.97) and (8.101), so that

$$H^{i-j}_{mnp\,klq} = (-1)^{m+k}\,H^{j-i}_{-klq\,-mnp} \qquad (8.76)$$

It can then be shown from Eq. (8.45) that T^{i-j} will also have this property, and that

$$\frac{1}{\overline{a}_{\nu}^{j}} T_{\mu\,\nu}^{i-j} - \sum_{\substack{j'=1 \\ j'\neq i}}^{N_S} \sum_{\nu'} T_{\mu\,\nu'}^{i-j'} H_{\nu'\,\nu}^{j'-j} = \delta_{ij}\,\delta_{\mu\nu} \tag{8.77}$$

This equation is now multiplied by $(T_{\mu\nu}^{i-j})^*$ and summed over j and ν. By employing the contraction properties of the translation matrices used in Eqs. (8.68) and (8.69), it follows that

$$\sum_{j=1}^{N_S} \sum_{\nu} \overline{b}_{\nu}^{j} \left|T_{\mu\,\nu}^{i-j}\right|^2 + \sum_{\nu} \left|T_{\mu\,\nu}^{i}\right|^2 = -\mathrm{Re}\, T_{\mu\,\mu}^{i-i} \tag{8.78}$$

This relation can be interpreted as a conservation-of-energy statement for an incident field that interacts solely with sphere i, i.e., for the hypothetical case when $\mathbf{E}_{inc,j} = 0$ for $j \neq i$. When multiplied by $2\pi\, I_{0,\lambda}\, d\lambda/\mathsf{k}^2$ and summed over μ, the term on the right would represent the orientation–averaged extinction power removed from the localized incident field at sphere i, e.g., including only the $j = i$ term in Eq. (8.74). The energy removed from the beam is either absorbed by the surrounding spheres (represented by the first term on the left-hand side) or scattered into all other directions (the second term).

A more concrete interpretation of Eq. (8.78) is obtained by multiplying the equation by $2\pi/\mathsf{k}^2\, \overline{b}_{\mu}^{i}$ and summing over all μ. By using the formula for $< C_{abs,i} >$ in Eq. (8.73) and rearranging, one obtains

$$\begin{aligned} \langle C_{abs,i} \rangle = \frac{2\pi}{\mathsf{k}^2} \sum_{\mu} \overline{b}_{\mu}^{i} \Bigg[-\mathrm{Re}\, T_{\mu\,\mu}^{i-i} \\ - \sum_{\nu} \overline{b}_{\nu}^{i} \left|T_{\mu\,\nu}^{i-i}\right|^2 - \sum_{\substack{j=1 \\ j\neq i}}^{N_S} \sum_{\nu} \overline{b}_{\nu}^{j} \left|T_{\mu\,\nu}^{i-j}\right|^2 \Bigg] \end{aligned} \tag{8.79}$$

Each of the terms appearing in this formula can be interpreted according to a radiant exchange relationship among the spheres. When multiplied by $4\pi\, I_{b,\lambda}(T_i)\, d\lambda$, the first and second terms on the right represent the source of emissive power from sphere i and the rate of reabsorption by sphere i; the net emissive power being the sum of the two. The third term accounts for the absorption of the net emissive power from i by all the other spheres in the ensemble. The net emissive power from a sphere, and the emissive exchange between a pair of spheres, can be used to define emission cross sections via

$$C_{emis,i} = -\frac{2\pi}{\mathsf{k}^2} \sum_{\mu} \overline{b}_{\mu}^{i} \left[\mathrm{Re}\, T_{\mu\,\mu}^{i-i} + \sum_{\nu} \overline{b}_{\nu}^{i} \left|T_{\mu\,\nu}^{i-i}\right|^2 \right] \tag{8.80}$$

$$C_{emis,i-j} = \frac{2\pi}{\mathsf{k}^2} \sum_{\mu} \overline{b}_{\mu}^{i} \sum_{\nu} \overline{b}_{\nu}^{j} \left|T_{\mu\,\nu}^{i-j}\right|^2 \tag{8.81}$$

The exchange cross section is defined so that net rate of monochromatic emissive transfer between a pair of spheres is given by

$$q_{i-j} = 4\pi \left(C_{emis,i-j}\, I_{b,\lambda}(T_i) - C_{emis,j-i}\, I_{b,\lambda}(T_j)\right)\, d\lambda \tag{8.82}$$

As per the symmetry property of the T^{i-j} matrix, the exchange cross section will automatically satisfy the required reciprocity condition of

$$C_{emis,i-j} = C_{emis,j-i} \tag{8.83}$$

When a sphere is nonabsorbing—which corresponds to $\overline{b} = 0$ for the sphere—the exchange cross section to the sphere will be zero. For a single, isolated sphere the self-interaction T matrix reduces to $T^{i-i}_{\mu\nu} \to \overline{a}^i_\mu \delta_{\mu\nu}$, and using

$$\overline{b}^i_\mu = -\frac{1}{\left|\overline{a}^i_\mu\right|^2} \left(\mathrm{Re}\,\overline{a}_\mu + \left|\overline{a}^i_\mu\right|^2\right) \tag{8.84}$$

it follows that

$$\begin{aligned} C_{emis,1}\big|_{N_S=1} &= \frac{2\pi}{k^2} \sum_\mu \left[-\mathrm{Re}\,\overline{a}^1_\mu - \left|\overline{a}^1_\mu\right|^2\right] \\ &= \frac{2\pi}{k^2} \sum_{n=1}^{L_1} \sum_{p=1}^{2} (2n+1) \left[-\mathrm{Re}\,\overline{a}^1_{np} - \left|\overline{a}^1_{np}\right|^2\right] \\ &= C_{ext,1} - C_{sca,1} = C_{abs,1} \end{aligned} \tag{8.85}$$

In other words, the emission cross section for an isolated sphere becomes identical to the absorption cross section for the sphere.

Unlike the case for the absorption cross sections of the individual spheres, the magnitude of $C_{emis,i}$ and $C_{emis,i-j}$, as well as the number of orders required for convergence of Eqs. (8.80) and (8.81), is strongly dependent on the separation distance between the spheres [24, 25].Indeed, as the gap between a pair of spheres shrinks to zero, Eqs. (8.80) and (8.81) individually will fail to converge, although their difference—which is the sphere absorption cross section per Eq. (8.79)—will converge. This behavior is a consequence of the isothermal-sphere assumption that formed the basis of the formulas; for two spheres each at uniform yet distinct temperatures, the radiative heat transfer between the spheres would become unbounded as the spheres approached contact. In this limit, however, the spheres could not maintain isothermal conditions as conduction heat transfer, within each sphere, would act to maintain temperature continuity at the point of contact.

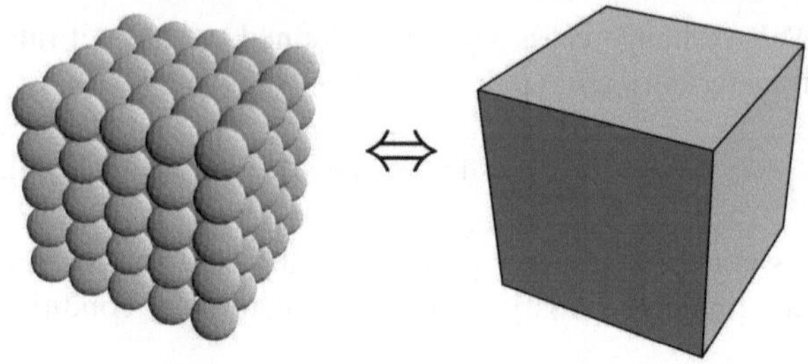

Fig. 8.6 Extensions of the multiple sphere formulation

8.5 Extensions and Future Challenges

The focus of the presentation so far has been on the *extension* of Mie theory to multiple sphere systems. In this sense, we have not examined a fundamentally new solution to a boundary value problem: the only real boundaries in the problem are at the sphere surfaces, and the continuity conditions at these points are provided by Mie theory. The general formulation provided by the multiple sphere solution, however, can be extended into topics well beyond that of light scattering by sphere clusters.

Figure 8.6 illustrates the two fundamental directions that such extensions can take. One path concerns the description of EM wave propagation in discretely inhomogeneous media; i.e., a medium consisting of particles dispersed in an otherwise continuous matrix. This situation can be viewed as the application of the multiple sphere solution to a semi–infinite system of spheres, and an objective of this application is to identify an effective propagation constant of the medium which describes the attenuation of the average (or coherent) field via absorption and scattering by the particles. In the diagram Fig. 8.6, this extension takes the situation on the left and models it as that on the right, i.e., representing an inhomogeneous medium as effectively homogeneous.

A robust theory exists for describing the propagation of EM waves in particulate media, and the reader is referred to the seminal works of Ishimaru [26], Varadan et al. [27], and Waterman and Pedersen [28] for the derivations and details. Most typically, the theoretical models address the reflection and propagation of a plane wave by a half space of particles that are either randomly distributed with a known concentration and pair distribution function, or arranged onto a lattice. Development of the models begins with the interaction equations for the sphere scattering coefficients, Eq. (8.10), applied to the semi–infinite expanse of particles. Closure of the equations is made possible by simplifying assumptions regarding the high–order correlations among the particle positions and scattered fields—such as the quasi–crystalline approximation—from which the configuration–averaged scattering coefficients can be represented by a functional, plane-wave model [29–31].

A simplified presentation of the key elements in an effective medium theory will be given in the following section. Before doing so, it will be noted that the second basic extension of the multiple sphere solution can be viewed as the reverse of the effective medium problem. The task here is to represent a homogeneous—albeit nonspherical—particle as an equivalent ensemble of "simple" particles with known properties; the point being that the multiple sphere formulation can be applied to the model particle to describe the scattering characteristics of the actual particle. This rationale forms the basis of the well–known discrete dipole approximation (DDA). Although the DDA was originally derived from an integral formulation of Maxwell's equations, the multiple sphere formulation described herein, when applied in the dipole limit (i.e., using only the first-order, TM coefficient in the field expansions) and when using a mixing rule formulation for the Mie coefficient $\overline{a}$, becomes equivalent to the DDA. The reader is referred to works by Draine and Flatau [32], and Yurkin and Hoekstra [33] for details on the derivation and implementation of the DDA. The calculation of the particle T matrix via a coupling of the present formulation with the DDA framework is presented in [34].

8.5.1 Application to Inhomogeneous Media

As opposed to applying Eq. (8.10) to a half space of particles, it is possible to identify the basic relations making up an effective medium theory via consideration of the equation for the sphere–sphere T^{i-j} matrix, Eq. (8.45), when applied to an infinite expanse of particles. Considering the case where origins i and j are not coincident, and assuming that the spheres are all identical so that $\overline{a}^i = \overline{a}$ (this constraint could be relaxed), the equations have

$$\frac{1}{\overline{a}_\mu} T^{i-j}_{\mu\,\nu} - \sum_{\substack{i'=1\\ i'\neq j,i}}^{\infty} \sum_{\mu'} H^{i-i'}_{\mu\,\mu'} T^{i'-j}_{\mu'\,\nu} - \sum_{\mu'} H^{i-j}_{\mu\,\mu'} T^{j-j}_{\mu'\,\nu} = 0 \tag{8.86}$$

Of interest here is identifying the configuration–averaged forms of the interaction matrices. A random dispersion of spheres will be considered, characterized by a volume fraction f and a pair correlation function $g(r_{i-j})$ which describes the relative probability of a sphere origin located a distance r_{i-j} from another sphere origin. Since the spheres are identical and do not overlap, $g = 0$ for $r_{i-j} < 2a$, and for a sufficiently large distance, denoted as r_c, the relative positions of the pair become uncorrelated so that $g \to 1$ for $r_{i-j} \geq r_c$.

For the infinite system of spheres, it can be assumed that 1) the averaged self–interaction matrix T^{j-j} becomes a diagonal and constant matrix, independent of position j, and 2) the averaged matrix T^{i-j} becomes a function solely of the distance vector between i and j. General relations for the average T^{j-j} and far-field T^{i-j} (i.e., for $r_{i-j} > r_c$) can be identified by examining Eq. (8.86) for the situation in which origins i and j become separated by at least $2r_c$, as illustrated in Fig. 8.7. Following the model of Ishimaru [26], the interaction matrix T^{i-j} is represented by a translation matrix evaluated for a medium with an effective propagation constant k_e, i.e.,

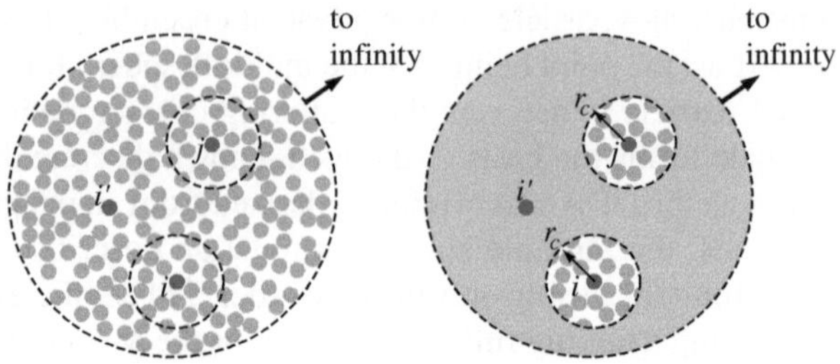

Fig. 8.7 Effective medium ansatz

$$T_{\mu\,\nu}^{i-j} \approx \alpha_\mu\, \mathscr{H}_{\mu\,\nu}^{i-j}\, \alpha_\nu, \quad \mathscr{H}_{\mu\,\nu}^{i-j} = H_{\mu\,\nu}(\mathsf{k}_e \mathbf{r}_{i-j}) \tag{8.87}$$

in which α denotes an as–of-yet undetermined coefficient. The sum over position i' in Eq. (8.86) is now split into three parts, corresponding to i' within the correlation distance of i, of j, or neither, and the properties of the translation matrices can be applied to convert the first two parts into sums involving spherically-symmetric functions. Omitting the subscripts on the matrices, and implying the usual matrix multiplication rules, the interaction term becomes

$$\begin{aligned} \sum_{\substack{i'=1 \\ i' \neq j,i}}^{\infty} H^{i-i'}\, T^{i'-j} = & \left(\sum_{r_{i-i'} < r_c} H^{i-i'}\, \alpha\, \mathscr{J}^{i'-i} \right) \mathscr{H}^{i-j}\, \alpha \\ & + H^{i-j} \left(\sum_{r_{j-i'} < r_c} J^{j-i'}\, T^{i'-j} \right) + \sum_{i' \in b} H^{i-i'}\, \alpha\, \mathscr{H}^{i'-j}\, \alpha \end{aligned} \tag{8.88}$$

in which $\mathscr{J}$ is the J translation matrix evaluated for a medium with propagation constant k_e, and b in the third sum denotes the background region, i.e., outside the zones surrounding i and j. This sum can now be represented by an integral over volume, the integrand of which will vanish at infinity. Since the translation matrices in the integrand both satisfy the scalar Helmholtz equation, the volume integral can be converted into a pair of surface integrals, by use of Green's second identity, over the two spherical surfaces surrounding the i and j zones. This gives

$$\sum_{i' \in b} H^{i-i'}\, \alpha\, \mathscr{H}^{i'-j}\, \alpha = -\left\langle H\, \alpha\, \mathscr{J} \right\rangle \mathscr{H}^{i-j}\, \alpha - H^{i-j}\, \langle J\, \alpha\, \mathscr{H} \rangle\, \alpha \tag{8.89}$$

in which

$$\langle H\,\alpha\,\mathscr{J}\rangle = \frac{3f\,r_c^2}{4\pi a^3(\mathsf{k}_e^2-\mathsf{k}^2)} \times \int_{4\pi}\left(\frac{\partial H(\mathsf{k}(-\mathbf{r}))}{\partial r}\,\alpha\,\mathscr{J}(\mathsf{k}_e\mathbf{r}) - H(\mathsf{k}(-\mathbf{r}))\,\alpha\,\frac{\partial\mathscr{J}(\mathsf{k}_e\mathbf{r})}{\partial r}\right)_{r=r_c} d\Omega \tag{8.90}$$

with $d\Omega = d(\cos\theta)\,d\phi$, and $-\mathbf{r} = (r, -\cos\theta, \phi+\pi)$. The analogous formula applies to $\langle J\,\alpha\,\mathscr{H}\rangle$. These quantities can be evaluated analytically by using the orthogonality properties of the translation matrix basis functions.

By substitution of Eqs. (8.87–8.89), Eq. (8.86) can be separated into two parts proportional to either $\mathscr{H}^{i-j}$ or H^{i-j}. Both of these parts must therefore equate to zero, as the result must be independent of r_{i-j}. The derived relations have the form

$$\alpha_\mu - \bar{a}_\mu \sum_\nu \left(\sum_{r_{0-i}<r_c} H^{0-i}_{\mu\,\nu}\,\alpha_\nu\,\mathscr{J}^{i-0}_{\nu\,\mu} + \langle H_{\mu\,\nu}\,\alpha_\nu\,\mathscr{J}_{\mu\,\mu}\rangle\right) = 0 \tag{8.91}$$

$$T^{j-j}_{\mu\,\mu} + \sum_\nu \left(\sum_{r_{j-i'}<r_c} J^{j-i'}_{\mu\,\nu}\,T^{i'-j}_{\nu\,\mu} - \langle J_{\mu\,\nu}\,\alpha_\nu\,\mathscr{H}_{\nu\,\mu}\rangle\,\alpha_\mu\right) = 0 \tag{8.92}$$

Equation (8.91) represents a homogeneous linear system for the α coefficients. Accordingly, the determinant of the system matrix must vanish, and this provides a formula for the effective propagation constant k_e. The inner sum in this relation can be evaluated as a finite sum, as done for a cubic lattice of spheres by Waterman and Pedersen [28], or via a volume integral involving the pair distribution function [27, 31]. Equation (8.92) can be viewed as a form of the Ewald–Oseen extinction theorem for the medium [31]. Referring to Eq. (8.74), this relation states the obvious point that the random–orientation cross sections of a sphere, surrounded by an infinite expanse of spheres, must be zero. On a practical level, Eq. (8.92) provides a relation between the self–interaction matrix T^{j-j} and the eigenvectors α of Eq. (8.91). A second relation, which would formally close the problem, would be obtained from the interaction matrix equations, Eq. (8.45), applied for $i = j$.

8.5.2 *Future Applications: Direct Simulations and the Bridge to the RTE*

Of course, many additional details would be needed to develop the basic concepts, presented above, into a working computational scheme.Analytical formulas for the effective propagation constant can be derived for spheres in the dipole limit [28, 31, 34]. However, calculation of k_e for sphere size parameters of order unity or greater—which involves multiple harmonic orders—can become a formidable numerical problem.

In this respect, an alternative and increasingly attractive route to calculating the effective radiative properties of inhomogeneous media is by direct simulations, or equivalently, numerical experiments. The basic concept is to apply the multiple sphere formulation, given by either Eq. (8.10) or (8.45), to systems containing a sufficiently large number of spheres so that the system, as a whole, can represent a radiative continuum. Such calculations have been made feasible by the development and availability of efficient codes for implementing the multiple sphere formulation, and—more importantly—the decreasing cost and increasing speed of computational hardware. A demonstration of a direct simulation approach to calculating the reflectance and transmittance properties of particle deposits was recently performed by the author, in which the modeled target consisted of a slab of spherical particles, with N_S ranging from 500 to 1,600, that was exposed to a Gaussian–profile incident beam [18]. More recently, Mishchenko et al. conducted sets of direct simulation calculations on systems containing on the order of 100–1,000 spheres to examine weak localization and coherent backscattering in discrete random media [19].

A practical objective of such calculations would be to use the exact, microscopic–level solution, provided by the multiple sphere formulation, as a benchmark—or replacement—to the phenomenological, macroscopic–level radiative transport equation in particulate media. This is especially relevant in situations such as particle deposits, paint pigments, composite materials, etc., for which the particle concentrations can become sufficiently high so that prediction of the bulk medium optical properties via the single–scattering formulations becomes suspect.

The microscopic–level detail of the fields in a particulate medium, which are accessible via direct simulations using the multiple sphere formulation, are also of interest in and of themselves. Specifically, resonant behavior of particle inclusions can lead to enhanced scattering and strong localization of the fields. Interest in this behavior relates to the possible manipulation of the EM field in periodic or random inhomogeneous media, with potential application in photonic devices (i.e., zero–threshold lasers, all–optical transistors, suppression of spontaneous emission) [35, 36].

8.6 Appendix

8.6.1 *Calculation of the Translation Matrix Elements*

Equations (8.33) and (8.34) provide a compact and relatively simple means of calculating the translation matrices. More computationally-efficient methods have been devised, such as recurrence relations for the matrix elements, yet in the overall scheme of things, the execution time required to evaluate the translation matrices will typically be small compared to that required to solve the interaction equations.

Calculation of the Bessel, Hankel, and Legendre functions that appear in the scalar harmonic ψ should pose no problem to those familiar with Lorenz/Mie theory. The Clebsch–Gordon coefficients, on the other hand, are not as common. These quantities appear most often in problems relating to angular momentum in quantum mechanics and the representation of the rotation group, and they are part of the intrinsic function set in *Mathematica*.In numerical codes they are most efficiently calculated using recurrence relations. The downwards recurrence is stable up to the maximum L values encountered in our codes, and is given by

$$C^{w-2}_{mn,kl} = a\left(b\, C^{w-1}_{mn,kl} - c\, C^{w}_{mn,kl}\right) \tag{8.93}$$

with

$$\begin{aligned} a &= 2(w-1)\left(\frac{(2w-3)}{(w-m-k-1)(w+m+k-1)(l-n+w-1)}\right)^{1/2} \\ &\quad\times\left(\frac{(2w-1)}{(n-l+w-1)(n+l-w+2)(n+l+w)}\right)^{1/2} \\ b &= \frac{(m-k)w(w-1)-(m+k)(n(n+1)-l(l+1))}{2w(w-1)} \\ c &= \frac{1}{2w}\left(\frac{(w+m+k)(w-m-k)(n-l+w)}{(w+w+1)}\right)^{1/2} \\ &\quad\times\left(\frac{(l-n+w)(n+l-w+1)(n+l+w+1)}{(w+w-1)}\right)^{1/2} \end{aligned} \tag{8.94}$$

The index w in Eq. (8.93) begins with $w = n+l$, and the Clebsch–Gordon coefficients for indices $w < w_s = \text{Max}(|n-l|, |m+k|)$ are implied to be zero. Starting values for the C coefficients are obtained from

$$C^{n+l}_{mn,kl} = \left(\frac{\binom{n+l+m+k}{l+k}\binom{n+l-m-k}{l-k}}{\binom{2(n+l)}{2l}}\right)^{1/2} \tag{8.95}$$

with

$$\binom{n}{l} = \frac{n!}{l!(n-l)!} \tag{8.96}$$

Symmetry and reflection properties of the J translation matrix are

$$J_{mnp\,klq}(x,y,z) = (-1)^{m+n+k+l+p+q} J_{-klq\,-mnp}(x,y,z) \tag{8.97}$$

$$J_{mnp\,klq}(-x,y,z) = (-1)^{p+q} J_{-mnp\,-klq}(x,y,z) \tag{8.98}$$

$$J_{mnp\,klq}(x,-y,z) = (-1)^{m+k+p+q} J_{-mnp\,-klq}(x,y,z) \tag{8.99}$$

$$J_{mnp\,klq}(x,y,-z) = (-1)^{m+n+k+l+p+q} J_{mnp\,klq}(x,y,z) \tag{8.100}$$

$$J_{mnp\,klq}(-x,-y,-z) = (-1)^{n+l+p+q} J_{mnp\,klq}(x,y,z) \tag{8.101}$$

The same relations will hold for H. Additional relations are

$$J_{mnp\,klq}(-x,y,z) = (-1)^{m+k+p+q} J^{*}_{mnp\,klq}(x,y,z) \tag{8.102}$$

$$J_{mnp\,klq}(x,-y,z) = (-1)^{p+q} J^{*}_{mnp\,klq}(x,y,z) \tag{8.103}$$

$$J_{mnp\,klq}(-x,-y,-z) = J^{*}_{klq\,mnp}(x,y,z) \tag{8.104}$$

In the three previous equations, it is understood that the conjugate operation does not apply to the Bessel function $j_w(\mathsf{k}r)$. This has no bearing for translation in a non-absorbing medium, for which k and $j_w(\mathsf{k}r)$ will be real. The three previous formulas can also be applied to the H translation matrix, yet again the conjugate operator does not apply to the Hankel function $h_w(\mathsf{k}r)$.

8.6.2 The Rotation-Axial Translation Operation

A time–intensive step during the iterative solution to Eq. (8.10) is the matrix–vector product,

$$a^{j,i}_{mnp} = \sum_{l=1}^{L} \sum_{k=-l}^{l} \sum_{q=1}^{2} H^{i-j}_{mnp\,klq}\, a^{j}_{klq} \tag{8.105}$$

For an arbitrary translation, the number of operations in this calculation, as performed directly, will scale (roughly) as L^4. This process can be made more efficient by the factorization of H into rotational and axial–translation matrices. The rotational matrices are diagonal in order, and the axial translation matrix are diagonal in degree. Consequently, the matrix-vector multiplication for the factored H will scale as L^3, and the time savings can become significant for large L.

The three–step procedure is

1. Rotate a^j towards i,

$$a^{j,1}_{mnp} = \sum_{k=-n}^{n} \mathscr{D}^{(n)}_{mk}(\cos\theta_{i-j})\, \exp(\mathrm{i}k\phi_{i-j})\, a^{j}_{knp} \tag{8.106}$$

2. Axial translation to i,

$$a_{mnp}^{j,2} = \sum_{l=1}^{L} \sum_{q=1}^{2} H_{mnp\,mlq}(\mathrm{k}r_{ij}, \theta = 0, \phi = 0)\, a_{mlq}^{j,1} \tag{8.107}$$

3. Inverse rotation,

$$a_{mnp}^{j,i} = \exp(-\mathrm{i}m\phi_{i-j}) \sum_{k=-n}^{n} \mathscr{D}_{-m-k}^{(n)}(\cos\theta_{i-j})\, a_{knp}^{j,2} \tag{8.108}$$

8.6.3 Generalized Spherical Functions

The generalized spherical functions $\mathscr{D}_{mk}^{(n)}(x)$ are defined by

$$\mathscr{D}_{mk}^{(n)}(x) = (-1)^{m+k} \left(\frac{(n-k)!(n+k)!}{(n-m)!(n+m)!} \right)^{1/2} \times \left(\frac{1+x}{2} \right)^{(m+k)/2} \left(\frac{1-x}{2} \right)^{(k-m)/2} P_{n-k}^{k-m,\,k+m}(x) \tag{8.109}$$

in which $P_n^{\alpha,\beta}$ is the Jacobi polynomial.

A stable recurrence formula for calculation of the $\mathscr{D}_{mk}^{(n)}$ functions starts with

$$\mathscr{D}_{0k}^{(n)}(x) = \left(\frac{(n-m)!}{(n+m)!} \right)^{1/2} P_n^k(x) \tag{8.110}$$

and, for $m = 1, 2, \ldots n$,

$$\mathscr{D}_{mk}^{(n)}(x) = \frac{1}{[(n-m+1)(n+m)]^{1/2}} \left([(n+k)(n-k+1)]^{1/2} \frac{1+x}{2} \mathscr{D}_{m-1k-1}^{(n)}(x) - [(n-k)(n+k+1)]^{1/2} \frac{1-x}{2} \mathscr{D}_{m-1k+1}^{(n)}(x) - k\,(1-x^2)^{1/2} \mathscr{D}_{m-1k}^{(n)}(x) \right) \tag{8.111}$$

and

$$\mathscr{D}_{-m-k}^{(n)}(x) = (-1)^{m+k} \mathscr{D}_{mk}^{(n)}(x) \tag{8.112}$$

References

1. G. Gouesbet, B. Maheu, G. Grehan, J. Opt. Soc. Am. A **5**, 1427 (1988)
2. A. Doicu, T. Wriedt, Appl. Opt. **13**, 2971 (1997)
3. S. Stein, Quart. Appl. Math. **19**, 15 (1961)
4. O. Cruzan, Quart. Appl. Math. **20**, 33 (1962)
5. C. Liang, Y. Lo, Radio Sci. **2**, 1481 (1967)
6. J.H. Brunning, Y.T. Lo, IEEE Trans. Antennas Propag. **AP-19**, 378 (1971)
7. F. Borghese, P. Denti, G. Toscano, O. Sindoni, Appl. Opt. **18**, 116 (1979)
8. F. Borghese, P. Denti, R. Saija, G. Toscano, O. Sindoni, J. Opt. Soc. Am. A **1**, 183 (1984)
9. K. Fuller, G. Kattawar, Opt. Lett. **13**, 90 (1988)
10. K. Fuller, G. Kattawar, Opt. Lett. **13**, 1063 (1988)
11. D.W. Mackowski, Proc. Roy. Soc. Lond. A **433**, 599 (1991)
12. D.W. Mackowski, J. Opt. Soc. Am. A **11**, 2851 (1994)
13. D.W. Mackowski, M.I. Mishchenko, J. Opt. Soc. Am. A **13**, 2266 (1996)
14. K.A. Fuller, D.W. Mackowski, in *Light Scattering by Nonspherical Particles: Theory, Measurements, and Applications*, ed. by M.I. Mishchenko, J.W. Hovenier, L.D. Travis, chap. 8, (Academic Press, New York, 2000), p. 226
15. F. Borghese, P. Denti, R. Saija, *Scattering from Model Nonspherical Particles* (Springer, Berlin, 2007)
16. M.I. Mishchenko, G. Videen, V.A. Babenko, N.G. Khlebtsov, T. Wriedt, J. Quant. Spectrosc. Radiat. Transf. **88**(1–3), 357 (2004)
17. M.I. Mishchenko, G. Videen, V.A. Babenko, N.G. Khlebtsov, T. Wriedt, J. Quant. Spectrosc. Radiat. Transf. **106**(1–3), 304 (2007)
18. D.W. Mackowski, ASME Conf. Proc. **2006**(47845), 205 (2006)
19. M. Mishchenko, L. Liu, D.W. Mackowski, B. Cairns, G. Videen, Opt. Express **15**, 2822 (2007)
20. P. Flatau, Opt. Express **12**, 3149 (2004)
21. K.A. Fuller, Appl. Opt. **30**(33), 4716 (1991)
22. Y. Xu, Appl. Opt. **36**, 9496 (1997)
23. C.F. Bohren, D.R. Huffman, *Absorption and Scattering of Light by Small Particles*, (Wiley, New York, 1983)
24. A. Narayanaswamy, G. Chen, Phys. Rev. B **77**(7), 075125 (2008)
25. D.W. Mackowski, M. Mishchenko, J. Heat Transf. **130**(11), 112702 (2008)
26. A. Ishimaru, *Wave Propagation and Scattering in Random Media.* vols. I and 2 (Academic Press, New York, 1978)
27. V.K. Varadan, V.N. Bringi, V.V. Varadan, A. Ishimaru, Radio Sci. **18**, 321 (1983)
28. P.C. Waterman, N.E. Pedersen, J. Appl. Phys. **59**, 2609 (1986)
29. L.L. Foldy, Phys. Rev. **67**(3–4), 107 (1945)
30. M. Lax, Phys. Rev. **85**(4), 621 (1952)
31. A. Doicu, T. Wriedt, Y.A. Eremin, *Light Scattering by Systems of Particles. Null-Field Method with Discrete Sources: Theory and Programs*, (Springer Science+Business Media, New York, 2006)
32. B.T. Draine, P.J. Flatau, J. Opt. Soc. Am. A **11**, 1491 (1994)
33. M.A. Yurkin, A.G. Hoekstra, J. Quant. Spectrosc. Radiat. Transf. **106**, 558 (2007)
34. D.W. Mackowski, J. Opt. Soc. Am. A **19**, 881 (2002)
35. H. Cao, J.Y. Xu, D.Z. Zhang, S.H. Chang, S.T. Ho, E.W. Seelig, X. Liu, R.P.H. Chang, Phys. Rev. Lett. **84**(24), 5584 (2000)
36. A. Yamilov, H. Cao, Phys. Rev. B **68**, 085111 (2003)

Index

W. Hergert and T. Wriedt (eds.), *The Mie Theory*, Springer Series in Optical Sciences 169, DOI: 10.1007/978-3-642-28738-1,

MIX
Papier aus verantwortungsvollen Quellen
Paper from responsible sources
FSC® C105338

If you have any concerns about our products,
you can contact us on
ProductSafety@springernature.com

In case Publisher is established outside the EU,
the EU authorized representative is:
Springer Nature Customer Service Center GmbH
Europaplatz 3, 69115 Heidelberg, Germany

Printed by Libri Plureos GmbH
in Hamburg, Germany